# METHODS OF MEASURING ENVIRONMENTAL PARAMETERS

# METHODS OF MEASURING ENVIRONMENTAL PARAMETERS

**YURIY POSUDIN**

National University of Life
and Environmental Sciences of Ukraine
Kiev, Ukraine

Published by John Wiley & Sons, Inc., Hoboken, New Jersey.
Published simultaneously in Canada.

For general information on our other products and services or for technical support, please contact our Customer Care Department within the United States at (800) 762-2974, outside the United States at (317) 572-3993 or fax (317) 572-4002.

Wiley also publishes its books in a variety of electronic formats. Some content that appears in print may not be available in electronic formats. For more information about Wiley products, visit our web site at www.wiley.com.

*Library of Congress Cataloging-in-Publication Data:*

Posudin, Yu.I. (Yuriy Ivanovich)
   Methods of measuring environmental parameters / Yuriy Posudin, National University of Life and Environmental Sciences of Ukraine, Kiev, Ukraine.
      pages cm
   Includes index.
   ISBN 978-1-118-68693-5 (cloth)
   1. Pollution–Measurement.   2. Environmental monitoring.   I. Title.
   TD193.P68 2014
   628.5028'7–dc23

                                                                                    2014008365

Printed in the United States of America

10  9  8  7  6  5  4  3  2  1

*To Professor Stanley J. Kays,*
*a good friend, gentleman, and my supervisor*

# CONTENTS

**PART III    HYDROGRAPHIC FACTORS**

**PART IV  EDAPHIC FACTORS**

## PART VI   PHYSICAL TYPES OF POLLUTION

### 26   Mechanical Vibration                                      345

### 27   Measurement of Vibration                                  348

### 28   Noise                                                     351

*Understanding that the world does not belong to any one nation or generation, and sharing a spirit of utmost urgency, we dedicate ourselves to undertake bold action to cherish and protect the environment of our planetary home.*

Al Gore
September 16, 1992
Sioux Fall, South Dakota

# PREFACE

The world population reached 7 billion October 31, 2011 and continues to grow at an unprecedented rate. The United Nations has projected the possible world population in 2300 using differing scenarios based on changes in fertility: declining growth (2.3 billion), average growth (9 billion), and high growth (36.4 billion).

Two primary processes impact population growth: (1) industrialization—the large-scale introduction of manufacturing, advanced technical enterprises, and other productive economic activities into an area, society, or country. There are typically significant social and economic changes during the transition between preindustrial and industrial societies; and (2) urbanization—the physical growth of urban areas as a result of rural migration into cities and concentration of large number of people in small areas, forming cities. The net effect is the emergence of megacities (cities with populations of over 1 million inhabitants) that are a characteristic feature of demographic growth. About 22 cities worldwide had populations that exceeded 10 million inhabitants in 2010.

The exponential growth of the world population is accompanied by increasing energy consumption, fertilizer use, and the concentration of pollutants in the environment (e.g., carbon dioxide, nitrogen dioxide, hydrogen sulfide, sulfur dioxide, hydrocarbons); a shortage of arable land; and the accumulation of radioactive waste. Mankind has reached or will reach its highest level of fossil fuel extraction in the near future: oil—in 2007, coal—in 2025, gas—in 2025. Electricity production by hydroelectric power stations will increase by 40% by 2050.

It is highly probable that armed conflicts around the world will increase due to the inequitable distribution of essential resources such as fossil fuels, fresh water, and other requisites. The rate of growth of the world population and the associated increase in energy consumption, lack of adequate food, deteriorating water and air quality due to the rapid rate of industrialization and urbanization, and the impact of biosphere pollution on climate, biogeochemical cycling, and the fauna and flora—all point toward an unprecedented environmental crisis in the immediate future.

The "greenhouse effect" is resulting in significant changes in the heat balance in the biosphere, which is precipitating potentially extreme climatic alterations. Increasing global temperature leads to melting of the polar ice, thermal expansion in the volume of the ocean water, elevated sea levels, and an appreciable reduction in the amount of dry land. Elevated temperature results in an increasing frequency of natural disasters such as severe floods, hurricanes, tsunamis, and drought. Changes in climate and sea level will also instigate extensive population migration.

Indications of this emerging crisis are natural and man-made environmental disasters that are accompanied by fatalities and irreversible harmful effects on the environment. Recent examples include the intense earthquakes and tsunamis in the Indian Ocean in 2004 (225,000 deaths), earthquakes in Pakistan in 2005 (74,500 deaths), China in 2008 (69,000 deaths), Haiti in 2010 (320,000 deaths); there were victims of hurricane "Katrina" in 2005, the most destructive hurricane in US history; the cyclone "Nargis" in 2005 (Burma or Myanmar); and the earthquake and tsunami in Japan in 2011.

In 2010, about one half of Europe was covered with dangerous ash as a result of the eruption of Eyjafjallajökull, a volcano in Iceland that paralyzed air travel for several weeks. The 2011 eruption of Grímsvötn, also in Iceland, similarly disrupted air travel in Northwestern Europe.

In 2003, an unexpected heat wave in Europe resulted in ~40,000 deaths; extremely high ambient temperatures in Russia in 2010 caused ~15,000 deaths and numerous fires.

Powerful floods occurred in Australia and Brazil in 2010–2011, in Amur district, Russia, in 2013.

Man-made disasters included an explosion at the AZF chemical factory in Toulouse (France) in 2001; an extremely serious accident on a British Petroleum oil well platform in the Gulf of Mexico in 2010 was the largest environmental disaster in US history; a fractured reservoir resulted in the release of toxic "red mud" in the Hungarian region of Veszprem in 2010; and the detection of dioxin in pig and poultry feed on farms in Germany in 2010.

It is impossible to ignore the harmful environmental impact of military activity and technology, and space exploration. A Ukrainian military tactical rocket struck a house in the town of Brovary, near Kiev, killing three people in 2000; a Ukrainian missile accidently shot down a Russian jet during military exercises in the Crimea in 2001, resulting in death of 78 people. There were a series of fires and explosions at military arsenals in Donetsk (2003), near the village of Novobohdanivka (2004) and in Lozova (2008) in Ukraine, that caused great losses and in Ulyanovsk (2009), Bashkortostan and Udmurtia (2011) in Russia.

There were also serious incidents in space. For example, the Russian "Cosmos-2251" and American "Iridium-33" satellites collided at an altitude of 790 miles over Siberia in 2009. In 2006, the wreckage of the Russian military satellite nearly struck a Latin American Airbus with 279 passengers on board. Tests of Chinese antisatellite weapons in 2007 led to the dispersal of considerable "space junk," adding to an estimated 2300 fragments and countless number of smaller items that greatly increase the possibility of a catastrophic loss of a manned spacecraft or satellites. Other accidents that resulted in debris in space include accidents or failures such as the cargo spaceship Progress M-12M (2011), the Russian interplanetary station "Phobos-Soil" that failed to reach Mars, and the Russian satellite "Meridian," that fell onto the street of Cosmonauts in the village Vahaytsevo in the Novosibirsk region. In addition to the tremendous number of debris fragments in space, each launch of a rocket is accompanied by the contamination of biosphere with dangerous fuel, pollutants and pieces of the spacecrafts.

We live in a very intense and harmful time from the point of view of damage to the environment and the resulting impacts on the biosphere. It is essential, therefore, that we utilize environmental monitoring and measurement of environmental parameters to obtain the essential information on changes in abiotic and biotic factors, and air, soil, and water quality. This may be accomplished using modern automated methods of measurement and remote sensing technologies. Understanding the nature and extent of environmental problems is essential for identifying and putting into action viable solutions.

# ACKNOWLEDGMENTS

The author of the textbook had the opportunity to cooperate with a number of scientists, researchers, and educators during his scientific and academic activities, namely:

Prof. E. Bazarov, Dr. G. Gerasimov, 1972–1975, Institute Radiotechnics and Electronics of the Academy of Sciences of USSR, Moscow, Frjazino, USSR; Prof. F. Lenci, Dr. G. Colombetti, 1979–1980, Institute of Biophysics CNR, Pisa, Italy; Prof. N. Massjuk, Dr. G. Lilitskaya, 1980–2010, M. G. Kholodny Institute of Botany of the National Academy of Sciences of Ukraine, Kiev, Ukraine; Prof. I. Lisker, 1992, Agrophysical Institute, St. Petersbourg, Russia; Prof. D.-P. Häder, 1993, Friedrich-Alexander-University Erlangen-Nuremberg, Germany; Dr. Chi N. Thai, 1996, Driftmier Engineering Center, University of Georgia, U.S.A.; Prof. H. K. Lichtenthaler, 1997, University of Karlsruhe, Germany; Prof. A. Flores-Moya, 2000, University of Malaga, Spain; Prof. Stanley J. Kays, 1996, 2000, 2008, Department of Horticulture, University of Georgia, U.S.A.; Prof. Hiroshi Kawai, 2002, Research Center for Inland Seas, Kobe University, Japan; Prof. C. Wiencke, Prof. D. Hanelt, 2003, Alfred-Wegener-Institute for Polar and Marine Research, Bremenhaven, Germany; Prof. Kyoichi Otsuki, Dr. Atsushi Kume, Dr. Tomo'omi Kumagai, 2007, Kyushu University, Toyama University, Japan; Dr. Gerry Dull, 2008, Department of Horticulture, University of Georgia, U.S.A.; Prof. Shunitz Tanaka, Dr. Kazuhiro Toyoda, 2010, Hokkaido University, Sapporo, Japan; Prof. R. Mnatsakanian, 2010, Central European University, Budapest, Hungary; Prof. Aimo Oikari, Prof. Timo Älander, 2010, Jyväskylä University, Finland.

These visits and collaboration undoubtedly influenced the forming of expanding horizons and world view of the author, who has a brilliant opportunity to express his sincere gratitude to all the above-mentioned supervisors, colleagues, and collaborators.

The author of this textbook, Fulbright Scholar in 1996, expresses his deep gratitude to the administration of the Fulbright Program in Ukraine, for organizational assistance to establish useful contacts with the US colleagues.

Special thanks to Prof. Stanley J. Kays for his moral support and encouragement, editorial revision, criticism, and suggestions during the preparation of the manuscript.

The author would like to thank the following reviewers of the manuscript: Prof. Motoyoshi Ikeda, The University of Hokkaido, Graduate School of Environmental Science, Sapporo, Japan; Dr. Atsushi Kume, Director of Ashoro Research Forest, Kyushu University, Ashoro, Hokkaido, Japan; Prof. Yuriy Masikevych, Head of Department of the Ecology and Law, Chernivtsi Faculty of the National Technical University "Kharkiv Polytechnic Institute," Chernivtsi, Ukraine; Prof. Yarema Tevtul', Chernivtsi National University, Chernivtsi, Ukraine; Anatoly Vid'machenko, Head, The Department for Physics of Solar System Bodies, Main Astronomical Observatory of the National Academy of Sciences of Ukraine, Kiev, Ukraine; Prof. Igor Yakymenko, Department of Biophysics, Bila Tserkva National Agrarian University, Bila Tserkva, Ukraine, as well as several anonymous reviewers, whose valuable opinions contributed greatly to the success of this manuscript.

Professor Stanley Kays (left) and author of the textbook (right) in Georgia

# ABOUT THE BOOK

The *main objective* of the textbook *"Methods of Measuring Environmental Parameters"* is to introduce students of Environmental Science and Engineering in current methods of environmental control and principles of devices used for measuring environmental parameters.

*Specific aims* of the textbook are:

1. a brief description of the main climatic (pressure, wind, temperature, humidity, precipitation, solar radiation), atmospheric, hydrographic, and edaphic factors;
2. assessment of abiotic factors, their effect on quality of atmosphere and indoor air, soil, water;
3. assessment of biotic factors, bioindication, and biomonitoring as perspective methods of observing the impact of external factors on ecosystems;
4. study the principal effects of environmental factors on living organisms, human health, and ecosystems;
5. review the basic methods and principles of modern instrumentation that can be applied for measuring environmental (mostly physical) parameters, with special emphasis on automated and remote sensing components of the environment;
6. comparative analysis of the advantages and disadvantages of the main methods of measurement presented in the textbook;
7. monitoring of student learning through practical exercises, tasks, problems, and tests.

The textbook has practical exercises, which allow a better understanding of the textbook content; the examples with solutions, and control exercises; questions and problems to provide self-testing the material presented in the textbook; constructive tests, for which there are no direct answers, but the student must find the answer himself. Examples of extreme environmental situations are given for the curious students. The list of references, electronic references and further reading is given at the end of each chapter. An appendix and a subject index are included at the end of the textbook.

The content of the textbook is based on course in *Methods of Measuring Environmental Parameters* given by author to students at National University of Kyiv-Mohyla Academy and undergraduates at National University of Life and Environmental Sciences of Ukraine, Kiev, Ukraine.

*Methods of Measuring Environmental Parameters* is intended to be a useful overview reference for professional ecologists, environmental scientists, meteorologists, climatologists, atmospheric physicists, aerobiologists, and soil and water managers. It can also serve as a suitable textbook for undergraduate and graduate students in these academic disciplines.

# ABOUT THE AUTHOR

Professor Yuriy Posudin, Doctor of Biological Sciences, National University of Life and Environmental Sciences of Ukraine, Kiev, Ukraine.

He studied at the Kiev State University (Radiophysical Faculty) 1964–1969, the Institute of Radiotechnique and Electronics, Moscow (1972–1975), and the Agrophysical Institute, St. Petersburg (1992).

Dr. Posudin's principal scientific interests are the investigation of photobiological reactions of algae and plants, and the non-destructive quality evaluation of agricultural and food products.

Academic duties of Dr. Posudin include lecturing of a cross-section of environmental topics, such as "Environmental Biophysics," "Methods of Measuring Environmental Parameters," and "Environmental Monitoring with Fundamentals of Metrology."

The main publications of Yuriy Posudin are:

Posudin, Yuri. 1998. *Lasers in Agriculture.* Science Publishers, Ltd, USA, p. 220.

Posudin, Yu. I. 2003. *Physics with Fundamentals of Biophysics.* Agrarna Nauka, Kyiv, p. 195.

Posudin, Yuriy. 2007. *Practical Spectroscopy in Agriculture and Food Science.* Science Publishers, Enfield, p. 196.

Posudin, Yu. I., Massjuk, N. P., and Lilitskaya, G. G. 2010. *Photomovement of Dunaliella Teod.* Vieweg + Teubner Research, p. 224.

Prof. Yuriy Posudin with the students of Environmental Sciences, Kiev, Ukraine.

# INTRODUCTION

## SOME PRINCIPAL DEFINITIONS

The existence and performance of a living organism depends on its environment.

*Environment* is a set of natural and human-altered abiotic and biotic factors that directly or indirectly affect an organism, population, or ecological community and influence its survival and development.

*Factor* is the reason or driving force of any process that occurs in the environment.

*Abiotic factors* are nonliving chemical and physical factors in the environment, which affect ecosystems.

Abiotic factors may be grouped into the following primary categories:

1. *Physical (climatic) factors*: pressure, wind, humidity, precipitation, temperature, solar radiation, ionizing radiation.
2. *Atmospheric factors* include the structure and composition of the atmosphere, the physical and chemical properties of the atmosphere that influence living organisms.
3. *Hydrographic factors* include the physical and chemical properties of water when it is a habitat for living organisms.

*Methods of Measuring Environmental Parameters*, First Edition. Yuriy Posudin.
© 2014 John Wiley & Sons, Inc. Published 2014 by John Wiley & Sons, Inc.

4. *Edaphic factors* include the structure and composition of the soil, a set of physical and chemical properties of soil that have ecological effects on living organisms.

*Biotic factors* in an ecosystem include all living factors.

An *ecosystem* is a community of living organisms in conjunction with the nonliving components of their environment, interacting as a system.

There are four spheres of the Earth: the *atmosphere* is the body of gaseous mass which surrounds our planet; the *hydrosphere* is composed of all of the water on or near the earth; the *biosphere* is composed of all living organisms; the *lithosphere* is the solid, rocky crust covering the entire planet.

*Environmental pollution* is the contamination of the physical and biological components of the ecosystem to such an extent that normal environmental processes are adversely affected.

*Measurement* is the process of experimentally obtaining one or more quantity values that can reasonably be attributed to a quantity.

*Measuring Instrument* is the device used for making measurements, alone or in conjunction with one or more supplementary devices.

*Parameter* is a value that characterizes any property of a process or phenomenon that occurs in the environment.

# PART I

## CLIMATIC FACTORS

# 1

# PRESSURE

## 1.1 DEFINITION OF PRESSURE

*Pressure* is a physical quantity that characterizes the intensity of normal force (perpendicular to the surface) with which one body acts on the surface of another. If the force exhibits a uniform distribution along the surface, the pressure $p$ is determined as the ratio of force to area:

$$p = \frac{F}{S}, \tag{1.1}$$

where $F$ is the magnitude of the normal force on the surface and $S$ is the area of this surface.

If the pressure is not uniform across the surface, the following expression defines the pressure at a specific point:

$$p = \sum_{\Delta S \to 0} \frac{\Delta F}{\Delta S} = \frac{dF}{dS}. \tag{1.2}$$

The SI unit for pressure is *pascal* (1 Pa $= 1$ N/m$^2$ $= 1$ kg/m·s$^2$). The following pressure units are also used:

1 Pa $= 9.87 \times 10^{-6}$ atm $= 7.5 \times 10^{-3}$ mmHg $= 0.000295$ inHg $= 7.5 \times 10^{-3}$ Torr
$= 10^{-5}$ bar $= 1.45 \times 10^{-4}$ psi;

*Methods of Measuring Environmental Parameters*, First Edition. Yuriy Posudin.
© 2014 John Wiley & Sons, Inc. Published 2014 by John Wiley & Sons, Inc.

1 bar = $10^5$ Pa = 0.987 atm = 750.06 mmHg = 29.53 inHg = 750.06 Torr = 14.504 psi;

1 atm = $1.01325 \times 10^5$ Pa = 1.01325 bar = 760 mmHg = 29.92 inHg = 760 Torr = 14.696 psi;

1 mmHg = 0.03937 inHg = 1 Torr = $1.3332 \times 10^{-3}$ bar = 133.32 Pa = $1.315 \times 10^{-3}$ atm = $19.337 \times 10^{-3}$ psi;

1 psi = 6894.76 Pa = $68.948 \times 10^{-3}$ bar = $68.046 \times 10^{-3}$ atm = 51.715 mmHg = 2.04 inHg = 51.715 Torr;

1 inHg = 25.4 mmHg = 3386.38816 Pa = 0.03342 atm = 25.4 Torr = 0.03386 bar = 0.49 psi.

A pressure of 1 Pa is small; the following non-SI metric units of pressure such as hectopascal (1 hPa = 100 Pa) or millibar (1 mbar = $10^{-3}$ bar) are used in meteorology and weather reports. The National Weather Service of the United States uses both inches of mercury (inHg) and hectopascals (hPa) or millibars (mbar). The pound per square inch (psi) is still popular in the United States and Canada.

## 1.2 ATMOSPHERIC PRESSURE

*Atmospheric pressure* is defined as the weight of a column of atmospheric air that acts on a given unit of surface area. Air is a mixture of gases, solids, and liquid particles. As a whole, atmospheric pressure depends on height; also it is characterized with horizontal distribution. The density and temperature of atmospheric air also depends on height.

The idea of a uniform distribution of atmospheric molecules within a volume of air is erroneous. Molecules are subject to the Earth's gravitational field; in addition, there is the effect of their thermal motion on the spatial distribution of the molecules. The combined action of the gravitational field and thermal motion leads to the state that is characterized by a decreasing gas concentration and pressure with increasing height.

Atmospheric air can be considered as an ideal gas, which can be described as follows:

$$p_A V_A = \frac{m_A}{M_A} R T_A, \tag{1.3}$$

where $p_A$ is the pressure of the air; $V_A$ is the volume of the air; $m_A$ is the mass of the air; $M_A$ is the molar mass of atmospheric air ($M_A = 0.029$ kg/mol for dry air, $M_A = 0.018$ kg/mol for water vapor); $R = 8.314$ J/K·mol is the molar gas constant; $T_A$ is the absolute temperature of the air.

Let us derive an expression for the pressure of an air in a container whose elementary volume is assumed to be in the shape of a cube (White, 2008). There are three forces that act on this volume. The force due to the pressure of the air contained on this volume is:

$$F_\downarrow = p_\downarrow S,$$ (1.4)

where $S$ is the base area of the elementary volume.Simultaneously the force of the gas, which is located below, acts in the opposite direction:

$$F_\uparrow = -p_\uparrow S.$$ (1.5)

The third force that acts on the elementary volume is the weight of the air:

$$P = \rho g V = \rho g S h,$$ (1.6)

where $\rho$ is the air density; $g$ is the acceleration due to gravity; $V$ is the elementary volume; $h$ is the edge length of the cube.The balance of these forces can be written as

$$F_\downarrow + F_\uparrow + P = p_\downarrow S - p_\uparrow S + \rho g S h = 0,$$ (1.7)

or

$$p_\downarrow - p_\uparrow = -\rho g h.$$ (1.8)

This equation can be written for infinitesimal changes of pressure in the differential form

$$dp = -\rho g dh.$$ (1.9)

The density $\rho_A$ of the atmospheric air can be expressed as

$$\rho_A = \frac{m_A}{V_A} = \frac{M_A p_A}{R T_A}.$$ (1.10)

The atmospheric pressure decreases with the change in altitude from $h$ to $h + dh$, according to Equation 1.9

$$\frac{dp_A}{dh} = -\rho_A g.$$ (1.11)

Combining (1.10) and (1.12), we can obtain the expression

$$\frac{dp_A}{p_A} = -\frac{M_A g}{R T_A} dh,$$ (1.12)

which by integrating this expression from the surface to the altitude $z$ leads to the *barometric formula*

$$p_A(z) = p_A(0) \exp[-(gM_A/RT_A)]z, \qquad (1.13)$$

where $p_A(0)$ is the pressure at sea level where the height $z$ is 0 (i.e., $z_0$); $g$ is the acceleration due to gravity (9.8 m/s$^2$); $M_A$ is the molar mass of the gas ($M_A = 0.029$ kg/mol for air); $R$ is the universal gas constant (8.314 J/K·mol); $T_A$ is the absolute temperature.

This equation indicates an exponential decrease in pressure with increasing elevation.

The meteorologists use the term "isobar"—a line connecting points of equal atmospheric pressure, which are depicted on weather maps. The rules for drawing isobars are

- Isobar lines may never cross or touch.
- Isobar lines may only pass through pressures of 1000 mbar ± 4 mbar.

In other words, allowable lines are 992, 996, 1000, 1004, 1008, and so on.

**Example** Find the atmospheric pressure at 10,000 m.

**Solution** The atmospheric pressure can be found as

$$p_A(z) = p_A(0) \cdot e^{[-(gM_A/RT_A)]z}$$
$$= 1.01325 \times 10^5 \text{Pa} \cdot e^{\left[-\frac{(0.029\,\text{kg/mol})\,(9.8\,\text{m/s}^2)\,(10^4\,\text{m})}{(8.314\,\text{m}^2\cdot\text{kg/s}^2\cdot\text{K}\cdot\text{mol})\,(223.25\,\text{K})}\right]}$$
$$= 1.01325 \times 10^5 \text{Pa} \cdot e^{-1.531} = 1.01325 \times 10^5 \text{ Pa} \cdot 0.2163 = 0.219 \times 10^5 \text{ Pa}.$$

### Exercises

1. Determine the atmospheric pressure at the altitude of the peak of Mount Everest (8848.82 m).
2. At what altitude is the atmospheric pressure reduced to 0.5 atm? The temperature is 260 K.
3. What is the temperature in Lhasa, Tibet (3650 m), if the atmospheric pressure is 652 mbar?

### Constructive Tests

1. To what altitude is the barometric formula valid?
2. What is the explanation for the exponential distribution of atmospheric pressure? Why are atmospheric molecules and particles not deposited on the Earth's surface due to the influence of gravity?

*Extreme Situations*

The highest barometric pressure ever recorded on Earth was 32.31 inHg (1094 hPa), measured in USSR, on December 31, 1968, in northern Siberia. The weather was clear and very cold at the time, with temperatures between 233.15 and 215.15 K.

The lowest pressure ever measured was 25.69 inHg (870 hPa), set on October 12, 1979, during Typhoon Tip in the western Pacific Ocean. The measurement was based on an instrumental observation made from a reconnaissance aircraft.

## 1.3   PHYSIOLOGICAL EFFECTS OF DECREASED AIR PRESSURE ON HUMAN ORGANISM

People who reach high altitudes suffer from *mountain sickness* (Folk, 1998): They feel changes in pulse and breathing rate, anorexia (an eating disorder due to loss of appetite), and a loss of body weight with increasing height. The principal symptoms of mountain sickness are dyspnea (difficult breathing), tachycardia (heart rate in excess of 100 beats per minute), malaise (vague body discomfort), nausea and vomiting, insomnia, and lassitude (a state or feeling of weariness, diminished energy, or listlessness). The prevalence of *mountain sickness* can be explained by the effect of altitude on the partial pressure of oxygen in the lung alveoli.

## 1.4   PHYSIOLOGICAL EFFECTS OF ALTITUDE ON ANIMALS

There are a number of animals that have adapted for living at high altitudes. The Lake Titicaca frog, likewise, lives at an altitude of 3812 m. The frog exhibits increased gas exchange due to extensive skin folds, and high hematocrit and erythrocyte concentrations. Thus, the capacity of the frog for transporting oxygen is substantially increased.

Mules are used at Aucanquilcha, a base camp for the International High Altitude Expedition, as a transport means in the 5250–6000 m altitude range. They demonstrate the ability to accurately assess their capacity for work and refuse to be pushed beyond a safe limit. Such animals may be mentioned as the vicuna (5000–6000 m), domestic sheep (up to 5250 m), and horses (up to 4600 m). Birds, however, hold the high altitude records: condors (7600 m), geese (8534 m), chough (9000 m), and griffon vulture (11,278 m).

## 1.5   EFFECTS OF ALTITUDE ON PLANTS

Altitudinal variation of climate induces morphological and physiological changes in plants and their canopy architecture. Often the plants maintain a compact or dwarf form with small, narrow or densely pubescent leaves. The ecological zone between 3230 and 3660 m is called an *alpine* area. Here, it is possible to find considerable changes in the quantitative and qualitative characteristics of the fauna. In addition, there are certain changes in climatic conditions that are related to the effects of pressure, wind, humidity and precipitation, temperature, radiation, and gas exchange, which in turn, also modify the fauna (Posudin, 2004).

## 1.6  VARIATION OF PRESSURE WITH DEPTH

Let us consider some liquid in a vessel. All points at the same depth feel the same pressure. Consider a cylinder of cross-sectional area $A$ and height $dy$ (Serway, 1990). The upward force that acts on the bottom of cylinder is $F_1 = pA$, and the downward force that acts on the top is $-F_2 = (p + dp)A$.

The weight of the cylinder is

$$dP = \rho g dV = \rho g A dy, \tag{1.14}$$

where $\rho$ is the density of the liquid.

In equilibrium, the resultant force is zero:

$$\sum F_y = pA - (p + dp)A - \rho g A dy, \tag{1.15}$$

Hence,

$$-dpS - \rho g S dy = 0, \tag{1.16}$$

or

$$\frac{dp}{dy} = -\rho g, \tag{1.17}$$

where the sign "minus" indicates that the increase of elevation corresponds to a decrease of pressure.

If $p_1$ and $p_2$ are the pressures at the levels $y_1$ and $y_2$ correspondingly, the following expression can be written as

$$p_2 - p_1 = -\rho g (y_2 - y_1). \tag{1.18}$$

If the vessel is open from the top, the pressure at the depth $d$ can be determined as

$$p = p_A + \rho g d, \tag{1.19}$$

where $p_A$ is atmospheric pressure; $\rho$ is the density of the liquid; $d$ is the depth; $g$ is the acceleration due to gravity.

This expression can be rewritten as

$$p = p_A + \frac{\rho g d A}{A} p_A + \frac{mg}{A} = p_A + \frac{P}{A}. \tag{1.20}$$

Thus, the absolute pressure $p$ at a depth $d$ below the surface of a liquid exceeds the atmospheric pressure by the value $\rho g d$, which corresponds to the pressure that is created by the weight of the liquid column of cross-sectional area $A$ and height $h$.

Thus, the pressure at the depth $d$ is determined for the opened vessel by Equation 1.19

$$p = p_A + \rho g d, \tag{1.21}$$

where $p_A \approx 1.01 \times 10^5$ Pa is the atmospheric pressure; $\rho$ is the density of the fluid; $g$ is the acceleration due to gravity.

In such a way, the absolute pressure $p$ at a depth $d$ below the surface of a liquid open to the atmosphere is greater than atmospheric pressure by an amount $\rho g d$.

**Example**   Calculate the pressure at the bottom of the Marianas Trench (depth 11,043 m). Assume the density of water is $1 \times 10^3$ kg/m$^3$ and take $p_A = 1.01 \times 10^5$ Pa.

**Solution**   Using formula (1.21), we can find the pressure:

$$p = p_A + \rho g d = 1.01 \times 10^5 \text{ Pa} + (1 \times 10^3 \text{ kg/m}^3)(9.8 \text{ m/s}^2)(11,043 \text{ m}) = 109 \text{ MPa}.$$

## 1.7   PHYSIOLOGICAL EFFECTS OF INCREASED PRESSURE ON HUMAN ORGANISM

*Pressure Problems.* The representative of some professions (Arabian sponge divers, Australian pearl divers, Japanese and Korean Ama) support their living by diving into the sea without any special equipment. When a diver descends, he is sensitive to the surrounding water pressure which increases at a rate of one atmosphere for each 10 m of descent (i.e., $\Delta p = 1$ atm for $\Delta d = 10$ m). The pressure of the surrounding water is transmitted to all internal parts of the body and the state of pressure equilibrium is established—the internal pressure of the body is equal to the surrounding pressure. During ascent and descent, the diver must support the same pressure of air in his lungs as that of the surrounding water. For example, if the diver goes to 30 m underwater by holding a full breath of air, he would have 4 atm in his lungs according to Equation 1.21. When he is ascending, the pressure at the water surface will be 1 atm, while the pressure in his lungs is 3 atm. This pressure difference can lead to the rupture of his lungs. Such a situation is called *pneumothorax*.

*Influence of Nitrogen.* The air we breathe consists of 79% of nitrogen and our blood is full of dissolved nitrogen. According to pneumothorax, the amount of dissolved gas in a liquid at constant temperature is directly proportional to the partial pressure of the gas. Liquids which are under high pressure can dissolve more gas than liquids which are under low pressure. The great pressure of air in the lungs provokes the formation of the bubbles into the blood vessels.

If bubbles are formed due to rapid decompression, the diver will suffer the effects of a painful disease called the *caisson disease* (from French *Caisse,* a chest) or the *bends*, which lead to neuralgic pains, paralysis, distress in breathing, and often collapse.

***Oxygen Poisoning.*** When a diver is under water, the tissues use a certain amount of dissolved oxygen from hemoglobin, the oxygen transport protein found in the erythrocytes of the blood. At sufficient depth the pressure is increased, this oxygen support by hemoglobin is stopped, the tissues remain saturated, and the convulsions induced by high-pressure oxygen can take place.

***Carbon Dioxide Influence.*** A great amount of carbon dioxide (more than 10%) at the depth can depress the cellular metabolism of the respiratory center; the diver develops lethargy and narcosis, and finally becomes unconscious.

## 1.8   PHYSIOLOGICAL EFFECTS OF PRESSURE ON DIVING ANIMALS

The deep-sea environment is characterized by a considerable high hydrostatic pressure which increases by approximately 1 MPa ($10^6$ Pa) for every 100 m of depth. The marine depths investigator Jacques Piccard observed through the illuminator of his bathyscaphe shrimp and fish at the depth about 10,912 m. Some invertebrates and bacteria have been found near the bottom of the Marianas Trench (depth 11,043 m, pressure about 110 MPa). There are a number of diving marine animals which can live under the pressure of tens or even hundreds of atmospheres. The depth range for different marine animals is 200–300 m (maximal depth is about 900 m)—fur seals, *Callorhinus*; 457 m—Weddell seal, *Leptonychotes weddelli*; 500–2000 m—deep-sea eel, *Synaphobranchus kaupi*; greater than 500 m—emperor penguin, *Aptenodytes fosteri*; below 1500 m—northern elephant seal, *Mirounga angustirostris*; 2000–3000 m—sperm whale, *Physeter catodon*; 7250 m—sea urchin, *Echinoidea*; 7360 m—sea star, *Asteroidea*; 8370 m—cuskeel fish, *Ophidiidae*; 10,190 m—sea cucumber, *Holothurioidea* (Folk, 1998).

What are the principal mechanisms of such depth adaptation of marine animals? The first peculiarity is related to the ability of these animals to exhale before diving: for example, cetaceans can exhale about 88% of its lung air with a single breathe while humans approximately 12%. Their lungs are collapsed quickly preventing atmospheric gases (nitrogen, oxygen) from entering the bloodstream. Another feature is the ability of diving animals to use oxygen from the blood which is characterized with high blood volume and high hematocrit (a number of blood cells). In such a way, the skeletal muscle tissue of these animals contains about 47% of overall body oxygen. The large amount of hemoglobin in the muscles is responsible for their deep red color. The diving animals have about 10 times more of myoglobin in the muscles than terrestrial animals. It is necessary to mention the sufficient ability of diving animals to decrease the heart rate during the diving in comparison to the sea surface situation: for instance, a muskrat has 320 beats per minute before dive and 34 beats per minute only at the depth; on land, seal has a heartbeat of 107 beats per minute, but at the depth its pulse decreases to a mean of 68 beats per minute. Such abilities of diving animals to collapse the lungs, to store myoglobin in their muscles, and to use oxygen located within the muscle cells and the blood can make it possible to escape the bends (Posudin, 2004).

## REFERENCES

Folk, G.E., Jr., Riedesel, M.L., and Thrift, D.L. 1998. *Principles of Integrative Environmental Physiology.* Austin and Winfield, San Francisco, CA.

Posudin, Y.I. 2004. *Physics with Fundamentals of Biophysics.* Agrarna Nauka, Kiev, Ukraine.

Serway, R.A. 1990. *Physics for Scientists and Engineers.* Harcourt Brace Jovanovich College Publishers, Orlando, FL.

White, F.M. 2008. *Pressure Distribution in a Fluid. Fluid Mechanics.* McGraw-Hill, New York.

# 2

# MEASUREMENT OF PRESSURE

## 2.1 MANOMETERS

A *manometer* is an instrument used for measuring the pressure of liquids and gases. There are two types of manometers for measuring atmospheric pressure: the *open-tube manometer (siphon manometer)* and *closed-tube manometer (cup manometer)*.

The open-tube manometer is a U-shaped tube containing a liquid; one end of the manometer is sealed and is deprived of the air; the open end is connected with the atmospheric air. Difference $h$ of liquid levels in two laps of the tube is calibrated in units of pressure (Figure 2.1). The pressure $p$ and atmospheric pressure $p_A$ are related as

$$p = p_A + \rho g h, \qquad (2.1)$$

where $\rho$ is the density of the liquid; $g$ is gravity acceleration; $h$ is the height of the liquid column.

The pressure $p$ is called *absolute pressure*, and the difference $p - p_A$ is called *gauge pressure*.

A closed-type manometer contains a vertical glass tube, sealed at the top and filled with a liquid. The lower end of the tube is immersed in a dish partially filled with the liquid. The pressure formed by a column of liquid in the tube is balanced by the atmospheric pressure (Figure 2.2).

Therefore $p_A = \rho g h$.

*Methods of Measuring Environmental Parameters*, First Edition. Yuriy Posudin.
© 2014 John Wiley & Sons, Inc. Published 2014 by John Wiley & Sons, Inc.

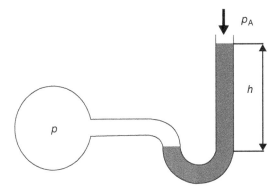

**FIGURE 2.1**    The open-tube manometer.

What about the liquids that can be used in manometers?

Evangelista Torricelli (1608–1647), a famous Italian physicist and mathematician, developed the first closed-type manometer in 1643. The choice of water for the liquid was based on its density: the density of water (4°C) is 1000 kg/m$^3$. Blaise Pascal (Rouen, France) repeated the experiment of Torricelli in 1646, but with wine,

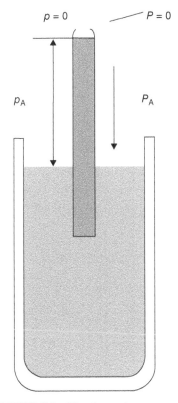

**FIGURE 2.2**    The close-tube manometer.

which was cheaper than water in France. The density of wine varies from 993.1 to 1116.4 kg/m$^3$ (density of dry wine is 10–13% less than the density of water). The height of tube in this experiment was 14 m.

Torricelli subsequently proposed using mercury as the liquid in manometers. The density of mercury is 13,600 kg/m$^3$. Consequently, pressure measured by mercury manometer is

$$p_A = \rho g h = (13.595 \times 10^3 \text{ kg/m}^3)(9.80665 \text{ m/s}^2)(0.76 \text{ m})$$
$$= 1.013 \times 10^5 \text{ Pa.}$$

(2.2)

A mercury manometer consists of a glass tube that is filled with mercury and immersed into a reservoir of mercury.

*The advantages of a mercury manometer:*

- It has a simple construction.
- This manometer provides a direct measurement; it does not require the calibration.
- Mercury manometer has high sensitivity and accuracy, which is 0.1 mbar; so, it can be used as a standard for calibrating aneroid barometers and altimeters.
- Mercury is denser than most other liquids and the column of mercury fell to about 760 mm (not 14 m as for water manometer).
- Mercury manometer has a much lower vapor pressure than that of water and does not evaporate easily; it is also easily available.
- At least it has a low purchase price.

*The disadvantages of a mercury manometer:*

- It is bulky and fragile; glass tube can be broken.
- Mercury devices must be perfectly installed vertically to the ground; the accuracy of the device depends on the vertical maintenance.
- Surface tension can provoke errors due to capillary rise and the shape of the meniscus.
- Its slow response makes it unsuitable for measuring fluctuating pressures.
- Mercury is a toxin that is dangerous for humans and wildlife. The European Union, however, has banned the use of mercury in manometers as part of a strategy to end the use of the highly toxic substance across the continent. This directive essentially stopped the production of new mercury barometers in Europe.

**Constructive Test**    Pascal repeated the experiment of Torricelli in 1646, using (like a true Frenchman) a red Bordeaux wine as the manometer's liquid. The density of the wine was 984 kg/m$^3$. What was the height of the wine column at normal atmospheric pressure?

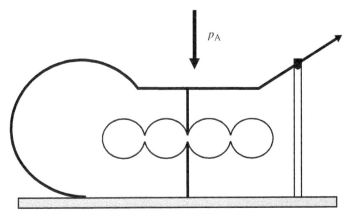

$p_A$

**FIGURE 2.3**   Aneroid barometer.

## 2.2   BAROMETERS

A *barometer* is an instrument which is used to measure changes in atmospheric pressure. There are two types of mechanical barometers: *aneroid* and *Bourdon tube*.

Aneroid barometer (Figure 2.3) consists of an aneroid capsule—a thin (0.2 mm), disk-shaped metallic box that is partially evacuated of gas, and is equipped with a spring. The deflection of the spring is proportional to the difference between the internal and external pressures. More sensitive aneroids contain up to 14 capsules.

Typically, atmospheric pressure is measured between 26.5 and 31.5 inHg.

The accuracy of an aneroid barometer depends on the range and varies from ±0.65 to ±1.00 mbar.

*The advantages of an aneroid barometer:*

- It is characterized by compactness, mechanical strength, and transportability.
- Aneroid capsules displacement can be easily converted into an electrical signal; these devices can be used for automated systems of pressure measurement.
- The measuring procedure is characterized with quick and easy handling of the instrument.

*The disadvantage of an aneroid barometer:*

- It has lower accuracy than mercury barometer due to temperature, hysteresis, and drift.

A *barograph* is a device for automatically recording changes in atmospheric pressure over a period of time (Figure 2.4).

*The advantages of a barograph:*

- It can be used as a tool for weather prediction. It is especially important at sea where it helps sailors and yachtsmen to plan their activity.

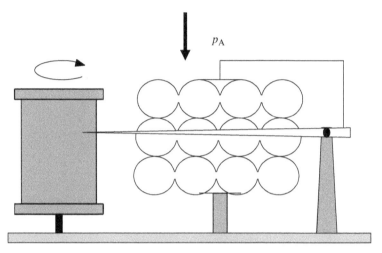

$p_A$

**FIGURE 2.4**   Barograph.

- The rate and history of change of barometric pressure can be an invaluable indicator of weather conditions.
- The time of action of barograph can last hours, days, or months.

*The disadvantage of a barograph:*

- The results of measurement do not provide the information about the root cause of pressure changes.

The second type, a Bourdon tube, (Figure 2.5) is equipped with a flattened curved tube that straightens under the influence of atmospheric pressure. The motion of the curved tube is transferred to a gear train and an indicating needle.

*The advantages of a Bourdon tube:*

- It is simple in construction and cheap.
- It can be available over a wide range of pressure.
- The device is characterized with high sensitivity, good repeatability, and good accuracy except at low pressures; accuracy is high especially at high pressure.

*The disadvantages of a Bourdon tube:*

- It is sensitive to shock and vibration.
- These devices respond slowly to changes in pressure and are subjected to hysteresis—a retardation of an effect when the forces acting upon a body are changed.
- The Bourdon tube is unsuitable for low pressure application.

Useful range: above $10^{-2}$ Torr (roughly 1 Pa); accuracy of the Bourdon tube is $\pm 2\%$ over the range.

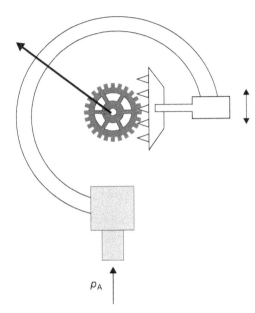

**FIGURE 2.5**   The Bourdon tube.

## 2.3   DIGITAL BAROMETRIC PRESSURE SENSOR

Recent technological advances have allowed producing a silicon capacitive absolute pressure sensor that consists of two silicon substrates fusion bonded together with a silicon dioxide ($SiO_2$) layer, forming a capacitor. The principle of operation is based on the change in capacitive of the pressure sensor in response to changes in atmospheric pressure. Pressure alters the distance between the silicon substrates that form the electrodes of the capacitor, resulting in a corresponding change in the capacitance (Figure 2.6).

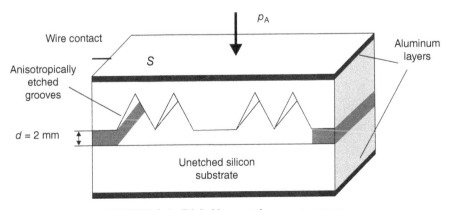

**FIGURE 2.6**   Digital barometric pressure sensor.

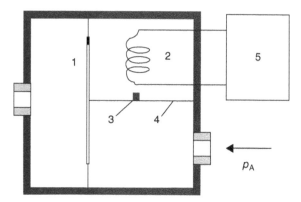

**FIGURE 2.7**    Vibrating wire sensor.

Digital barometric pressure sensors (model PTB210, Vaisala, Finland) have a pressure range between 500 and 1100 hPa, a temperature range of −40°C to +60°C, a total accuracy between ±0.15 and 0.35 hPa, a weight of 110 g, and a size of 122 mm (http://www.usitt.ecs.soton.ac.uk/capacitive.shtml).

## 2.4    VIBRATING WIRE SENSOR

It is known that the natural frequency of a stretched string depends on the string's tension force:

$$F = \frac{1}{2L}\sqrt{\frac{T}{\mu}}$$
(2.3)

where $F$ is the fundamental resonance frequency of the string (Hz); $L$ is the string length (m); $T$ is the string tension (N); $\mu$ is the unit mass of string (kg/m).

The vibration of the diaphragm (1) under the influence of varying pressure is converted into electromagnetic oscillations of the coil (2) during the movement of a magnet (3) connected to the wire (4) (Figure 2.7). The electromagnetic oscillations are recorded by the system of registration (5).

This device can be used for differential, absolute, or gauge measurements.

## 2.5    CAPACITIVE PRESSURE SENSOR

This type of sensor uses a thin diaphragm that is made of metal or quartz with a sputtered metal surface (sputtering yields a smoother, more uniform film deposit). This forms two condensers, the capacitance of which is changed in response to applied pressure. Both these condensers form the electrical bridge with the other two condensers, C1 and C2 (Figure 2.8).

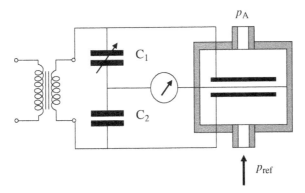

**FIGURE 2.8**   Capacitive pressure sensor.

It is known that the capacitance C of a parallel plate capacitor is determined as:

$$C = \frac{\varepsilon \varepsilon_0 A}{d}, \tag{2.4}$$

where $\varepsilon$ is the dielectric permittivity of the medium between the plates of a capacitor; $\varepsilon_0$ is the dielectric constant ($8854 \times 10^{-12}$ $C^2/N \cdot m^2$); A is the area of the plates; and d the distance between them.

Changes in capacitance (which can reach a few percent of the original capacity) lead to a frequency change in the output signal, the scale of which is calibrated in units of pressure.

*The advantages of a capacitive pressure sensor:*

- This sensor is robust, it has no moving parts.
- Cost of the sensor is low.
- Small mechanical dimensions.
- Has very high sensitivity.
- Operating temperatures up to 250°C (480°F).

*The disadvantage of a capacitive pressure sensor:*

- It needs complex electronics.

**Constructive Tests**

1. Compare the accuracy of the abovementioned devices.
2. What types of the abovementioned devices can be used for automated measurement of atmospheric pressure? Explain your opinion.

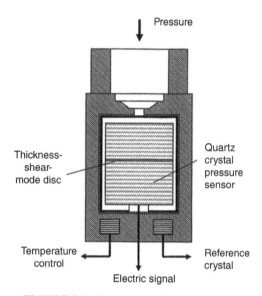

**FIGURE 2.9**    Quartz crystal pressure sensor.

## 2.6    MEASUREMENT OF PRESSURE AT DEPTH

Measuring the pressure at sea depths makes it possible to approach the understanding of the processes that take place as a result of ocean currents, underwater oscillatory phenomena, and earthquakes and tsunami.

Quartzdyne's Pressure Sensor contains a quartz resonator which can change its frequency in response to applied pressure (Figure 2.9).

A thickness-shear-mode disc resonator divides the central portion of the hollow cylinder. This pressure produces internal compressive stress in the resonator and corresponding changes of the vibrating frequency of the sensor. The principle of operation of the sensor is based on the piezoelectric effect.

*Piezoelectricity* is the ability of some materials (notably crystals and certain ceramics) to generate an electric field or electric potential in response to applied mechanical stress. Quartz is a typical piezoelectric crystal.

**Example**    Crystal quartz has a thickness of 0.25 cm. Determine the voltage that occurs on the surface of the crystal as a result of applying pressure 345 N/m², where the sensitivity of the crystal is 0.055 V m/N.

*Solution*    The voltage measured on the surfaces of piezoelectric crystal is calculated as follows:

$$U = vdp, \tag{2.5}$$

where $v$ is the sensitivity of the crystal; $d$ is the thickness of the crystal.

Using Equation 2.5, we obtain $U = vdp = (0.055 \text{ V·m/N})·(0.0025 \text{ m})·(345 \text{ N/m}^2)$ = 0.047 V.

**Control Exercise**    Determine the piezoelectric constant of quartz if the crystal of area of 1 $cm^2$ creates the charge $7.7625 \times 10^{-14}$ C under the pressure 345 $N/m^2$. *Answer:* $2.25 \times 10^{-12}$ C/N.

*The advantages of a quartz pressure sensor:*

- This device is characterized by long operating life in battery-powered applications.
- It provides stability, shock resistance, and high resolution with low power consumption.

## QUESTIONS AND PROBLEMS

1. Explain why the atmospheric pressure decreases with increasing the height above the Earth's surface.

2. Explain the barometric formula.

3. What is the isobar?

4. Write the formula of absolute pressure versus the depth below the surface of a liquid open to the atmosphere.

5. What is the difference between barometer and barograph?

6. Explain the principle of operation and advantages of digital barometric pressure sensor.

## FURTHER READING

Benedict, R.P. 1984. *Fundamentals of Temperature, Pressure, and Flow Measurements*. John Wiley & Sons, New York.

Meteorological Office. 1981. *Handbook of Meteorological InstrumentI*, Volume 1 Measurement of Atmospheric Pressure, 2nd edition. Stationery Office Books.

## ELECTRONIC REFERENCES

Capacitive pressure sensor, Williams C.D.H. Introduction to Sensors, http://newton.ex.ac.uk/teaching/CDHW/Sensors/ (accessed February 5, 2013).

Digital Barometric Pressure Sensor, http://www.usitt.ecs.soton.ac.uk/capacitive.shtml (accessed February 8, 2013).

European Union bans sale of toxic mercury thermometers, http://www.cbc.ca/news/health/story/2007/07/11/mercury-ban.html (accessed February 10, 2013).

Manometers Advantages and Limitations—M. Subramanian, http://www.msubbu.in/ln/fm/Unit-I/ManometersLimit.htm (accessed February 14, 2013).

Micromachined Silicon Capacitive Sensors, http://www.usitt.ecs.soton.ac.uk/capacitive.shtml (accessed February 15, 2013).

Nurbey, G. 2011-03-27. *What textbooks concealed*—Amazing Physics, http://www.modernlib.ru/books/nurbey_gulia/udivitelnaya (accessed February 15, 2013).

Posudin, Y. 2011. *Environmental Biophysics*. Fukuoka-Kiev, http://www.ekmair.ukma.kiev.ua/handle/123456789/951 (accessed February 19, 2013).

Quartzdyne's Pressure Sensor, http://www.quartzdyne.com/quartz.php (accessed February 21, 2013).

Vaisala BAROCAP® Digital Barometer PTB210, http://www.vaisala.com/en/products/pressure/Pages/default.aspx (accessed February 25, 2013).

Vibrating Wire Sensor, http://instrumenttoolbox.blogspot.com/2011/02/vibrating-wire-sensors.html (accessed February 26, 2013).

Williams C.D.H. Introduction to Sensors, http://newton.ex.ac.uk/teaching/CDHW/Sensors/ (accessed February 28, 2013).

# PRACTICAL EXERCISE **1**

# ANALYSIS OF OBSERVED DATA: THEORY OF ERRORS

## 1  APPROXIMATION OF DATA

### 1.1  Rules for Dealing with Significant Numbers

If a student measures the area ($S$) of a square with a ruler and one edge ($a$) of the square is $205 \pm 1$ mm, only the first two digits (i.e., 20) are *significant*. The final digit is *doubtful*—it can be either 6 or 4. The calculation of the area of the square gives:

$$S = a^2 = 205^2 = 42025 \ \text{mm}^2$$

The actual area is between $206^2 = 42436$ mm$^2$ and $204^2 = 41616$ mm$^2$. Here only the first digit (4) is significant and the other digits are doubtful. This example demonstrates that the *precision of the final result depends only on the precision of measurements; it is not possible to increase the precision of the answer by increasing the precision of calculations.* In such a way, the precision of a result is determined by the number of significant figures used to express the number. The principal rules for dealing with significant numbers are as follows (Wood, 1992):

1. *The most significant number* is always the left-most nonzero digit, regardless of where the decimal point is found.

| e.g., | 2̲53 | 0.02̲15 |
|---|---|---|

---

*Methods of Measuring Environmental Parameters*, First Edition. Yuriy Posudin.
© 2014 John Wiley & Sons, Inc. Published 2014 by John Wiley & Sons, Inc.

2. *The least significant number* is the right-most nonzero digit if there is no decimal point.

| e.g., | 45<u>3</u>7 | 5<u>3</u>00 |
|---|---|---|

3. *The least significant number* is the right-most digit, whether zero or not, if there is a decimal point.

| e.g., | 21.3<u>4</u> | 1.34<u>0</u> |
|---|---|---|

*All digits between the least and most significant digits are counted as significant numbers.*

e.g., *Three significant numbers*: 245; 24,500; 24.5; 2.45; 0.245; 0.0245; 0.00245.

*Four significant numbers*: 11.35; 5608; 0.05638; 2.590; 8.342 × 10⁴.

### 1.2  The Precision of the Measurement During Multiplication or Division

Rule: *The result of multiplication or division should have as many significant numbers as the least precise of its factors.*

| e.g., | 13.56 | 2.1315 |
|---|---|---|
| | ×4.56 | ×0.0114 |
| | 61.8336 ≈ 61.8 | 0.029841 ≈ 0.0298 |

### 1.3  The Precision of the Measurement During Addition or Subtraction

Rule: *The result of addition or subtraction should not have significant numbers in the least digit orders which are absent in at least in one of the summands.*

| e.g., | 16.28 | 5.32 |
|---|---|---|
| | +0.514 | −1.2 |
| | +42.6 | 4.12 ≈ 4.1 |
| | 59.394 ≈ 59.4 | |

### 1.4  The Precision of the Measurement During Raising to a Power or Extracting a Root

Rule: The result of raising to a power or extracting a root should have as many significant numbers as in the initial digit which is measured.

| e.g., | $3.2^3 = 10.24 \approx 10$ | $\sqrt{25} = 5.0$ |
|---|---|---|

## 2  THEORY OF ERRORS

### 2.1  Types of Errors

If the measurements are physically derived using measuring tools, they are referred to as *direct*; if derived using a formula, they are *indirect*. For example, determining the length of an object with a metric ruler is a direct measurement, while determining the moment of inertia using the formula $I = mr^2$ is an indirect measurement.

The measured value for a quantity always differs from the true value for the quantity. The reasons for the difference are *instrumental measurement errors* (due to the imperfection of the measuring instrument) and *personal errors* (due to measuring errors by the individual). Another method for classifying errors is based on their properties. An error is *systematic* if it is constant over several measurements or *random* if it changes during the measurements. The following definitions more precisely describe these terms (Posudin, 2004):

> *Error*—the difference between the determined value of a physical quantity and the true value.

The foregoing classification of measurement errors is based on the cause of the errors.

*Systematic errors* are caused by the imperfection of measuring methods and inaccuracy of instruments. These errors remain constant or change in a regular fashion in repeated measurements of one and the same quantity.

*Random errors* appear due to human factors and other accidental causes that vary randomly in repeated measurements of the same value; they include errors owing to incorrect reading of the tenth graduation of an instrument scale, small changes of the measurement conditions, asymmetric placement of the indicator mark, etc. These errors are changing in an irregular fashion in repeated measurements of one and the same quantity.

## 2.2  Errors in Direct Measurements

Let $x_1, x_2, x_3, \ldots x_n$ denote the raw data derived from experimental observations. *The arithmetic mean* ($\langle x \rangle$) is the sum of observations divided by the number ($n$) of observations:

$$\langle x \rangle = \frac{x_1 + x_2 + x_3 + \ldots + x_n}{n} = \frac{\sum\limits_{i=1}^{n} x_i}{n} \tag{P1.1}$$

The difference between a data point and the mean is referred to as a *deviation* ($\delta_i$):

$$\delta_i = x_i - \langle x \rangle \tag{P1.2}$$

*Dispersion* or *variance* ($\sigma^2$) is defined as the sum of the squared deviations divided by $n - 1$:

$$\sigma^2 = \frac{\sum\limits_{i=1}^{n} \delta_i^2}{n - 1} \tag{P1.3}$$

*Sample standard deviation* ($\sigma$) is defined as the square root of the dispersion by the following formula:

$$\sigma = \sqrt{\frac{\sum\limits_{i=1}^{n} \delta_i^2}{n-1}} \tag{P1.4}$$

*Confidence interval* is the difference between the largest and smallest observations in a sample.

*Confidence interval of systematic error* ($\Delta_c$) is the smallest division of the measuring instrument.

*Confidence interval of random error* ($\overset{0}{\Delta}$) is defined by the following formula:

$$\overset{0}{\Delta} = t\sqrt{\frac{\sum\limits_{i=1}^{n} \delta_i^2}{n(n-1)}} \tag{P1.5}$$

The Student's coefficient ($t$) is a criterion of reliability of results and is determined from Table P1.1 using the number ($n$) of observations and confidence probability ($P$) desired.

*Confidence interval of total error* $\Delta$ is

$$\Delta = \overset{0}{\Delta} + \Delta_c \tag{P1.6}$$

**TABLE P1.1    The Student's coefficient**

| N | P = 0.90 | P = 0.95 | P = 0.98 | P = 0.99 | P = 0.999 |
|---|----------|----------|----------|----------|-----------|
| 3 | 2.92 | 4.30 | 6.97 | 9.93 | 31.60 |
| 4 | 2.35 | 3.18 | 4.54 | 5.84 | 12.94 |
| 5 | 2.13 | 2.78 | 3.75 | 4.60 | 8.61 |
| 6 | 2.02 | 2.57 | 3.37 | 4.03 | 6.86 |
| 7 | 1.94 | 2.45 | 3.14 | 3.71 | 5.96 |
| 8 | 1.90 | 2.37 | 3.00 | 3.50 | 5.41 |
| 9 | 1.86 | 2.31 | 2.90 | 3.36 | 5.04 |
| 10 | 1.83 | 2.26 | 2.82 | 3.25 | 4.78 |
| 11 | 1.81 | 2.23 | 2.76 | 3.17 | 4.59 |
| 12 | 1.80 | 2.20 | 2.72 | 3.11 | 4.44 |
| 13 | 1.78 | 2.18 | 2.68 | 3.06 | 4.32 |
| 14 | 1.77 | 2.16 | 2.65 | 3.01 | 4.22 |
| 15 | 1.76 | 2.15 | 2.62 | 2.98 | 4.14 |
| 20 | 1.73 | 2.09 | 2.54 | 2.86 | 3.88 |
| $\infty$ | 1.64 | 1.96 | 2.33 | 2.58 | 3.29 |

*Relative error* ($\varepsilon$) is the confidence interval of the total error as a percentage of the mean:

$$\varepsilon = \frac{\Delta}{\langle x \rangle} 100\% \qquad (P1.7)$$

## 2.3   Errors in Indirect Measurements

Suppose the volume (V) of a cylinder has been determined by the following formula:

$$V = \pi \frac{D^2}{4} h \qquad (P1.8)$$

where $D$ is the diameter and $h$ the height of the cylinder.

The logarithm of the previous formula (2.8) is

$$\ln V = \ln \pi + 2 \ln D - \ln 4 + \ln h \qquad (P1.9)$$

The differentiation of expression (2.9) leads to the following:

$$d \ln V = d \ln \pi + 2d \ln D - d \ln 4 + d \ln h \qquad (P1.10)$$

Take into account that $d \ln x = \frac{dx}{x}$:

$$\frac{dV}{V} + \frac{d\pi}{\pi} + 2\frac{dD}{D} - \frac{d4}{4} + \frac{dh}{h} \qquad (P1.11)$$

The differentials are substituted by the confidence intervals with a "+" sign and the symbols of the values by their means (where $\frac{d4}{4} = 0$):

$$\frac{\Delta V}{V} = \frac{\Delta \pi}{\pi} + 2\frac{\Delta D}{D} + \frac{\Delta h}{h} \qquad (P1.12)$$

or

$$\varepsilon_V = \varepsilon_\pi + 2\varepsilon_D + \varepsilon_h \qquad (P1.13)$$

The error of a tabular value is determined as one half of the last significant summand. For example, error ($\Delta \pi$) of tabular value $\pi = 3.14$ is $\Delta \pi = 0.005$; while the error of tabular value $\pi = 3.141$ is $\Delta \pi = 0.0005$, etc.

Errors of direct measurements $\Delta D$ and $\Delta h$ are determined according to Section 2.2.

The confidence interval of the total error ($\Delta$) is

$$\Delta_V = \varepsilon_V \cdot \langle V \rangle \qquad (P1.14)$$

and the arithmetic mean ($\langle V \rangle$) can be determined as follows:

$$\langle V \rangle = \pi \frac{\langle D \rangle^2}{4} \langle h \rangle \qquad\qquad (P1.15)$$

**Exercise**    How many significant numbers are there for each of the following?

| | | | | | |
|---|---|---|---|---|---|
| 0.045; | 24.25; | 6583; | 2.0; | 0.00087; | 9000 |

**Exercise**    Multiply the following and indicate the significant numbers.

| | | |
|---|---|---|
| 5.3456 | 254.7 | 6.43 |
| ×0.0134 | ×6.43 | ×0.78 |
| ? | ? | ? |

**Exercise**    Add the following and indicate the significant numbers.

| | | |
|---|---|---|
| 234.5 | 17.456 | 84.234 |
| +34.794 | +435.7 | +3.17 |
| +65.34 | +15.05 | ? |
| +5.13 | ? | |
| ? | | |

**Exercise**    Subtract the following and indicate the significant numbers.

| | | |
|---|---|---|
| 93.173 | 76.1643 | 486.3 |
| −5.14 | −5.032 | −6.2349 |
| ? | ? | ? |

**Exercise**    Calculate the following and indicate the significant numbers.

| | | |
|---|---|---|
| $\pi(3.74)^2 = ?$ | $(6.213 \times 10^{-4})^2 = ?$ | $\sqrt{2.567} = ?$ |

**Example**    The results of measuring wind speed velocity by anemometer (accuracy is 0.5 m/s) are presented in the following Table.

Wind speed in m/s

| № | 1 | 2 | 3 | 4 | 5 | 6 | 7 | 8 | 9 | 10 |
|---|---|---|---|---|---|---|---|---|---|---|
| $v$, м/c | 22.5 | 2.0 | 1.5 | 1.0 | 2.5 | 3.5 | 1.5 | 1.0 | 0.5 | 0.5 |

Find relative error of measurement of wind speed by anemometer.

**Solution**    The arithmetic mean $\langle x \rangle$ is

$$\langle x \rangle = \frac{2.5 + 2.0 + 1.5 + 1.0 + 2.5 + 3.5 + 1.5 + 1.0 + 0.5 + 0.5}{10} = 1.7 \, \text{m/s}.$$

The values of deviation $\delta_i$ are

$$\begin{aligned}
\delta_1 &= |2.5 - 1.7| = 0.8 \, \text{m/s}; & \delta_6 &= |3.5 - 1.7| = 1.8 \, \text{m/s}; \\
\delta_2 &= |2.0 - 1.7| = 0.3 \, \text{m/s}; & \delta_7 &= |1.5 - 1.7| = 0.2 \, \text{m/s}; \\
\delta_3 &= |1.5 - 1.7| = 0.2 \, \text{m/s}; & \delta_8 &= |1.0 - 1.7| = 0.7 \, \text{m/s}; \\
\delta_4 &= |1.0 - 1.7| = 0.7 \, \text{m/s}; & \delta_9 &= |0.5 - 1.7| = 1.2 \, \text{m/s}; \\
\delta_5 &= |2.5 - 1.7| = 0.8 \, \text{m/s}; & \delta_{10} &= |0.5 - 1.7| = 1.2 \, \text{m/s}.
\end{aligned}$$

*Dispersion* or *variance* $\sigma^2$ is calculated as

$$\sigma^2 = \frac{0.8^2 + 0.3^2 + 0.2^2 + 0.7^2 + 0.8^2 + 1.8^2 + 0.2^2 + 0.7^2 + 1.2^2 + 1.2^2}{10} = 8.6 \, \text{m/s}.$$

Confidence interval of random error $\overset{0}{\Delta}$ is determined by the formula (2.5):

$$\overset{0}{\Delta} = t \sqrt{\frac{\sum\limits_{i=1}^{n} \delta_i^2}{n(n-1)}} = 2.26 \sqrt{\frac{8.6}{10 \cdot 9}} = 0.7 \, \text{m/s}.$$

Here we have determined the Student's coefficient $t = 2.26$ from the Table P1.1 for $n = 10$ and $P = 0.95$.

Confidence interval of systematic error $\Delta_c$ for our anemometer is $\Delta_c = 0.5$ m/s.
Confidence interval of total error $\Delta$ is

$$\Delta = 0.7 + 0.5 = 1.2 \, \text{m/s}.$$

Relative error $\varepsilon$ is

$$\varepsilon = \frac{1.2}{1.7} = 0.7 = 70\%.$$

**Control Exercise**    The systolic blood pressure $p$ (mmHg) of three students was measured 10 times giving the following data:

| Variant | $p$ (mmHg) | | |
|---|---|---|---|
| 1 | 73 | 84 | 91 |
| 2 | 72 | 66 | 80 |
| 3 | 87 | 88 | 90 |
| 4 | 94 | 104 | 117 |
| 5 | 100 | 99 | 95 |
| 6 | 78 | 64 | 69 |
| 7 | 85 | 95 | 112 |
| 8 | 87 | 92 | 101 |
| 9 | 101 | 114 | 107 |
| 10 | 95 | 81 | 84 |

Elaborate the results of measurement of each student. Determine the following values: arithmetic mean; deviation; dispersion; sample standard deviation; confidence intervals of random, systematic, and total errors; and relative error for the blood pressure measurements (confidence interval of systematic error $\Delta_c = 1$; confidence probability $P = 0.95$).

Use the rules of data approximation.

**Control Exercise**    A student measured the stem elongation rate ($l$ = mm/day) of three plants for 10 days grown under the same conditions and obtained the following data:

| Variant | $l$ (mm/day) | | |
|---|---|---|---|
| 1 | 64 | 81 | 68 |
| 2 | 69 | 73 | 78 |
| 3 | 64 | 75 | 70 |
| 4 | 66 | 71 | 54 |
| 5 | 56 | 78 | 68 |
| 6 | 92 | 79 | 74 |
| 7 | 75 | 68 | 73 |
| 8 | 74 | 77 | 69 |
| 9 | 69 | 62 | 79 |
| 10 | 61 | 68 | 70 |

Elaborate the results of measurement of each student. Determine the following values: arithmetic mean; deviation; dispersion; sample standard deviation; confidence intervals of random, systematic, and total errors; and relative error of the stem elongation measurements (confidence interval of systematic error $\Delta_c = 1$; confidence probability $P = 0.95$).

Use the rules of data approximation.

## REFERENCES

Posudin, Y.I. 2004. *Physics with Fundamentals of Biophysics*. Agrarna Nauka, Kiev, Ukraine.

Wood, R.M. 1992. *Experiments for an Introductory Physics Course*, 2nd edition. Department of Physics and Astronomy, University of Georgia, CRC, Athens, GA.

## ELECTRONIC REFERENCE

JCGM 200. 2012. International vocabulary of metrology—Basic and general concepts and associated terms (VIM) 3rd edition 2008 (version with minor corrections) http://www.bipm.org/utils/common/documents/jcgm/JCGM_200_2012.pdf (accessed January 7, 2013).

# 3

# WIND

## 3.1 DEFINITION OF WIND

*Wind* is the motion of air relative to the ground. It is a vector quantity that is characterized by both numerical and directional properties.

## 3.2 FORCES THAT CREATE WIND

Wind is induced by a series of forces (Henry and Heike, 1996):

1. A *pressure gradient force* that arises from the difference in pressure across a surface which induces air motion from areas of high pressure to areas of low pressure.
2. *Gravity* induces the acceleration of air downward at a rate of 9.8 m/s$^2$. This force is always directed perpendicular to the Earth's surface but does not participate in the formation of horizontal winds.
3. *Friction*, the resistance that the air particles encounter during contact with surface of the ground, can be defined as

$$F = -\mu v, \tag{3.1}$$

where $\mu$ is the coefficient of friction which depends on the type of surface; $v$ is the wind velocity. Force of friction is proportional to the velocity of wind and has an opposite direction.

*Methods of Measuring Environmental Parameters*, First Edition. Yuriy Posudin.
© 2014 John Wiley & Sons, Inc. Published 2014 by John Wiley & Sons, Inc.

4. *Coriolis force* is induced by the rotation of the earth; it deflects the wind to the right or to the left depending on the hemisphere. Coriolis force depends on the wind speed and is expressed by

$$F_k = 2\rho v \omega \sin \varphi, \tag{3.2}$$

where $\rho$ is the density of air; $v$ is the wind velocity; $\omega$ is the rate of rotation of the earth (i.e., $7.3 \times 10^{-5}$ rad/s); $\varphi$ is the latitude.

5. *Centripetal force* is a force acting on a body in curvilinear motion that is directed toward the center of curvature or axis of rotation. It causes a change in the direction of the wind but not a change in the velocity. Centripetal force is calculated as

$$F_B = \frac{mv^2}{r}, \tag{3.3}$$

where $r$ is the radius of curvature of the trajectory.

## 3.3  PARAMETERS OF WIND

Winds are characterized by their speed, uniformity of speed (gustiness), and direction.

*Wind speed* is measured in meters per second (m/s), kilometers per hour (km/h, kph), miles per hour (m/h, mph), and sometimes in knots (nautical miles per hour = 1 international knot = 1.852 kilometers per hour).

Wind speed unit conversion can be presented as follows:

1 knot = 1.2 mph = 0.5 m/s = 1.7 ft/s = 1.8 km/h;
1 mph = 0.9 knots = 0.4 m/s = 1.5 ft/s = 1.6 km/h;
1 m/s = 2.2 mph = 1.9 knots = 3.3 ft/s = 3.6 km/h;
1 ft/s = 0.7 mph = 0.6 knots = 0.3 m/s = 1.1 km/h;
1 km/h = 0.6 mph = 0.5 knots = 0.3 m/s = 0.9 ft/s.

**Constructive Test**    The knot is a unit of speed equal to one nautical mile (1.852 km) per hour. Explain the origin (etymology) of this term. Why is this unit so called?

Intense winds often have special names, such as *gales, hurricanes, tornados,* and *typhoons*.

A *gale* is a very strong wind that has a speed from 51 to 102 km/h (32–63 mph) or a force of 7–10 on the Beaufort scale.

A *hurricane* is a tropical storm with wind speed of 119–257 km/h (74–160 mph) that usually originates in the equatorial regions of the Atlantic Ocean or eastern regions of the Pacific Ocean and is accompanied by lightening, thunder, rain, and flooding. These speed values correspond to a force 12 on the Beaufort scale.

A *tornado* is a violently rotating column of air ranging in width from several meters to more than a kilometer that is characterized by destructively high speeds (up to 177 km/h or 110 mph) and emerges from cumulonimbus thunderstorm clouds. Tornados can travel on the ground from a very short distance to up to a number of kilometers before dissipating. They occur most frequently in the United States.

A *typhoon* is a violent tropical cyclone occurring in the Western Pacific or Indian oceans.

Wind speed is estimated by the *Beaufort wind force scale*. It was devised in 1805 by Sir Francis Beaufort and is based on an estimation of wind strength without instruments but on the visual observation of the environmental effects that were produced by the wind. This scale contains the 12 Beaufort categories.

For example, category 0 corresponds to a calm situation, when the smoke rises vertically; category 4 characterizes moderate breeze, when the small branches begin to move; category 8 means fresh gale, when small branches are broken from trees and cars are veered on road; category 12 is estimated as hurricane force which provokes extreme destruction, such as damage of mobile homes and poorly constructed sheds and barns.

The *Saffir–Simpson Hurricane Scale* is used for estimating the intensity of a hurricane. The scale is based on 5-level rating of the potential property damage and flooding: 1 (minimal) 119–153 km/h, the destruction of trees, mobile homes, partial flooding of coastal areas; 2 (moderate) 154–177 km/h, significant damage to vegetation, tearing trees, flooding coastal roads; 3 (extensive) 178–279 km/h, the destruction of small houses, large trees blown down; 4 (extreme) 210–249 km/h, the destruction of roofs, windows, complete destruction of mobile homes, flooding up to 10 km; 5 (catastrophic) >249 km/h, the destruction of buildings, industrial plants, the need for evacuation in the area of 8–16 km.

The *Fujita Tornado Scale* (*F*-Scale) is a scale for rating tornado intensity, based primarily on an estimation of the damage to vegetation, buildings, and other man-made structures. This scale uses the following categories: $F_0$ (gale); $F_1$ (moderate); $F_2$ (significant); $F_3$ (severe); $F_4$ (devastating); $F_5$ (incredible); $F_6$ (inconceivable).

**Constructive Test**   L. Frank Baum wrote in 1901 a children's novel *The Wonderful Wizard of Oz* where he described a Kansas farm girl Dorothy Gale who lived with her aunt Em and uncle Henry in their Kansas farm home.

Dorothy and her pet dog Toto stayed at home alone when the tornado broke. Dorothy discovered that she, her dog, and the house have been swept away and were soaring through the core of the tornado.

Find in the internet a Fujita Tornado Scale and estimate the category and corresponding wind speed of tornado, which hit Dorothy and Toto.

*Wind direction* is the direction from which the wind is blowing. It is usually reported in cardinal directions or in azimuth degrees.

A *gust of wind* is a strong, abrupt rush of wind. The change of wind speed is $\Delta v = \pm 3$ m/s at $v = 5$ to 10 m/s; $\Delta v = \pm 5$ to 7 m/s at $v = 11$ to 15 m/s.

### *Extreme situations*

*A wind speed of 318 mph (509 km/h) inside a tornado was recorded in Oklahoma City on May 3, 1999—the fastest wind speed ever recorded on Earth.*

*The fastest non-tornado wind ever recorded was 231 mph (370 km/h), measured at Mount Washington in New Hampshire in 1934.*

## 3.4   EFFECT OF WIND ON LIVING ORGANISMS

*Insects Response to Wind.* The ability of flying insects to demonstrate the responsive movement toward or away from such an external stimulus as wind is called *anemotaxis*.

Flying insects attempt to fly directly along wind direction when they are in contact with a pheromone plume and to fly back and forth perpendicular to the wind when this contact is lost. *Pheromones* (from Greek *pherein*—to carry or transfer, and *hormōn*—to excite or stimulate) are the molecules that are transported by wind and used for communication between animals. In addition, flying insects can control their position by observing the ground surface below; if the ground is moving directly underneath, the insect does not change orientation; if the ground moves to the left, the insect also turns to the left to keep direct relation with wind orientation.

The beetles can provide a spatial orientation due to the *Johnston organ*—a number of sensory cells on the antennae.

Honeybee evaluates the wind speed and direction by the sensitive long hairs that are located on the head and wings and act as aerodynamic sensors of air-current direction or flight speed relative to air.

The effect of wind on the sensor hairs of locust is accompanied with the increasing frequency of generation of electric pulses from 50–70 Hz to 245 Hz.

A fly has receptors on the antennae, which are sensitive to wind and help the insect to change its position relative to the wind direction and speed.

*Plant Response to Wind.* In addition to the stimulation of nastic movements in plants, the wind is involved in heat and mass transfer, changes the boundary layer resistance, and the rate of evaporation. Wind can also induce significant asymmetry in plant architecture by either direct damage (breaking of stems or foliage) or indirect damage via materials such as salt and sand transported by it. A very important agricultural problem is lodging of cereals which can cause a decrease in harvestable yield due to poor light penetration to the canopy, damaged conducting system, and weakening of photosynthetic activity of the plant. Also noteworthy is that wind mediates the deposition and dispersion of soil, plant pollen, seeds, spores, and droplets of agrochemical substances. It is necessary to mention that wind plays an important role in the formation of soil erosion.

## REFERENCE

Henry, J.G. and Heike, G.W. 1996. *Environmental Science and Engineering*. Prentice-Hall, Inc., New Jersey.

# 4

# MEASUREMENT OF WIND PARAMETERS

*Anemometer* (from the Greek *anemos*—wind) is a device for measuring wind speed either directly, through estimation of the rotation of cups or a windmill, or via measuring the propagation speed of ultrasound or light signals.

## 4.1 CUP ANEMOMETER

The *cup anemometer* consists of three or four semispherical cups which are mounted one on each end of horizontal arms, which lie at equal angles to each other (Figure 4.1). The axis of rotation of the cups is vertical and the velocity of rotation of the cups is proportional to the wind speed. The number of turns of the cups during a certain time period makes it possible to determine the average wind speed.

For example, Model 014A Met One Wind Speed Sensor (Campbell Scientific, Inc., USA.) is characterized by the following parameters: threshold 0.45 m/s (1 mph); calibrated range 0–45 m/s (0–100 mph); accuracy 1.5% or 0.11 m/s (0.25 mph); temperature range from 50°C to −70°C.

If $C_1$ and $C_2$ are clutch coefficients of concave and convex surfaces of anemometer cups with air, respectively, the forces acting on the opposing cups are defined by the expression:

$$F_y = \frac{1}{2}C_1\rho S(v - v_t)^2, \tag{4.1}$$

$$F_o = \frac{1}{2}C_2\rho S(v + v_t)^2, \tag{4.2}$$

*Methods of Measuring Environmental Parameters*, First Edition. Yuriy Posudin.
© 2014 John Wiley & Sons, Inc. Published 2014 by John Wiley & Sons, Inc.

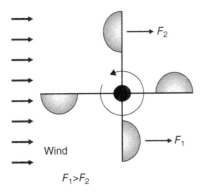

Wind

$F_1 > F_2$

**FIGURE 4.1**   Cup anemometer.

where $v$ is the wind speed; $v_t$ is the tangential speed of the cup that rotates; $\rho$ is the air density; $A$ is the cross-sectional area of the cup.

The rotating speed of the system that is in equilibrium with air flow provides equality of the forces $F_y$ and $F_o$:

$$C_1(v - v_t)^2 = C_2(v + v_t)^2. \tag{4.3}$$

The solution of this equation for $v_t$ leads to the following expression:

$$v_t = v\left(\frac{1 - \sqrt{C_1 C_2}}{C_1 - C_2}\right). \tag{4.4}$$

*The advantages of a cup anemometer:*

- This device is characterized by a linear dependence of $v_t$ on the wind speed $v$, if to take into account that clutch coefficients are constant (this condition is satisfied for cup anemometer).
- Threshold sensitivity of cup anemometer varies from a range of 90 mm/s to 2.24 m/s.
- Anemometers of this type are portable, simple, and sensitive.
- The cup anemometers have quite low cost.

*The disadvantages of a cup anemometer:*

- Moving parts of such an anemometer wear out.
- Slow to react to gusts.
- Not sensitive to wind speeds of fractions of a meter per second.

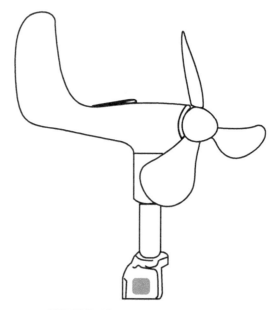

**FIGURE 4.2**   Windmill anemometer.

## 4.2   WINDMILL ANEMOMETER

A *windmill anemometer (aerovane)* combines a three- or four-blade propeller and a tail that orients the propeller toward the direction of the wind (Figure 4.2). Such an anemometer can be combined with a rotating vane to estimate wind speed and direction simultaneously.

The relation of rotation velocity of blade $v_t$ and wind speed $v$ is described as:

$$U_t = \frac{v}{ktg\theta},\tag{4.5}$$

where $\theta$ is the propeller blade angle relative to the axis of rotation; $k$ is the coefficient that depends on the design of the propeller ($k \cong 1$). If we choose the angle of the blade $\theta = 45°$, then $tg\theta = 1$ and the tangential velocity of the blade is approximately equal to the speed of the wind. Maximum sensitivity of the propeller anemometer is 1.1 m/s. These instruments are used for measuring wind speeds up to 90 m/s.

*The advantages of a propeller anemometer:*

- This anemometer is simple and compact; it has small weight (the propeller can be made of plastic).
- It is characterized by three times greater speed than the cup anemometer and the capability of measuring weak air currents and the turbulence.
- It is more responsive to gusts than the cup anemometer.
- The propeller anemometer has a relatively low cost.

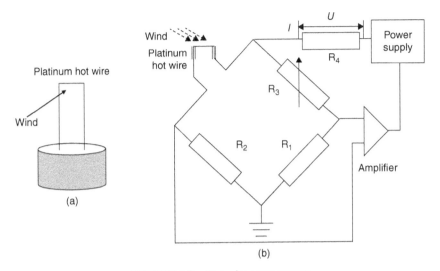

**FIGURE 4.3**    Hot-wire anemometer.

*The disadvantages of a propeller anemometer:*

- Moving parts of such an anemometer wear out.
- It must be oriented into the wind.

## 4.3   HOT-WIRE ANEMOMETER

A *hot-wire anemometer* consists of a thin tungsten or platinum wire (about several micrometers) that is heated by passing an electric current through it to a temperature above its surroundings (Figure 4.3a). Wind flowing across cools the wire. The resistance of the wire is dependent upon its temperature, making it possible to estimate the wind speed. Hot-wire anemometers characteristically are highly sensitive.

L.V. King described in 1914 the behavior of a long cylinder located in free flow; the heat losses $Q$ from such a cylinder can be described by the following equation (King, 1914):

$$Q = (a + b\,U^{1/2})\Delta T^{n}, \tag{4.6}$$

where $a$ and $b$ are constants; $U$ is the velocity of the flow; $\Delta T = T - T_{a}$ is the temperature difference; and $T$ and $T_{a}$ the temperatures of the wire and air flow, respectively; $n$ is the coefficient close to 1.

The heat $Q$ dissipated in a resistor can be expressed as:

$$Q = I^{2}R, \tag{4.7}$$

where $R$ is the resistance of a conductor.

Combining Equations 4.6 and 4.7, we get:

$$I^2 R = (a + bU^{1/2})(T - T_a)^n. \tag{4.8}$$

Using the expression:

$$R_2 = R_1 [1 + \alpha(T_2 - T_1) + \beta(T_2 - T_1)^2 + ...], \tag{4.9}$$

where $\alpha$ and $\beta$ are the temperature coefficients of resistivity and neglecting the members of the second order, we obtain:

$$R_2 I^2 \cong (a' + b' U^{1/2})(R_2 - R_1), \tag{4.10}$$

where $a'$ and $b'$ are new constants.

Typical scheme of hot-wire anemometer is shown in Figure 4.3b.

*The advantages of a hot-wire anemometer:*

- It has extremely high frequency response and high spatial resolution (measures the flow in a precise location).
- It can be used for the detailed study of turbulent flows with rapid velocity fluctuations.

*The disadvantage of a hot-wire anemometer:*

- Costly, orientation sensitive, fragile, and "wire" can accumulate debris in a dirty flow.

## 4.4   SONIC ANEMOMETER

*Ultrasonic waves* are longitudinal waves with frequencies over 20,000 Hz, which is about the upper limit of human hearing. The primary unique features of ultrasound are its *high energy* and *rectilinearity of propagation*.

An *ultrasonic anemometer* is based on the dependence of the propagation speed of ultrasound in air on the direction of wind. It usually consists of two–three pairs of transducers; the path length between them is 10–50 cm. Each pair of transducers includes a generator and ultrasound receiver oriented at different angles to each other. Ultrasound propagates more quickly in the wind direction and more slowly in the opposite direction (Figure 4.4).

An ultrasonic anemometer detects the phase shifting of sound which depends on the orientation of the transducers relative to the wind direction.

Typical parameters of commercial ultrasound anemometer (Kaijo Denki DA-600-3TV, Kaijo Cooperation, Japan) are range 0–30 m/s; resolution 0.005 m/s.

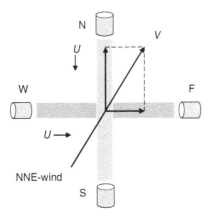

**FIGURE 4.4**    Sonic anemometer: $U$, velocity of ultrasound; $V$, velocity of wind. Ultrasound propagates from North to South and from West to East. The wind propagates in the NNE direction.

*The advantages of an ultrasonic anemometer:*

- The anemometers of this type are rather accurate and respond quickly to wind-speed fluctuations; they can be used for turbulence measurements.
- They are insensitive to icing.
- The lack of moving parts makes them the preferred instruments for wind measurement in weather stations in remote locations.
- There is less chance of these instruments becoming damaged and needing repairs.

*The disadvantages of an ultrasonic anemometer:*

- There is the dependence of the speed of sound on temperature, humidity, and atmospheric pressure which requires calibration of the instruments.
- These anemometers provide the distortion of the flow itself by the structure supporting the transducers.
- These anemometers demonstrate lower accuracy in precipitation, as raindrops can alter the speed of sound.
- Electronic equipment increases the complexity and cost of this type of anemometer.

## 4.5    REMOTE WIND SENSING

*Remote sensing* wind parameters in the atmosphere at very high elevations are based on the use of radiosondes, radars, sodars, lidars, satellites, and rockets.

### 4.5.1  Radiosonde

A *radiosonde* is a device that can be used for measuring certain wind parameters and transmitting the information to a terrestrial receiver; it also contains sensors which measure temperature, humidity, and pressure at different levels.

Estimating the horizontal position of a radiosonde balloon in relation to the terrestrial point where it was released makes it possible to calculate the average wind speed and direction.

A radiosonde is tracked usually by radar which determines the direction of the signal and turns in the direction of the radiosonde. These data are sent to a computer, visualized, and processed. Modern radiosondes can determine the wind speed and direction via GPS (Global Positioning System).

Radiosondes are flown on balloons made of natural or synthetic rubber; it can reach the altitude 20 km during the 90–120 minutes of flight.

A radiosonde that is designed to measure wind speed and direction only is called a *rawinsonde*.

The National Weather Service launches radiosondes from 92 stations in North America and the Pacific Islands twice daily.

### 4.5.2  Radar

A *radar* (**RA**dio **D**etection **A**nd **R**anging) is an object-detection system that uses radiowaves to determine the range, altitude, direction, or speed of objects.

This type of remote sensing technology involves the use of electromagnetic waves in the range from 0.1 cm to 2 m (corresponding to frequencies from 100 MHz to 50,000 MHz). The registration of the time of arrival of radar signal can make it possible to determine a backscattering object's position. The angle of arrival at a receiving station can be determined by the use of a directional antenna. Microwave location makes it possible to determine the position, motion, and nature of distant objects.

*The advantages of a radar anemometer:*

- Not sensitive to acoustic noise.
- Better operates in turbulent atmosphere.
- Moderate cost.

*The disadvantages of a radar anemometer:*

- Effect of birds and aircrafts.
- Effect of precipitation.
- Limitation in altitude (not less than 50 m).
- Interference with the ground sources of electromagnetic radiation.

### 4.5.3  Sodar

A *sodar* (**SO**und **D**etection **A**nd **R**anging) is an ultrasonic anemometer that can be used for remote measurement of wind at various heights above the ground. It can be used as a wind profiler to measure the scattering of ultrasound waves by atmospheric turbulence.

*The advantages of an ultrasound anemometer:*

- Such systems can measure the *wind profile*—detecting the wind speed and direction at various altitudes with an accuracy within 5%.
- Continuous regime of sampling.
- Fine vertical resolution.
- Sampling below 100 m.

*The disadvantages of an ultrasound anemometer:*

- The speed of propagation of ultrasound depends on temperature, humidity, atmospheric pressure, and precipitation.
- Such devices require appropriate calibration.
- The electronic equipment increases the cost of devices of this type.

### 4.5.4  Lidar

A *lidar* (**LI**ght **D**etection and **R**anging) is an optical remote sensing device that can measure the distance to an object by illuminating the object with laser radiation. The principle of operation of lidar is based on the estimation of scattering of laser radiation by aerosols, dust, water droplets, particles of dirt, pollutants, or salt crystals that travel at the speed of wind. Such laser systems can be used for measuring and evaluating the speed and direction of wind and air turbulence at high altitude. Fiber laser systems that were developed during recent years are characterized by extremely high ($10^{-12}$) sensitivity.

*The advantages of a laser anemometer:*

- Laser beam does not disturb the flow being measured.
- This anemometer can be used for accurate measurements in unsteady and turbulent flows where the velocity is fluctuating with time.
- No calibration required.
- Wide velocity range (from zero to supersonic).
- One, two, or three velocity components can be measured simultaneously.
- Measurement distance varies from centimeters to meters.
- Flow reversals can be measured.
- High spatial and temporal resolution.

*The disadvantages of a laser anemometer:*

- The anemometer of this type is expensive (typically $40,000 for a simple system).
- It requires a transparent medium through which the laser beam can pass. Limited range in precipitations, fog, clouds, and aerosols.

All these systems (sodar, radar, and lidar) can utilize the Doppler effect during measurement of wind speed.

### 4.5.5  Doppler Effect

There is a well-known train-spotter's observation that the pitch of a train's whistle changes as it passes the observer. As the train approaches, the observer hears a note which is higher than the true note and on passing, the pitch quickly falls to a lower note than the true pitch. Doppler in 1842 was the first to give an explanation for the phenomenon which has been entitled the *Doppler effect*.

When either the source or the receiver of a propagating wave moves, there is usually a change in frequency called the *Doppler shift*. Such a shift can be used to determine the velocity of the target along the line to it, which is to say its *approach velocity*. If the transmitter generates a frequency of $f_0$, the velocity of the sound is $v_s$, and the approach velocity is a much smaller value ($v_a$), the received frequency is approximated by:

$$f = f_0 \left( 1 + \frac{2v_a}{v_s} \right). \tag{4.11}$$

The frequency shift is proportional to the approach velocity:

$$\Delta f = f - f_0 = \frac{2f_0}{v_s} v_a. \tag{4.12}$$

Vaisala LAP®-3000 Lower Atmosphere Wind Profiler (Finland) is a Doppler radar which provides vertical profiles of horizontal wind speed and direction, and vertical wind velocity up to an altitude of 3 km above ground level.

**Example**    What is the Doppler frequency shift received by a bat, if the insect is immobile with respect to the bat's motion with the velocity equal to 5 m/s. The frequency of the bat's sound is 60 kHz.

**Solution**    We can insert the values into the Doppler shift Equation 4.12:

$$\Delta f = \frac{2f_0}{v_s} v_a = \frac{2 \times 60 \times 10^3 \text{ Hz}}{340 \text{ m/s}} 5 \text{ m/s} = 1.76 \text{ kHz}$$

### 4.5.6  Satellite and Rocket Remote Sensing

This technique makes it possible to construct a map of the winds on the Earth's surface, to examine the flow of air in the atmosphere, to study wind erosion risk, and dust emission–deposition.

The geostationary weather satellites are widely used for detection and monitoring of wind extreme situations: hurricanes, their spatial distribution, and temporal behavior. The combination of observing systems such as weather satellites, rockets, reconnaissance aircrafts, coastal radars, meteorological stations makes it possible to reach accurate results.

The American Space Agency (NASA) launched, in 2012, five rockets at intervals of 80 seconds to study high-speed air flow in the upper atmosphere. The project was named ATREX (**A**nomalous **T**ransport **R**ocket **EX**periment). The test center was located on the Wallops Island in Virginia.

At an altitude of about 80 km, missiles threw a special reagent (trimethylaluminum) that reacts with oxygen. This reaction is accompanied by luminescence. It was possible to observe high-speed streams (hundreds of kilometers per hour) at altitudes of 100–110 km, almost on the border with the space. Traditional methods to study these flows are limited because of the very low air density at these altitudes.

## 4.6  MEASUREMENT OF WIND DIRECTION

*Wind vanes* are usually used to monitor wind direction as they always point into the wind. Measured clockwise from the true north, an easterly wind is designated as 90°, a southerly wind as 180°, a westerly wind as 270°, and a northerly wind as 360°. The wind recorded as 0° is only during calm conditions.

A *wind sock* is a striped open bag of conical shape that indicates wind direction and relative speed at airports.

The transfer of wind direction information in modern instruments utilizes a *selsyn*—a system that consists of a generator and a motor connected by wire in such a way that angular rotation or position in the generator is reproduced simultaneously in the motor. The simplest selsyn consists of a three-phase rotor and a stator (the stationary part of a rotor system) whose winding is formed by three coils with axes oriented at 120° (Figure 4.5). Two such devices are electrically connected to each other—a stator with a stator and a rotor with a rotor. An alternating voltage is applied to the rotors. Under these conditions, the rotor of a selsyn generator induces currents in the three coils of the stator, and a corresponding magnetic field in the selsyn motor, causing rotation of the rotor of selsyn motor which is associated with the indicator. The accuracy of determining the wind direction is ±3°.

To determine the dominant wind direction, a *wind rose* is used—a vector diagram that describes the speed and direction of the wind at a location based on long-term observations. A wind rose looks like a polygon, with rays diverging from the center of the chart in different directions that are proportional to the frequency of winds in the area (Figure 4.6).

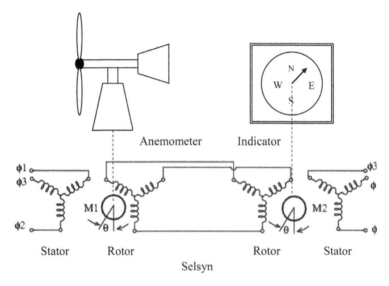

**FIGURE 4.5** Transfer of information on the wind direction through the selsyn system.

Simultaneous measurement of velocity and direction of air movement is realized by *anemorumbometer*. Amount of rotations of propeller is transformed into a sequence of electrical pulses. The frequency of these pulses is proportional to wind speed, and phase shift depends on the direction.

International Meteorological Organization requires from instruments for measuring wind direction the range of wind speeds from 0.5 to 50 m/s with a resolution of ±2° to ±5°.

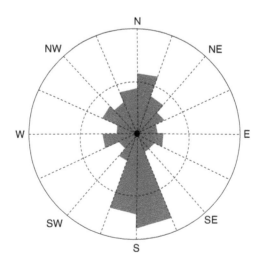

**FIGURE 4.6** Wind rose.

## 4.7   CYCLONE ASSESSMENT

National Oceanic and Atmospheric Administration (NOAA) provides the estimation of the activity of individual tropical cyclones and entire tropical cyclone seasons through the *Accumulated Cyclone Energy (ACE)* that is given by

$$ACE = 10^{-4} \sum v_{max}^2, \tag{4.13}$$

where $v_{max}$ is estimated sustained wind speed in knots.

The ACE of a season is calculated by summing the squares of the estimated maximum sustained velocity of every active tropical storm (wind speed 35 knots or higher), at 6-hour intervals.

The NOAA proposes the following system categories:

- *Above-normal season*: an ACE value above 103 (115% of the current median), provided at least two of the following three parameters exceed the long-term average: number of tropical storms (10), hurricanes (6), and major hurricanes (2).
- *Near-normal season*: neither above normal nor below normal
- *Below-normal season*: an ACE value below 66 (74% of the current median)

### Extreme Situations

*The highest ever ACE estimated for a single storm in the Atlantic is 73.6, for Hurricane San Ciriaco in 1899.*

## REFERENCE

King, L.V. 1914. On the convection of heat from small cylinders in a stream of fluid: determination of the convection constants of small platinum wires with applications to hot-wire anemometry. *Phil. Trans. Roy. Soc. (London). A*, 214:373–432.

# PRACTICAL EXERCISE 2

# MODELING THE VARIATION IN WIND SPEED

## 1 MODELING VARIATION IN WIND SPEED NEAR THE GROUND

The theoretical determination of the average wind-speed profile in a turbulent boundary layer was developed by Prandtl (1920). Fluctuations in horizontal velocity are associated with the fluctuations of velocity in the vertical direction according to this theory.

The equation that describes the change in wind speed with height is

$$v(z) = \frac{v^*}{0.4} \ln \frac{z}{z_m}, \tag{P2.1}$$

where $v^*$ is the friction velocity (which is a constant for the lowest surface layer, that is, 50–100 m in the atmosphere); $z_m$ is called the "roughness parameter" of the surface; $z$ is the height; 0.4 is the von Karman constant.

**Example** Calculate the dependence of wind speed on the height above the Earth's surface if the friction velocity is $v^* = 0.35$ m/s and the roughness parameter is $z_m = 0.005$ m.

*Methods of Measuring Environmental Parameters*, First Edition. Yuriy Posudin.
© 2014 John Wiley & Sons, Inc. Published 2014 by John Wiley & Sons, Inc.

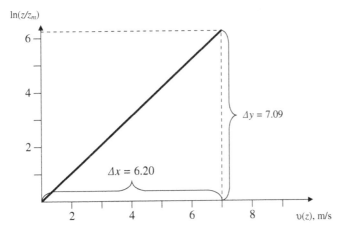

**FIGURE P2.1**   Wind-speed profile near ground: graph of the dependence $\ln(z/z_m) = f[v(z)]$.

***Solution***   Construct a graph $\ln z = f[v(z)]$ for the range of heights 0.005–6 m.

| $z$, m | $\ln z$ | $\ln(z/z_m)$ | $v(z) = 0.875$ $\ln(z/z_m)$, m/s |
|---|---|---|---|
| 6 | 1.79 | 7.09 | 6.20 |
| 4 | 1.39 | 6.68 | 5.85 |
| 3 | 1.10 | 6.40 | 5.60 |
| 2 | 0.69 | 5.99 | 5.24 |
| 1 | 0 | 5.30 | 4.64 |
| 0.5 | −0.69 | 4.60 | 4.03 |
| 0.1 | −2.3 | 2.99 | 2.62 |
| 0.05 | −2.99 | 2.30 | 2.01 |
| 0.01 | −4.61 | 0.69 | 0.61 |
| 0.005 | −5.30 | 0 | 0 |

The value $v^*$ can be found from the graph $\ln(z/z_m) = f[v(z)]$ as the slope of dependence $\ln(z/z_m)$ versus $v(z)$ (Figure P2.1):

$$\Delta y / \Delta x = \ln(z/z_m)/v(z) = 7.09/6.20 = 1.14 = 0.4/v^*.$$

Therefore

$$v^* = 0.4/1.14 = 0.35 \, \text{m/s}.$$

The value of $z_m$ can be found from the graph $\ln z = f[v(z)]$: the intercept with ordinate axis, where $v(z) = 0$, is $\ln z_m$ (Figure P2.2):

$$\ln z_m = -5, 3.$$

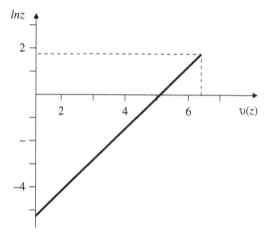

**FIGURE P2.2**    Wind-speed profile near ground: graph of the dependence $\ln z = f[v(z)]$.

Hence,

$$z_m = e^{-5.3} = 0.005.$$

## 2  MODELING THE VARIATION IN WIND SPEED ABOVE A PLANT CANOPY

When calculating the wind above a plant canopy, we assume that the wind speed at the terrestrial surface is zero and increases with height up to constant value. The equation that describes the change of wind speed with height is

$$v(z) = \frac{v^*}{0.4} \ln \frac{z - d}{z_m}, \tag{P2.2}$$

where $d$ is the zero plane displacement. Equation P2.2 is valid for $z \geq z_m - d$.

**Example**    Estimate the friction velocity if the wind speed measured 2 m above a potato crop 60 cm high is 4 m/s, the roughness parameter is $z_m = 0.04$ m, and the zero plane displacement is 1.2 m.

**Solution**    Using Equation P2.2 gives:

$$v(z) = \frac{v^*}{0.4} \ln \frac{2.6 - 1.2}{0.04} = \frac{v^*}{0.4} \ln 35 = v^* \cdot 8.9 = 4 \, \text{m/s}.$$

Hence,

$$v^* = \frac{4 M \cdot c^{-1}}{8.9} = 0.45 \, \text{m/s}.$$

**Example**  Plot graphs of the dependence $\ln[(z-d)/z_m] = f[v(z)]$, $\ln(z-d) = f[v(z)]$ and $z = f[v(z)]$ for the following parameters: roughness parameter is $z_m = 0.2$ m; zero plane displacement is 1.4 m; friction velocity $v^* = 0.9$ m/s. The range in the height change is 1.6–6 m. Graphically find the values for zero plane displacement $d$, roughness parameter $z_m$, and friction velocity $v^*$.

**Solution**  Substituting the numerical data into Equation P2.2 beginning from $z = 6$ m:

$$v(z) = \frac{v^*}{0.4} \ln \frac{z-d}{z_m} = \frac{0.9 \, \text{m/s}}{0.4 \, \text{m/s}} \ln \frac{6-1.4}{0.2} = 2.25 \ln 23 = 2.25 \cdot 3.135 = 7.05 \, \text{m/s}.$$

This gives $z - d = 6 - 1.4 = 4.6$ m; $\ln(z-d) = \ln 4.6 = 1.53$;

$$\ln[(z-d)/z_m] = \ln(4.6/0.2) = 3.1.$$

Similar calculations can be done for $z = 4$ m; 3 m; 2 m; 1.75 m; 1.6 m; the results have been entered in the following table:

| $z$, m | $v(z)$ | $z-d$, m | $\ln(z-d)$ | $\ln[(z-d)/z_m]$ |
|---|---|---|---|---|
| 6 | $2.25 \cdot 3.135 = 7.05$ | $6-1.4 = 4.6$ | $\ln 4.6 = 1.53$ | $\ln(4.6/0.2) = 3.1$ |
| 4 | 5.76 | 2.6 | 0.96 | 2.56 |
| 3 | 4.69 | 1.6 | 0.47 | 2.08 |
| 2 | 2.47 | 0.6 | −0.51 | 1.1 |
| 1.75 | 1.26 | 0.35 | −1.05 | 0.56 |
| 1.6 | $x = 0$ | 0.2 | −1.61 | 0 |

Calculate the dependence $\ln[(z-d)/z_m] = f[v(z)]$. The slope of this graph makes it possible to find the friction velocity $v^*$ (Figure P2.3):

$$v^* = \frac{0.4 v(z)}{\ln \frac{z-d}{z_m}} = \frac{0.4}{\frac{\Delta y}{\Delta x}} = \frac{0.4}{0.44} = 0.9 \, \text{m/s}.$$

Calculate the dependence $\ln(z-d) = f[v(z)]$ (Figure P2.4).

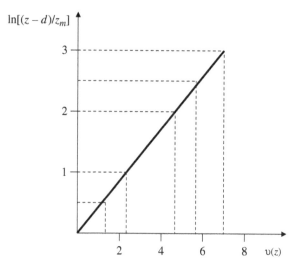

**FIGURE P2.3**   Dependence $\ln[(z-d)/z_m] = f[\upsilon(z)]$.

To find the value $\ln(z-d)$ that corresponds to $\upsilon(z) = 0$:

$$0 = \frac{\upsilon^*}{0.4}\ln\frac{z-d}{z_m} = \frac{0.9\,\text{м/c}}{0.4\,\text{м/c}}[\ln(z-d) - \ln z_m] = 2.25[\ln(z-d) - \ln z_m].$$

Hence, $\ln(z-d) = \ln z_m = -1{,}6$; $z_m = 0.2$ m. Wind-speed profile $z = f[\upsilon(z)]$ above plant canopy is illustrated in Figure P2.5.

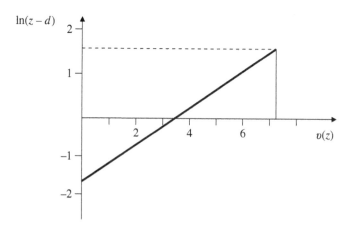

**FIGURE P2.4**   Dependence $\ln(z-d) = f[\upsilon(z)]$.

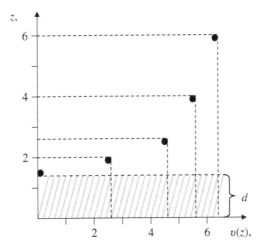

**FIGURE P2.5**    Wind-speed profile $z = f[v(z)]$ above plant canopy.

**Control Exercise**    Plot graphs of the dependences $\ln[(z - d)/z_m] = f[v(z)]$, $\ln(z - d) = f[v(z)]$, and $z = f[v(z)]$ for the surface with parameters that correspond to a certain variant.

Calculate from the graph the values for the roughness parameter $z_m$ and the friction velocity $v^*$.

| No. of variant | Type of the Surface | $d,m$ | $z_m$, m | $v^*$, m/s |
|---|---|---|---|---|
| 1 | Ice | 0.0006 | $10^{-3}$ | 0.5 |
| 2 | Water surface | $4 \times 10^{-3}$ | $6 \times 10^{-4}$ | 0.06 |
| 3 | Smooth desert | $10^{-4}$ | $10^{-4}$ | 0.05 |
| 4 | Coniferous forest | 7 | 1.1 | 0.5 |
| 5 | Cotton 1.3 m tall | 0.8 | $1.3 \times 10^{-1}$ | 0.4 |
| 6 | Citrus orchard | 2.5 | $4 \times 10^{-1}$ | 0.4 |
| 7 | Village | 3.25 | $5 \times 10^{-1}$ | 0.45 |
| 8 | Town | 20 | 1.75 | 0.5 |
| 9 | Grass 50 cm high | 0.65 | 0.1 | 0.2 |
| 10 | Calm sea surface | 0.03 | $5 \times 10^{-3}$ | 0.08 |
| 11 | Snow surface | 0.15 | 0.02 | 0.07 |
| 12 | Soil | 0.4 | 0.06 | 0.25 |

## QUESTIONS AND PROBLEMS

**1.** Name the main parameters of the wind.

**2.** Explain the primary causes of wind.

3. Check the dimensions of the left and right sides of the equation of the Coriolis force $F_k = 2\rho v \omega \sin\varphi$.

4. Make a comparative analysis of cup and propeller anemometers.

5. Explain the principle of action of a thermoanemometer.

6. What is the principle of operation of an ultrasonic anemometer?

7. Explain how the Doppler shift is used.

## REFERENCE

Prandtl, L. 1920. *Theory of lifting surfaces.* NACA TN 9.

## FURTHER READING

Campbell, G.S. and Norman, J.M. 1998. *Environmental Biophysics*, 2nd edition. Springer Verlag, New York.

Guyot, G. 1998. *Physics of the Environment and Climate*. John Wiley & Sons, Chichester.

Meteorological Office. 1981. *Handbook of Meteorological Instruments*, Volume 4 Measurement of Surface Wind, 2nd edition. Her Majesty's Stationery Office.

Monteith, J.L. and Unsworth, M. 1990. *Environmental Physics*. 2nd edition. Edward Arnold, London.

## ELECTRONIC REFERENCES

Articles tagged as: weather vane. November 9, 2011. Weather Vanes Are Ancient Tools, http://tamutimes.tamu.edu/tag/weather-vane/ (accessed January 12, 2013).

Posudin, Y. 2011. Environmental Biophysics. Fukuoka-Kiev, http://www.ekmair.ukma.kiev.ua/handle/123456789/951 (accessed January 24, 2013).

Prandtl, L. Theory of lifting surfaces. UNT Digital Library, http://digital.library.unt.edu/ark:/67531/metadc53693/ (accessed April 27, 2013).

Unlimited access to wind Speed and direction data, http://www.3tier.com/en/package_detail/wind-prospecting-tools/ (accessed February 1, 2013).

Windsock, http://en.wikipedia.org/wiki/Windsock (accessed February 10, 2013).

# 5

# TEMPERATURE

## 5.1 DEFINITION OF TEMPERATURE

*Temperature* is a physical quantity that characterizes the state of thermodynamic equilibrium of a macroscopic system. Temperature characterizes how hot or cold a body is.

Quantitative measurement of temperature is possible using a thermometer or temperature scales.

The *thermometer* is a device for the quantitative estimation of the temperature of a system.

## 5.2 TEMPERATURE SCALES

There are three temperature scales in use today: Fahrenheit, Celsius, and Kelvin.

Daniel Gabriel Fahrenheit used the following reference points in the *Fahrenheit scale* in 1724: the temperature of a mixture of water, ice, and ammonium chloride as the zero point (0°F); the temperature of the same mixture but without salt (30°F); the temperature of the human body (96°F). The freezing point of water was selected as 32°F and the boiling point as 212°F. The interval between the two points was divided into 180 parts—the *degree Fahrenheit* (°F).

*Methods of Measuring Environmental Parameters*, First Edition. Yuriy Posudin.
© 2014 John Wiley & Sons, Inc. Published 2014 by John Wiley & Sons, Inc.

Anders Celsius created in 1742 the *Celsius temperature scale* which uses such reference points as 0°C for the freezing point of water and 100°C for the boiling point of water under a pressure of one standard atmosphere. The interval between these two points was estimated as the *degree Celsius* (°C).

Lord Kelvin (William Thompson) proposed in 1848 the *Kelvin temperature scale* (K). This temperature was based on two reference points: absolute zero—the lowest possible temperature in the universe below which temperatures do not exist, and the triple point of water (the intersection on a phase diagram where the three-phase equilibrium point consists of ice, liquid, and vapor). The zero point is defined as 0 K and corresponds to −273.15°C, while by definition the triple point of water (273.16 K) and corresponds to 0.01°C.

*Conversion formulas* for the interconversion of temperatures expressed on the Celsius, Kelvin, or Fahrenheit scales are as follows:

$$\left.\begin{array}{l} °C = K - 273.15; \\ K = °C + 273.15; \\ °C = (°F - 32)5/9; \\ °F = 9/5°C + 32; \\ K = (°F - 32)5/9 + 273.15; \\ °F = (K - 273.15)9/5 + 32. \end{array}\right\} \quad (5.1)$$

**Example**    A saucepan of milk was heated from 20°C to 80°C. What is the change in its temperature on the Kelvin scale and on the Fahrenheit scale?

**Solution**    From Equation 5.1, we can calculate that the change in temperature on the Celsius scale is equal to the change in temperature on the Kelvin scale:

$$\Delta T_K = \Delta T_C = 80 - 20 = 60°C = 60 \text{ K}$$

The change in temperature on the Fahrenheit scale is

$$\Delta T_F = \frac{9}{5}\Delta T_C = \frac{9}{5}(80 - 20) = 108°F.$$

**Exercise**    You are flying by airplane from Kiev to Atlanta; the stewardess informs that the temperature in Atlanta is 86°F. Express this temperature on the Celsius scale.

**Exercise**    Convert −30°C into degrees on the Fahrenheit scale.

## 5.3   ATMOSPHERIC TEMPERATURE

The Earth's atmosphere consists of the following layers: troposphere, stratosphere, mesosphere, and thermosphere.

In the *troposphere* (from the Greek word *tropos*, meaning "mixed"), which is in about 10 km altitude, there is vertical mixing of air due to heating of the Earth's surface by solar radiation. The troposphere temperature decreases with height due to the movement of air in a horizontal direction.

The *stratosphere* (from New Latin *stratum*—"a spreading out") occupies the region between 10 and 50 km. The temperature of this layer increases with altitude. Here ozone ($O_3$), which absorbs short-wave solar radiation, plays a critical role in heating the stratosphere.

The *mesosphere* (from the Greek *mesos*, meaning "middle") occupies the region 50–85 km above the surface of the earth. The mesosphere's temperature decreases with altitude due to significant mixing of air by wind, the speed of which can reach 150 m/s.

The *thermosphere* (from the Greek *thermos*, meaning "temperature") is found in the 85–640 km region and the *exosphere* (from the Greek *exo*, meaning "outside"), in the 500–1000 km region. They are characterized by the temperature increasing with height due to the high kinetic energy of the gas molecules which absorb high-energy solar radiation. The temperature of the thermospheric particles can reach 2000°C, although there is a great decrease in the mass of the air (i.e., rarefaction).

## 5.4   SOIL TEMPERATURE

Typically soil temperature changes with the depth and time of day; the latter is characterized by a sinusoidal diurnal variation.

The following equation describes changes in soil temperature as a function of depth $z$ and time $t$ when the mean daily soil surface temperature $\langle T \rangle$ is known:

$$T(z, t) = \langle T \rangle + A(0) \exp(-z/D) \sin[\omega(t - 8) - z/D], \qquad (5.2)$$

where $\langle T \rangle$ is the mean daily soil surface temperature; $A(0)$ is the amplitude of the temperature fluctuations at the surface; $D$ is the damping depth ($D = 0.1$ m for moist soil and $D = 0.03$–$0.06$ m for dry soil).

The range in soil temperature variation at a particular depth is given as

$$T(z, t) = \langle T \rangle \pm A(0) \exp(-z/D), \qquad (5.3)$$

where the sign "+" corresponds to the maximum temperature and sign "−" gives the minimum temperature.

The damping depth $D$ is determined as

$$D = \frac{z_1 - z_2}{\ln A_2 - \ln A_2},$$    (5.4)

where $z_1$ and $z_2$ are two different soil depths, $A_1 = T(z_1) - \langle T \rangle$ and $A_2 = T(z_2) - \langle T \rangle$ are the corresponding amplitudes of the temperature waves.

## 5.5  TEMPERATURE OF WATER RESERVOIRS

The thermal regime in water reservoirs such as lakes is determined by the high thermal capacity of water in comparison with soil. The thermal capacity buffers against sharp changes in the temperature of the lake water resulting in more stable thermal conditions. In lakes and slow-flowing rivers, the surface layers of water are heated by solar radiation more quickly than deeper water and mixing of the upper layers is facilitated by the wind. Warmer water has a lower density and overlays the colder, dense layers. Such separation of the lake into several zones based on temperature is called *thermal stratification*.

The upper warm layer is called the *epilimnion*, while the lower cold layer is called the *hypolimnion*. Between these two layers is a third layer—so called the *metalimnion* which is characterized by strong vertical temperature gradients. Thermal stratification influences biological productivity and water quality.

## 5.6  HEAT FLUX

*Heat flux* is the amount of heat transferred in a unit of time. The unit of measurement of heat flow is J/s.

The *density of heat flux* is the amount of heat that is transferred per unit time per unit surface area. The unit of measurement of heat flux density is $J/m^2 \cdot s$ or $W/m^2$.

Heat that induces a temperature change in an object is called *sensible heat*. For example, the thermal energy transferred between the surface and air due to the difference in temperature between them, or due to the vertical temperature gradient, is sensible heat. We feel sensible heat transfer as a change in the temperature of the air. Heat is transferred into the air by conduction or convection, the latter being the more efficient way of transferring sensible heat.

If heat is transferred upward from a surface that is warmer than the air, it is called *positive sensible heat transfer*; conversely, if the air is warmer than the surface, it is called *negative sensible heat transfer.*

The amount of energy in the form of heat released or absorbed by a substance during the transition from one physical state to another is called *latent* or *hidden* heat. "Latent" is from the Latin *latere*—to hide.

Possible phase transitions are solid → liquid → gas. For example, the transition from solid to liquid is called melting; from liquid to gaseous state is called evaporation, from gas to liquid is called condensation.

For example, the heat used during evaporation is called a *positive latent heat flux*. The energy is removed from the surface and the temperature of the surface is decreased.

The rate of heat loss LE, when water vapor is transferred from the surface to the atmosphere through evaporation, depends on the temperature of the body and the relative humidity of the air and is given by the product of the density of water $\rho_w$ (kg/m$^3$), the latent heat of vaporization $\lambda$ (J/kg), and the evaporation rate $E$—the amount of water loss per unit area per unit time (m$^3$/m$^2$·s = m/s):

$$LE = \rho_w \lambda E. \tag{5.5}$$

The latent heat of vaporization of water $\lambda$ is 2.454 MJ/kg at 20°C. Here LE is measured in J/m$^2$·s.

If $E$ is measured in kg/s, LE is measured in J/s, and the last equation becomes

$$LE = \lambda E. \tag{5.6}$$

## 5.7    EFFECT OF TEMPERATURE ON LIVING ORGANISMS

*Thermoregulation* is the process of reflectory changes in heat production and heat transfer of organism to keep its temperature within certain boundaries, even when the surrounding temperature is very different.

### 5.7.1    Heat Production

*Vasoconstriction and Vasodilation.* Heat is produced in all body tissues, but particularly in the core organs of a resting animal and in muscles during activity. Heat must be distributed around the body, principally by the blood or other body fluids, and dissipated at the surface according to need. The simplest form of thermal control is to manage the rate and volume of the flow of blood to the surface relative to the core. It is possible to distinguish *vasoconstriction*—the narrowing of surface blood vessels in response to cold temperature, keeping heat in the core, and *vasodilation*—the expansion of surface blood vessels in response to warm temperatures. For example, elephants use their large ears for thermoregulation, to cool themselves in the hot equatorial sun. As the elephant flaps its ears, blood vessels in the ears are cooled. The cooled blood circulates throughout the elephant and helps regulate the overall body temperature. Another example, the fennec (*Fennecus zerda*), is the smallest of the wild canid species. The fennec has incredibly large ears, which measure up to 15 cm in length, making it extremely sensitive to sound. It also assists in thermoregulation, which is essential in its North African desert habitat.

*Huddling and Aggregation.* Endothermic animals in polar regions, such as penguins and some seals, cluster together when the temperature drops substantially (e.g., −50°C), typical in an Antarctic winter and when the wind speeds are over 160 km/h. Penguins form much denser huddles than seals with up to several thousand (up to 6000) individuals, which can have a biomass of 100 tons. Clustering together saves as much as 80% of the heat loss that would occur from an isolated bird.

*Insulation.* The fur of large polar animals, commonly 30–70 mm thick, is often so effective that the complete thermal gradient from air temperature to normal body temperature is contained across the fur and skin. Fur in boreal mammals varies with the season. Winter coats of animals are thicker and provide better thermal insulation than summer coats. Arctic foxes and the stoat are classic examples of sub-Arctic tundra species that grow a very dense and highly camouflaged winter coat. Insulation is the reciprocal of the total heat flux per unit area per unit of temperature difference, and therefore, has units of $°C/m^2 \cdot W$. Insulation may be provided by internal air sacs and fat layers, or externally by cuticular bristles of chitin or by various hairs and other structures made of keratin. Internal insulators are much less effective; a layer of blubber 60–70 cm thick has about the same insulating properties as 2 cm of good mammalian fur.

The typical values of the insulation ($°C/m^2 \cdot W$) of some biological materials are human tissue, 14; fat, 38; cattle fur, 50; sheep wool, 102; husky dog fur, 157; lynx fur, 170; still air, 270.

*Animal Coloration.* There is an apparent paradox in animal coloration, in that the polar bear and many other polar animals are white, when black fur would appear to be beneficial in maximizing the absorption of solar radiation. The explanation for this is that white pelage makes the animals less conspicuous against a bright ice or snow. An additional reason is that for polar bears, the air space reflects visible light so that the hair appears white (it is actually translucent) and the hairs act like optical fibers which permit ultraviolet radiation to pass from the hair tips to the bear's skin, facilitating radiative heating.

For example, laboratory experiments demonstrated that a South American frog, *Bokermannohyla alvarengai*, kept in the dark or at lower (20°C) temperature had darker skin color while the frogs kept in the light or higher (30°C) temperature had skin color of a lighter hue.

The dark-colored feathers on the back surface of penguins provide the warming of these birds by solar radiation.

*Shivering.* Shivering is a high-frequency reflex producing oscillatory contractions of skeletal muscles. The adenosine triphosphate (ATP) hydrolyzed to provide energy for contractions is producing minimal physical work but substantial heat output. The intensity of shivering is linearly related to oxygen consumption. The animals are regulating the extent of their shivering to compensate for cold by generating internal heat.

Several snakes, such as the Indian python (*Python* spp.), use similar shivering mechanisms to achieve endothermic warming when they are incubating eggs. The female wraps her body tightly round the clutch and produces a low frequency but powerful shivering response, raising her body temperature and thus the egg temperature to around 30–33°C, often 7–8°C above ambient, thus facilitating more rapid development of the young.

Certain moths, beetles, dragonflies, flies, wasps, and many bees also shiver, doing so to warm their muscles before flying.

*Basking.* This mechanism is particularly effective for small ectotherms. The temperature of moderately sized insects can be raised at least 15°C above ambient by basking alone. Basking postures are also important. Several species of flowers (e.g., *Dryas integrifolia, Papaver radicatum*) are shaped like bowls and rotate via phototropism such that their corolla always point toward the sun. Their shape acts as a parabolic reflector which concentrates radiation into the center of the flower, raising the flower's temperature by 5–8°C above ambient. Mosquitoes, hoverflies, blow flies, and dance flies use these flowers as basking sites.

*Burrowing.* The internal temperature of air in the burrow is kept constant.

So, the burrow of deer mice in the Nevada region has constant internal temperature about 26°C while the temperature of external air varies from 16°C to 44°C.

*Migration.* There are many endothermic examples of polar animals (e.g., caribou, reindeer, polar bears, lemmings, and polar birds) that migrate to avoid very low temperatures and to find areas suitable for foraging. Long-range migration is not a very common strategy for ectotherms. The monarch butterfly (*Danaus plexippus*) migrates from Canada and the northern United States each year to overwinter in Mexico. Colias butterflies also migrate seasonally from northern Scandinavia to the southern Baltic regions.

### 5.7.2 Heat Transfer

Heat transfer is the process whereby heat moves from living organism to environment by radiation, conduction, convection, or a combination of these methods. Heat transfer can appear only if a temperature difference exists, and then only in the direction of decreasing temperature. Heat transfer can be realized through various mechanisms, such as thermal conduction, thermal convection, thermal radiation, or evaporation.

*Thermal Conductivity.* This process can be viewed on an atomic level as an exchange of kinetic energy between molecules, where the less energetic particles gain energy by colliding with more energetic particles. The conduction of heat occurs only if there is a difference in temperature between two parts of the conducting medium. If we consider a sheet of material of cross-section area $S$, thickness $dx$, and temperature difference $dT$, the *Law of Heat Conduction* is

$$Q_c = -kS\frac{dT}{dx} \qquad (5.7)$$

where the proportionality constant $k$ is called the *thermal conductivity* of the material, and $\frac{dT}{dx}$ is the *temperature gradient*—the variation in temperature with position. The minus in this *Fourier's equation* denotes the fact that heat flows in the direction of decreasing temperature. The typical values of thermal conductivity (W/m·K) of air and water at different temperatures are air (−10°C) 0.0237; air (0°C) 0.0243; air

(10°C) 0.0250; air (20°C) 0.0257; air (30°C) 0.0264; air (40°C) 0.0270; air (50°C) 0.0277; water (0°C) 0.565; water (20°C) 0.599; water (40°C) 0.627.

**Example**    When a pig lies on concrete, the animal's belly and the concrete on which it lies are in contact. The temperature of the concrete's surface approximates that of the pig's belly surface. Assume that the 8-cm-thick concrete slab is on ground with a temperature of 0°C, that the belly–floor contact area is 3000 cm², the body temperature of the pig is 30°C, and the thermal conductivity of the concrete is 2.43 W/m·K.

Estimate the conductive heat transfer under steady-state conditions.

**Solution**    Conductive heat transfer flux from the belly to the concrete can be calculated as

$$Q_c = -kS\frac{\Delta T}{\Delta x}$$
$$= -2.43\,\frac{W}{mK}\cdot 3000\times 10^{-4}\,m^2\frac{(0-30)\,K}{8\times 10^{-2}\,m}$$
$$= -2.43\times 3\times 10^{-1}\frac{(-30)}{8\times 10^{-2}}$$
$$= 273.37\;J/s.$$

**Convection.** Heat transferred by the movement of a heated substance is said to have been transferred by *convection*. When the movement results from differences in density, it is referred to as *natural convection*. When the heated substance is forced to move by a fan or pump, the process is called *forced convection*.

Consider the convective heat exchange of objects with various shapes. The amount of heat conducted across the boundary layer and convected away from the flat plate (e.g., the surface of a leaf per unit time and area) by forced convection is

$$J_C = -2k_{air}\frac{(T_{leaf}-T_{air})}{\delta} \tag{5.8}$$

where $k_{air}$ is the thermal conductivity coefficient of air; $T_{leaf}$ is the leaf temperature; $T_{air}$ is the temperature of the air outside a boundary layer of thickness $\delta$. This can be expressed as

$$\delta\;(mm) = 4.0\sqrt{\frac{L(m)}{v\;(m/s)}} \tag{5.9}$$

where $L$ is the mean length of the leaf; $v$ is the ambient wind speed; the factor 4.0 has dimensions in $m/s^{1/2}$.

In the case of a cylinder (e.g., an animal), the heat flux density is

$$J_C = -2k_{air}\frac{(T_{surf}-T_{air})}{r\ln\left(\frac{r+\delta}{r}\right)} \tag{5.10}$$

where $r$ is the cylinder radius; $T_{surf}$ surface temperature; $\delta$ is calculated as:

$$\delta \text{ (mm)} = 5.8\sqrt{\frac{D(m)}{v \text{ (m/s)}}} \tag{5.11}$$

where $D$ is the cylinder diameter.

It is possible to use the following relation for objects of irregular shape which is known as *Newton's Law of Cooling*. It describes the rate of heat loss $J_C$ in W/m² per unit surface area of a body in a cool air stream as:

$$J_C = k_c(T_{surf} - T_{air}) \tag{5.12}$$

where $k_c$ is the convection coefficient with units W/m²·deg.

**Example**  Calculate the heat flux density conducted across the boundary layer and convected away from the surface of a sheep if the body of the animal approximates a cylinder with a radius of 60 cm, the surface temperature $T_{surf}$ is 38°C, the temperature of the air $T_{air}$ is 20°C, and the ambient wind speed $v$ is 80 cm/s.

**Solution**  The average of the boundary layer ($\delta$) is

$$\delta = 5.8 \cdot \sqrt{\frac{D}{v}} = 5.8 \cdot \sqrt{\frac{0.6}{0.8}} = 5 \text{ mm} = 5 \times 10^{-3} \text{ m}.$$

Using Equation 5.12 and $k_{air}$ (0.0257 W/m·K at 20°C), calculate the heat flux density conducted across the boundary layer.

$$J_C = -2k_{air}\frac{(T_{surf} - T_{air})}{r \ln\left(\frac{r+\delta}{r}\right)} = \frac{0.0257(38-20)}{0.3 \cdot \ln\left(\frac{0.3+0.005}{0.3}\right)} = 93 \text{ W/m}^2.$$

*Radiation.* Two bodies at different temperatures will exchange heat even when there is no possibility of exchange by conduction or convection; the transfer of heat takes place by *radiation*.

The rate at which an object emits radiant energy is proportional to the fourth power of its absolute temperature. This is known as *Stefan–Boltzmann's law* which is expressed in equation form as

$$R = \sigma A \varepsilon T^4 \tag{5.13}$$

where $R$ is the power radiated by the body (W); $\sigma = 5.7051 \times 10^{-8}$ W/m²·K⁴ is a constant; $A$ is the surface area of the object, $\varepsilon$ is a constant called the *emissivity*; $T$ is the temperature in kelvin.

When an object is in equilibrium with its surrounding, it radiates and absorbs energy at the same time:

$$R_n = Q_a - Q_e = A\sigma\left(aT_s^4 - \varepsilon T_e^4\right) \tag{5.14}$$

where $a$ is a constant called the *absorptivity.*

The typical values of emissivity for farm animals and humans is $\varepsilon = 0.95$.

**Example**  If a brown cow ($a = 0.80$; $\varepsilon = 0.95$) has an effective radiant-surface area of 4 m² and a radiant-surface temperature averaging 27°C, and the average temperature of the environment is −3°C, calculate the net flux of thermal radiant energy between the animal and its environment.

**Solution**  A net heat flux from animal to environment via thermal radiation can be determined from Equation 5.14:

$$R_n = Q_a - Q_e = A\sigma\left(aT_s^4 - \varepsilon T_e^4\right)$$
$$= 4\ \text{m}^2 \times 5.7051 \times 10^{-8}\ \text{W/m}^2 \cdot \text{K}^4(0.80 \times 300^4 - 0.95 \times 270^4) = 632\ \text{W}.$$

*Evaporation.* Evaporative loss of water is an excellent way for animals to dissipate heat. The rate of evaporation depends not only on the surface temperature, but also on the difference in water vapor density between the animal's surface and the environment, and on the resistance to water loss from the surface. Typical values for evaporative water loss are 7.0–20.9 W/m².

### Extreme Situations

*The coldest place is Antarctica, where the temperature can reach −89.2°C; the highest temperature is about 80°C in desert areas.*

*The lower limit for aquatic animals is determined by the freezing point of sea water at −1.86°C. The wood frog, Rana sylvatica, is capable to withstand temperatures as low as −8°C, with 65% of its body water converted to ice. Terrestrial large polar animals and birds can tolerate the temperatures up to about −60°C.*

*The upper thermal limit for animals and plants is about 50°C and for thermophilic bacteria about 90°C.*

*The sensitivity of a rattle snake to rapid changes of temperature is 0.002°C.*

*Evidence exists of prokaryotes living at temperatures as high as 155°C.*

*For land animals, the upper limit is about 50°C for certain insects and reptiles.*

*Certain large polar mammals and birds can tolerate an ambient temperature of −60°C.*

# 6

# MEASUREMENT OF TEMPERATURE

## 6.1 LIQUID-IN-GLASS THERMOMETERS

The use of *liquid-in-glass thermometers* is based on the expansion of liquids in response to temperature. It is known that the volume of a material (solid or liquid) changes with temperature: the change in volume $\Delta V$ at a constant pressure is proportional to the original volume $V$ and the change in temperature $\Delta T$ as $\Delta V = \beta V \Delta T$, where $\beta$ is the coefficient of volume expansion. In a liquid-in-glass thermometer, the liquid rises inside a capillary column because the coefficient of volume expansion of liquids is much higher than that of glass: $1.81 \times 10^{-4}$ $(°C)^{-1}$—mercury; $10.6 \times 10^{-4}$ $(°C)^{-1}$—alcohol; $9.16 \times 10^{-4}$ $(°C)^{-1}$—toluene; $2.5 \times 10^{-5}$ $(°C)^{-1}$—glass.

A liquid-in-glass thermometer consists of a *bulb*—the reservoir for containing the liquid; a *stem*—the glass tube having a capillary bore along which the liquid moves with changes of temperature; a *scale*—an engraved or etched scale with well-defined, narrow lines graduated in degrees Celsius.

Liquid-in-glass thermometers are employed in meteorological and oceanographic applications.

The most widely used liquid in this type of thermometer is mercury. It is a liquid material that does not deteriorate or adhere to the glass. The temperature range of mercury thermometers is from $-38.83°C$ to $+356.7°C$. For measurement of low temperatures, liquids such as alcohol or toluene are used: the melting point of alcohol is $-114°C$, and $-95.1°C$ for toluene. However, these liquids are chemically less stable than mercury and can decompose in sunlight.

*Methods of Measuring Environmental Parameters*, First Edition. Yuriy Posudin.
© 2014 John Wiley & Sons, Inc. Published 2014 by John Wiley & Sons, Inc.

*The advantages of a liquid-in-glass thermometer:*

- They are portable, durable, handy, and convenient to use.
- It is easily calibrated.
- These thermometers are cheaper than other types of temperature-measurement devices.
- They demonstrate compatibility with most environments.
- They have a wide range of temperatures which can be measured (the range of which depends upon the liquid selected).
- They do not need power supply or batteries and can be used in areas where there are problems of electricity.

*The disadvantages of a liquid-in-glass thermometer:*

- The display is harder to read.
- They are very weak and delicate, they must be handled with extra care because they can be easily broken; the liquid element contained in a glass thermometer (especially mercury) may be dangerous or risky to health.
- They cannot provide digital and automated results.
- These types of thermometers can be affected by the environmental temperature.
- Reading temperature via liquid-in-glass thermometers requires excellent eyesight.

**Example**    Mercury thermometer (Figure 6.1) has a capillary of diameter 0.003 cm and a flask of diameter 0.20 cm. Find the change of the height $h$ of the mercury column in the capillary when the temperature is changed to 35°C. Thermal expansion of glass is negligible.

*Solution*    The volume of the bulb can be found from the expression

$$V = \frac{4}{3}\pi R^3 = \frac{4}{3}\pi \left(\frac{D}{2}\right)^3 = \frac{4}{3} \times 3.14 \left(\frac{0.20}{2}\right)^3 = 4.2 \times 10^{-3} \text{ cm}^3.$$

The change in volume $\Delta V$ at a constant pressure is proportional to the original volume $V$ and the change in temperature $\Delta T$ as:

$$\Delta V = \beta V \Delta T = (1.82 \times 10^{-4})(4.2 \times 10^{-3})(35) = 267.54 \times 10^{-7} \text{ cm}^3.$$

The volume of capillary $V_{\text{cap}}$ is

$$V_{\text{cap}} = \frac{\pi D^2}{4} \cdot h.$$

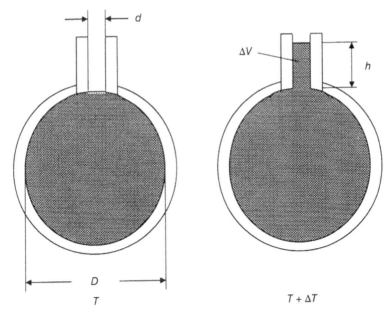

**FIGURE 6.1**    Mercury thermometer.

The change of the height of the mercury column in the capillary is

$$h = \frac{V_{cap} \cdot 4}{\pi D^2} = \frac{\Delta V \cdot 4}{\pi D^2} = \frac{4\beta V \Delta T}{\pi D^2} = \frac{4 \times 267.54 \times 10^{-7}}{3.14 \times 0.003^2} = 3.79 \text{ cm.}$$

## 6.2   BIMETALLIC THERMOMETER

A *bimetallic thermometer* uses the thermal expansion of solids. By using a bimetallic strip consisting of two metals with differing coefficients of expansion, the difference in their expansion rates in response to temperature is measured. Typically metal pairs such as steel–copper or steel–nickel are used. One end of the strip is fixed while the other acts as the indicator (Figure 6.2). Since the two metals have different rates of expansion, a change in temperature results in bending of the bimetallic strip. The displacement $\Delta X$ of the free end depends on the change of temperature $\Delta t$ as $\Delta X = K \Delta t$, where $K$ is the coefficient of proportionality. This device has a linear displacement response to temperature.

*The advantages of a bimetallic thermometer:*

- They do not need power supply.
- They can be used to 500°C.
- They are easily calibrated.
- They have durable and simple construction of less weight.
- They are quite inexpensive.

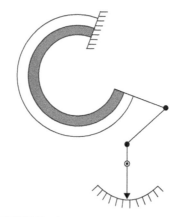

**FIGURE 6.2**    Bimetallic thermometer.

*The disadvantages of a bimetallic thermometer:*

- The accuracy is not very high.
- Requires frequent calibration.
- These thermometers are not suitable for very low temperatures because the expansion of metals tend to be too similar.

## 6.3    RESISTANCE THERMOMETER

A *resistance thermometer* is based on the change in the electrical resistance of some materials with changing temperature. There are two main types of these thermometers—metallic devices which are called *thermoresistors* and semiconductor devices called *thermistors*.

The resistance of a metal is linearly related to the temperature according to the following equation:

$$R = R_0(1 + \alpha \Delta T), \tag{6.1}$$

where $R$ is the resistance at temperature $T$; $R_0$ is the resistance at temperature $T_0$; $\alpha$ is the thermal coefficient of metal; $\Delta T = T - T_0$ (where $T_0$ is usually taken to be 20°C).

The most widely used material for thermoresistors is platinum. Platinum resistors consist of a fine wire (diameter less than 0.1 mm) covered with a layer of glass or ceramic (Figure 6.3). Platinum is highly stable and resistant to corrosion and the action of aggressive chemicals. The resistance of platinum to changes is about 0.3% for each temperature change of 1 K and the temperature range is from −50°C to 550°C.

Copper is characterized with a very linear resistance–temperature relationship, but copper can be oxidized at moderate temperatures and cannot be used over 150°C (302°F).

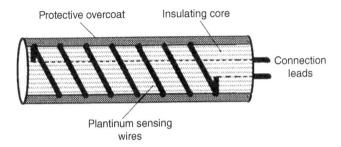

**FIGURE 6.3**    Platinum resistor.

The resistance of semiconductors decreases with increasing temperature as follows:

$$R = ae^{b/T}, \tag{6.2}$$

where $a$ and $b$ are constants depending on the semiconductor material and $T$ is the temperature in K.

Thermistors are fabricated from oxides of various metals (nickel, manganese, cobalt, and copper). Their size can reach 0.2 mm and their accuracy of temperature measurement is typically $\pm 1°C$ within a temperature range from 50°C to 100°C.

*The advantages of a resistance thermometer:*

- They are characterized by linearity over a wide operating range, high accuracy, stability at high temperature, low drift, and suitability for precise measurements.
- They have rapid response.
- They have a wide temperature operating range.
- Display is easy to read.

*The disadvantages of a resistance thermometer:*

- Platinum resistance thermometers are less sensitive to small temperature changes; they have a slower response than thermometers with thermistors.
- Platinum resistance thermometers have common sources of errors which are related to calibration, hysteresis, lead wires, self heating, stem conduction.
- They are expensive.
- They can be affected by shock and vibration.

## 6.4   THERMOCOUPLES

*Thermocouples* are thermoelectric devices consisting of two different metals welded together at one end so that a potential difference generated between the junction end and tail (or reference) end which can be used as a measure of the temperature

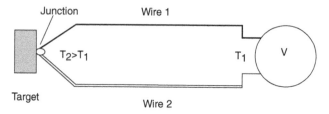

**FIGURE 6.4**    Principle of operation of thermocouple.

difference between junction end ($T_2$) and tail end which is held at ambient temperature $T_1$ (Figure 6.4).

Typical thermocouples consist of such metals or alloys as copper–constantan, chromel–alumel, chromel–constantan, iron–constantan, platinum–platinum/rhodium (10%).

*The advantages of thermocouples:*

- They have a fast response to temperature changes
- These devices are inexpensive.
- They can measure a wide range of temperatures (e.g., chromel–alumel from $-200°C$ to $+1200°C$; iron–copper from $-40°C$ to $+750°C$; copper–constantan from $-200°C$ to $+350°C$.
- They have the ability to be brought into direct contact with the material they are measuring.
- The display is easy to read.

*The disadvantages of thermocouples:*

- The primary disadvantage of thermocouples is their low accuracy which is typically greater than $1°C$.
- They are vulnerable to corrosion.
- They require the calibration of thermocouples—the process which is tedious and difficult.
- The measuring process (especially signal disposal) is complex; it provides many sources of errors.

## 6.5    OPTICAL PYROMETRY

*Optical pyrometry* is a non-contact measurement of the temperature of an object by estimating its emissivity—the ratio of energy radiated by the material to energy radiated by a black body at the same temperature. An instrument that measures the temperature by means of pyrometry is called an *optical pyrometer* and consists of a radiation source, calibrated lamp, filter with a narrow band of transmission, and a detector (Figure 6.5a).

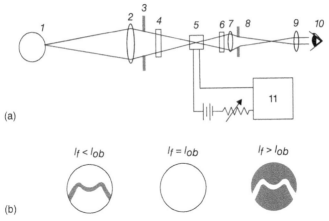

**(a)**

$I_f < I_{ob}$      $I_f = I_{ob}$      $I_f > I_{ob}$

**(b)**

**FIGURE 6.5**  Optical pyrometry. (a) Design of optical pyrometer. 1, source of light; 2, lens; 3, diaphragm; 4, filter; 5, incandescent lamp; 6, red filter; 7, lens; 8, diaphragm; 9, ocular; 10, observer; 11, readout system. (b) Image of an internal lamp filament in the eyepiece.

The measurement is based on comparing the brightness of the object $I_{ob}$ being measured with the brightness of the wire filament $I_f$ of the lamp. The brightness and temperature of the wire is regulated by adjusting the electric current that passes through the wire and the temperature is read when the brightness of the object and filament are equal (Figure 6.5b).

*The advantages of an optical pyrometer:*

- It has a very high accuracy.
- The absence of any direct contact between the optical pyrometer and the object.
- The distance between both of them is not at all a problem; thus, the device can be used for remote sensing.

*The disadvantages of an optical pyrometer:*

- The device can be used only for measuring a minimum temperature of 700°C.
- The device cannot be used for obtaining continuous values of temperatures at small intervals since it is manually operated.

## 6.6  INFRARED THERMOMETERS

Thermal radiation is electromagnetic radiation generated by the thermal motion of charged particles in matter. A matter with a temperature greater than absolute zero emits thermal radiation. The rate at which an object emits radiant energy is proportional to the fourth power of its absolute temperature. This is known as *Stefan's law* and is expressed in equation form as:

$$P = \sigma A e T^4,$$

(6.3)

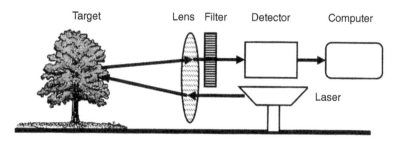

**FIGURE 6.6**    Block diagram of an infrared laser thermometer.

where $P$ is the power radiated by the body in watts (or joules per second); $\sigma$ is a constant equal to $5.6696 \times 10^{-8}$ W/m$^2 \cdot$K$^4$; $A$ is the surface area of the object in square meters; $e$ is a constant called the *emissivity*; $T$ is temperature in Kelvin. The value of $e$ can vary between zero and unity, depending on the properties of the surface.

The temperature range of an infrared thermometer is from $-50$ to $+500°$C.

An infrared laser thermometer contains an optical system that directs the laser radiation onto the target and focuses the infrared radiation flux through a lens and filter on a photodetector (Figure 6.6). The detector is a thermal sensor that converts infrared radiation into an electrical signal that is proportional to the intensity of the radiation. A filter makes it possible to utilize only the desired wavelength. Infrared thermometers have a microcomputer that calculates the temperature from the radiation signal and power source.

*The advantages of an infrared thermometer:*

- Accuracy of infrared thermometers is usually less than $1°$.
- They can be used for remote measurement.
- The device is characterized by quick response, versatility, portability, non-invasiveness.
- They have no effect on the measured object.

*The disadvantages of an infrared thermometer:*

- They are very expensive (approximately $240.00).
- Can be temporarily affected by environmental factors (frost, moisture, dust, fog, smoke, rapid changes in ambient temperature).
- They can only measure the surface temperature.
- They cannot "see" through glass, liquids, or other transparent surfaces.

## 6.7    HEAT FLUX MEASUREMENT

A heat-flux sensor is intended for automated measurement of heat flux. This type of device is designed to measure the heat transferred by conduction, convection, or radiation. It generates an electrical signal proportional to the total heat flux through the surface of the sensor.

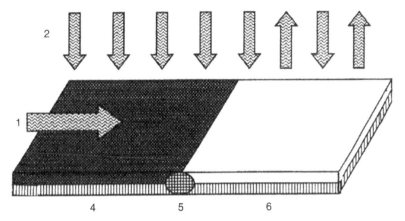

**FIGURE 6.7**     Principle of operation of heat flux sensor RC01. 1, convective heat; 2, radiative heat; 3, heat flux sensor with black-body detector, sensitive to convective and radiative heat flux effects; 4, temperature sensor (for sensor temperature measurement); 5, heat flux sensor with gold plated detector, convective heat effects only.

The heat flux sensor (such as an RC01 radiation/convection (radcon) sensor, Hukseflux Thermal Sensors) consists of two detectors, the surface of one is covered with gold and is sensitive only to convective flux, while the surface of the other detector is black and is designed to record both convective and radiative heat fluxes (Figure 6.7).

Under the surface of the detector is a thermocouple made of chromel–alumel alloy (chromel contains 90% nickel and 10% chromium; alumel contains 95% nickel, 2% magnesium, 2% aluminum, and 1% silicon). The dimensions of the sensor are 22 × 10 × 3 mm and of the entire device is 65 × 65 × 13 mm.

Another type of heat flux sensor (such as a HFP01, Hukseflux Thermal Sensors) contains a thermocouple (copper–constantan) that is placed in a ceramic–plastic cage (Figure 6.8).

This thermocouple measures the temperature difference generated at the boundary planes of the sensor. The output signal, measured in millivolts, is proportional to the local heat flux passing through the sensor.

*The advantages of a heat flux sensor:*

- Important characteristics of the sensor are the linearity of the temperature to the output signal and the temperature range for the RC01 sensor is from −200°C to +1250°C.
- This device which is equipped with thermopiles that have high stability, good signal-noise ratio.

*The disadvantages of a heat flux sensor:*

- The temperature dependence of the thermocouple provokes the temperature dependence and the corresponding non-linearity of the heat flux sensor.
- It demonstrates low sensitivity of the output.

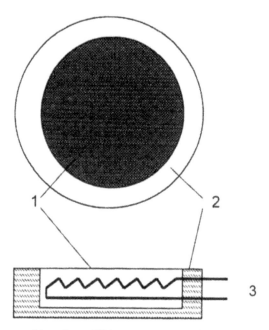

**FIGURE 6.8**    Sensor of heat flow HFP01. 1, sensor area; 2, guard of ceramic–plastic composite; 3, cable, standard length is 5 m.

## 6.8    METHOD OF SCINTILLOMETRY

The propagation of optical radiation in the atmosphere can be accompanied by a distortion of its characteristics—intensity, polarization, and phase. The primary causes of these distortions are scattering and absorption of radiation by atmospheric gases and particles, which attenuate the energy of the beam. Diffusion is the most important factor affecting the intensity of radiation and the most significant mechanism on which the radiation interacts with the atmosphere forming small fluctuations in the refractive index $n$ of the air. These turbulent refractive index fluctuations induce corresponding fluctuations of optical radiation, known as *scintillation* (Meijninger, 2003; Kleissl et al., 2007).

A beam of radiation is transmitted over a path and the fluctuations in the radiation intensity at the receiver are analyzed to determine the variation in refractive index along the path and, as a result, the turbulent characteristics of the atmosphere.

A *scintillometer* is an optical device that consists of a transmitter (radiation source) and a receiver (Figure 6.9). The transmitter generates optical radiation that undergoes fluctuations during its passage through the atmospheric boundary layer. These fluctuations are registered by a receiver (Odhiambo and Savage, 2009). Since the fluctuations are induced by temperature fluctuations, it is possible to record the fluxes of sensible and latent heat along the propagation path of the optical beam, which varies from 100 to 5000 m.

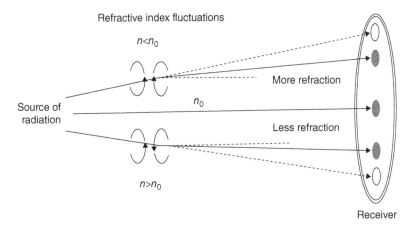

Refractive index fluctuations

$n < n_0$

More refraction

Source of radiation

$n_0$

Less refraction

$n > n_0$

Receiver

**FIGURE 6.9**    Diagram illustrating the principle of scintillometry. The transmitter generates optical radiation passing through a turbulent atmospheric layer, and the receiver analyzes the intensity fluctuations that characterize the turbulent eddies.

If the aperture of the scintillometer is large enough and it exceeds the size of large eddies, the receiver will be able to average the signal received throughout the aperture. This means that small fluctuations in the optical signal, which are presented by dark and bright spots will compensate with each other within the aperture of the scintillometer.

*The advantages of a scintillometer:*

- The sensible and latent heat fluxes and evapotranspiration can be measured.
- Sensible heat fluxes over distances from 100 m to 4.5 km can be derived.
- Limited power is required; therefore a simple solar-charged battery can be sufficient.
- The built-in data logger stores several months of measurements and results.
- All the scintillometers (Kipp & Zonen) are designed to offer easy installation and low-maintenance operation.

## REFERENCES

Kleissl, J., Gomez, J., Hong, S.-H., Hendrickx, J.M.H., Rahn, T., and Defoor, W.L. 2007. *Large Aperture Scintillometer Intercomparison Study.* Boundary Layer Meteorology. Manuscript No. 335.

Meijninger, W.M.L. 2003. *Surface Fluxes over Natural Landscapes Using Scintillometry.* Wageningen University and Research Centrum, Wageningen.

Odhiambo, G.O. and Savage, M.J. 2009. Surface layer scintillometry for estimating the sensible heat flux component of the surface energy balance. *S. Afr. J. Sci.* 105:N.5–6.

# PRACTICAL EXERCISE 3

# MODELING VERTICAL CHANGES IN AIR TEMPERATURE

## 1   MEASUREMENT OF TEMPERATURE ABOVE UNIFORM SURFACE

The profile of temperature above uniform surface within stationary conditions is described by the equation (Campbell and Norman, 1998):

$$t(z) = t_0 - \frac{H}{0.4\rho c v^*} \ln \frac{z-d}{z_H}, \tag{P3.1}$$

where $t(z)$ is the mean air temperature at height $z$; $t_0$ is the apparent aerodynamic surface temperature; $H$ is the sensible heat flux from the surface to the air; $\rho c$ is the volumetric specific heat of air (1200 J/m$^3\cdot$°C at 20°C and sea level); $d$ is the zero plane displacement; $v^*$ is the friction velocity; $z_H$ is a roughness parameter for heat transfer.

For a flat smooth surface, $d = 0$. The uniform surface of vegetation is characterized by the following relations:

$$z_H \cong 0.02\,h; \;\; d \cong 0.6\,h, \tag{P3.2}$$

where $h$ is the canopy height.

---

*Methods of Measuring Environmental Parameters*, First Edition. Yuriy Posudin.
© 2014 John Wiley & Sons, Inc. Published 2014 by John Wiley & Sons, Inc.

**Example**   The results of the temperature measurement above a 15-cm high crop on a clear day are presented in the following table.

| Height $z$, m | 0.2 | 0.5 | 1.0 | 1.5 |
|---|---|---|---|---|
| Temperature $t$, °C | 27 | 25 | 23 | 21 |

Find the aerodynamic surface temperature $t_0$.

**Solution**   Application of Equation P3.2 makes it possible to determine the roughness parameter and the zero plane displacement: $z_H = 0.003$ m; $d = 0.09$ m.

The following values of $(z - d)/z_H$ and $\ln(z - d)/z_H$ can therefore be obtained:

| Height $z$, m | 0.2 | 0.5 | 1.0 | 1.5 |
|---|---|---|---|---|
| Temperature $t$, °C | 27 | 25 | 23 | 21 |
| $(z - d)/z_H$ | 36.7 | 136.7 | 303.3 | 470 |
| $\ln[(z - d)/z_H]$ | 3.60 | 4.92 | 5.71 | 6.15 |

It is possible to determine the aerodynamic surface temperature $T_0$ from the graph $\ln[(z - d)/z_H] = f[t(z)]$ through the extrapolation of a graph line to zero (Figure P3.1). The intercept is at $t_0 = 37$°C.

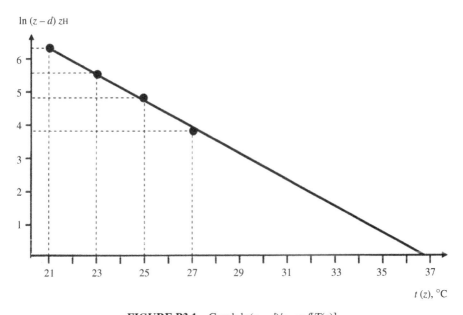

**FIGURE P3.1**   Graph $\ln(z - d)/z_H = f[T(z)]$

**Example** The mean temperature at a height of 3 m above the soil surface is 4°C and at a height 1 m the temperature is 2°C. If the crop below the point where these temperatures are measured is 70 cm tall, will the crop experience a temperature below freezing?

### Algebraic Solution

Let us get into account that the value $H/(0,4c\rho\upsilon^*)$ is constant for all heights. Then Equation P3.1 can be written for each height as follows:

$$4 = t_0 - A \ln \frac{3 - 0.6 \times 0.7}{0.02 \times 0.7} = t_0 - 5.216 A;$$

$$2 = t_0 - A \ln \frac{1 - 0.6 \times 0.7}{0.02 \times 0.7} = t_0 - 3.724 A,$$

where $A = H/(0.4c\rho\upsilon^*)$.

Subtracting the second equation from the first, and solving for $A$ gives:

$$4 = t_0 - 5.216 A;$$

$$2 = t_0 - 3.724 A;$$

$$4 + 5.216 A = 2 + 3.724 A;$$

$$2 = -1.492 A;$$

$$A = -1.34 \,°C.$$

The aerodynamic temperature is

$$t_0 = 2 + 3.724(-1.34°C) = -2.99°C.$$

The temperature at the height $h = 0.7$ m is

$$t(0.7) = -2.99 + 1.34 \ln \frac{0.7 - 0.6 \times 0.7}{0.02 \times 0.7} = 1.025°C.$$

So the top of the canopy is not below the freezing temperature.

### Graphic Solution

Fill a table for three temperatures.

| Height $z$, m | 3 | 1 | 0.7 |
|---|---|---|---|
| Temperature $t$, °C | 4 | 2 | ? |
| $(z - d)/z_H$ | 184.29 | 41.43 | 20 |
| $\ln(z - d)/z_H$ | 5.216 | 3.724 | 2.995 |

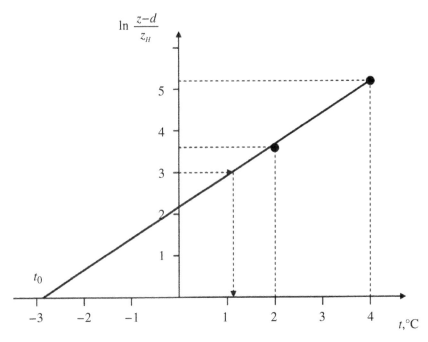

**FIGURE P3.2**   Graph $\ln(z-d)/z_H = f(t)$

The graph $\ln(z-d)/z_H = f(t)$ is presented in Figure P3.2:
It is possible to determine the mean temperature at a height 0.7 m above the soil surface: $t(0.7) = 1.025°C$. The intercept is at $-2.99°C$, which is the aerodynamic surface temperature $t_0$.

**Control Exercise**   The following temperatures were measured at the various heights (see the following table).

| Height $z$, m | Mean temperature of the air, °C |
|---|---|
| 6.4 | 31.08 |
| 3.2 | 31.72 |
| 1.6 | 32.37 |
| 0.8 | 33.05 |
| 0.4 | 33.80 |
| 0.2 | 34.91 |
| 0.1 | 36.91 |

Perform the following tasks:

1. Construct a graph $\ln(z - d)/z_H = f(<t>)$, assuming $h = 0.15$ m.
2. Find the aerodynamic temperature $t_0$.

## 2    MEASUREMENT OF SENSIBLE HEAT FLUX

Equation P3.1 can be written for two temperatures as

$$t(z_1) = t_0 - A \ln \frac{z_1 - d}{z_H};$$ (P3.3)

$$t(z_2) = t_0 - A \ln \frac{z_2 - d}{z_H};$$ (P3.4)

Excluding aerodynamic temperature from Equations P3.3 and P3.4 we can get the expression:

$$t(z_1) - t(z_2) = A \left[ \ln \frac{z_1 - d}{z_H} - \ln \frac{z_2 - d}{z_H} \right];$$ (P3.5)

or

$$t(z_1) - t(z_2) = A \ln \frac{z_1 - d}{z_2 - d}.$$ (P3.6)

Hence we can determine the sensible heat flux $H$ from the surface to the air:

$$H = A \cdot 0.4 \, c\rho v^* = [T(z_1) - T(z_2)] / \ln \frac{z_1 - d}{z_2 - d} \cdot 0.4 \, c\rho v^*$$ (P3.7)

Let us consider the result of measuring temperature at two different heights: $t_1 = 3°C$ at the height $z_1 = 2$ m; $t_2 = 1°C$ at the height $z_2 = 1$ m. Assume the canopy is 50-cm tall; the volumetric specific heat of air $c\rho = 1200$ J/m$^3 \cdot °$C; the friction velocity is $v^* = 0.15$ m/s.

Substituting the numerical data into Equation P3.7 gives

$$1 - 3 = A \ln \frac{2 - 0.6 \cdot 0.5}{1 - 0.6 \cdot 0.5} = A \ln \frac{1.7}{0.7} = A2.43.$$

Hence,

$$A = H/(0.4 c\rho v^*) = -2.25 \, °C.$$

So, the sensible heat flux $H$ from the surface to the air is

$$H = 0.4 \times 1280 \times 0.15 \, (-2.25) = -173 \ \text{W/m}^2.$$

## QUESTIONS AND PROBLEMS

**1.** Describe the change in atmospheric temperature with altitude.

**2.** What is the change of temperature in soil and water?

**3.** What is the principle of operation of a bimetallic thermometer?

**4.** Name the benefits of a platinum thermistor.

**5.** Explain the principle of operation of a thermoelectric thermometer.

**6.** Explain the difference between sensible and latent heat.

**7.** Define heat flux and heat flux density.

**8.** How does a heat flow sensor work?

**9.** Explain the operating principle of scintillometry.

## REFERENCE

Campbell, G.S. and Norman, J.M. 1998. *Environmental Biophysics*. 2nd edition. Springer, New York.

## FURTHER READING

Benedict, R.P. 1969. *Fundamentals of Temperature, Pressure, and Flow Measurements*. John Wiley & Sons, New York.

Deacon, E.L. 1969. Physical processes near the surface of the earth. In: *World Survey of Climatology. Vol. 2. General Climatology* (ed. H.E. Landsberg). Elsevier, Amsterdam.

Gates, D.M. 1980. *Biophysical Ecology*. Dover Publications, Inc., Mineola, NY.

Guyot, G. 1998. *Physics of the Environment and Climate*. John Willey & Sons, Chichester.

Montheith, J.L. and Unsworth, M. 1990. *Principles of Environmental Physics*, 2nd edition. Edward Arnold, London.

Willmer, P., Stone, G., and Johnston, I. 2000. *Environmental Physiology of Animals*. Blackwell Science Ltd.

## ELECTRONIC REFERENCES

AMRL. In-Focus Newsletter Spring 2011. -Knake M., The Anatomy of a Liquid-in-Glass Thermometer by Maria Knake, http://www.amrl.net/AmrlSitefinity/default/aboutus/newsletter/Newsletter_Spring2011/8.aspx (accessed March 29, 2013).

DialTempTM, *Bi-Metal Stem Thermometers*, 3" Heads, http://www.omega.com/ppt/pptsc_lg.asp?ref=A_R_DIALTEMP&Nav= (accessed March 30, 2013).

Posudin, Y. 2011. *Environmental Biophysics*. Fukuoka-Kiev, http://www.ekmair.ukma.kiev.ua/handle/123456789/951 (accessed March 31, 2013).

# 7

# HUMIDITY

## 7.1 DEFINITION OF HUMIDITY

Water can exist in the atmosphere in three phases: gaseous, liquid, and solid. *Vapor* is the gaseous phase of a substance that is normally a solid or liquid. The concentration of water molecules in the gaseous phase increases during evaporation. An equilibrium is established when the number of molecules escaping from the liquid equals the number being recaptured by the liquid.

*Humidity* is the amount of water vapor in the air.

## 7.2 PARAMETERS OF HUMIDITY

*Absolute humidity* ($a$) is the mass ($m$) of water vapor (grams) per unit volume ($V$) of moist air ($m^3$): $a = m/V$.

*Partial pressure* ($e$) is the pressure exerted by one gas in a mixture of gases.

The saturation vapor pressure ($E$) is the static pressure of a vapor when the vapor phase of the atmosphere is in equilibrium with the liquid phase of atmosphere. The saturation vapor pressure of the atmosphere depends on the absolute temperature;

$$\lg E = 9.4 - \frac{2345}{T}, \tag{7.1}$$

where $E$ is the saturation vapor pressure (millibars or hectopascals) and $T$ is the absolute temperature (kelvin).

*Methods of Measuring Environmental Parameters*, First Edition. Yuriy Posudin.

*Relative humidity* (*r*) is the ratio of the actual partial vapor pressure of water to the saturation vapor pressure at a given temperature: $r = \frac{e}{E} \cdot 100$.

*Vapor deficit* (*d*) is the difference in vapor pressure between saturated and ambient air: $d = E - e$.

*Dew point* ($T_d$) is the temperature to which, if unsaturated air is cooled, the water vapor becomes saturated.

## 7.3    EFFECT OF HUMIDITY ON LIVING ORGANISMS

### 7.3.1    Effect of Humidity on Human Organism

The human organism participates in such heat exchange processes as evaporation (through perspiration), heat convection, and thermal radiation. The first process depends strongly on the humidity: under high levels of air humidity, the effectiveness of sweating to cool the human body is reduced due to the limitation of the evaporation of perspiration from the human surface and the process of maintaining body temperature is rather embarrassing. The combination of high temperature and high humidity of the atmosphere provokes the restriction of heat exchange between blood flows in the body and surrounding air through conduction; this situation causes hyperpyrexia that is an excessive elevation of body temperature greater than or equal to 41.1°C (106°F). The quantity of blood that reaches the internal organs of the body is decreased and this shortage of the blood leads to various negative consequences such as heat stroke (or hyperthermia), an acute condition that occurs when the body produces or absorbs more heat than it can dissipate.

The optimal level of relative humidity for an average person in the home ranges between 30% and 65–70%, but the recommended interval is between 30% and 50%.

### 7.3.2    Effect of Humidity on Microorganisms

The survival of airborne microflora in indoor air depends on the relative humidity. The activity of infectious bacteria and viruses is minimized if the range of indoor relative humidity is between 40% and 70%. Allergenic mites and fungal populations stop their activity under relative humidity less than 50%, while these populations reach the maximum at 80%. Mold spores and dust mites can provoke such human diseases as allergy and asthma.

### 7.3.3    Effect of Humidity on Animals

Relative humidity plays an important role in the viability of terrestrial organisms. The air that surrounds the animal maintains less water than its own body. Loss of water by living organism is realized by the final products of metabolism. Supply of water to the organism is provided during feeding and drinking. All living organisms have their own ability to adapt to air humidity. Humidity also determines the distribution of terrestrial animals which remain to be "aquatic organisms" due to their ability to maintain water balance, such as amphibian, terrestrial crustacean, nematode, molluscs, and so on. Usually they inhabit the areas with the relative humidity near 100%.

### 7.3.4   Effect of Humidity on Plants

Relative humidity determines the rate of transpiration of plants. The fact is that the substomatal leaf space is usually close to saturation level while the vapor pressure of atmospheric air depends on either the relative humidity or temperature of air. If the relative humidity of the surrounding air rises, the transpiration rate falls because the water evaporates easier into drier air than into saturated air. Quite the reverse, a low level of relative humidity induces the diffusion of water from substomatal leaf space into outer air.

# 8

# MEASUREMENT OF AIR HUMIDITY

## 8.1  HYGROMETERS

The instrument that is intended for measuring atmosphere humidity is called a
*hygrometer.* There are several types of hygrometers—psychrometer, hair hygrometers, capacitive hygrometers, condensation or dew point hygrometers, electrolytic
hygrometers, and radiation absorption hygrometers. The following critiques the principles of psychrometer and hygrometers that can be used for the automated measurement of atmospheric humidity.

## 8.2  ASSMANN PSYCHROMETER

The Assmann psychrometer is an instrument used to determine air humidity. The
Assmann psychrometer was invented in the late nineteenth century by Adolph Richard
Aßmann (1845–1918).

Figure 8.1 shows the design of an aspiration psychrometer.

An Assmann psychrometer consists of two thermometers—the first is a dry-bulb
thermometer that measures the air temperature $T_a$, and the second has a muslin jacket
that is located in a reservoir of water and measures the wet-bulb temperature $T_w$.

The psychrometer has a fan that draws air through the tubes; the fan is driven by
a clockwork mechanism to ensure a consistent speed. The ventilating speed at the
bulbs averages 3 m/s.

*Methods of Measuring Environmental Parameters*, First Edition. Yuriy Posudin.
© 2014 John Wiley & Sons, Inc. Published 2014 by John Wiley & Sons, Inc.

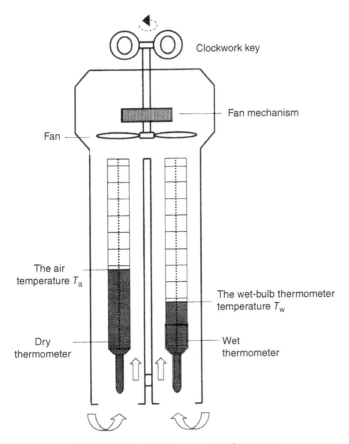

**FIGURE 8.1**    An Assmann psychrometer.

The wet-bulb temperature is lower than that of the air due to evaporation of water from the muslin.

The relationship between the two thermometer readings can be written as a *psychrometer equation:*

$$e = E_1 - A(T_a - T_w)p_A \tag{8.1}$$

where $e$ is the water vapor pressure; $E_1$ the saturated vapor pressure at the temperature of the wet-bulb thermometer; $A$ the psychrometric coefficient ($6.62 \times 10^{-4}$ K); $T_a$ the air temperature, $T_w$ the wet-bulb thermometer temperature; and $p_A$ the atmospheric pressure (mmHg or Pa).

The psychrometric coefficient $A$ depends on the ventilation rate: the value $A$ decreases from $13.0 \times 10^{-4}$ K for ventilation rate 0.12 m/s to $6.7 \times 10^{-4}$ K for ventilation rate 4.00 m/s.

Besides, the errors of the Assmann psychrometer are negligible if the air temperature is higher than 10°C; it is necessary to take into account the errors at low temperatures.

**Exercise**    Use Equation 8.1 and plot the graph of dependence of the saturation vapor pressure of the atmosphere on the absolute temperature.

Automated measurement of atmospheric humidity utilizes an Assmann psychrometer located in a *Stevenson screen* or instrument shelter that is enclosed to shield the meteorological instruments against precipitation and direct heat radiation from outside sources, while still allowing air to circulate freely around them. This system is equipped with a water pump, electric contact thermometers, electric fan, and data lodger.

*The advantages of an Assmann psychrometer:*

- High accuracy (corrected accuracy can be better than ±0.1°C).
- Can be used as a reference instrument for other types of psychrometers.
- Wet- and dry-bulb thermometers are cheap, robust, and reliable.

*The disadvantages of an Assmann psychrometer:*

- Moving air must be used.
- The errors are high at low temperatures.
- The muslin of the wet-bulb thermometer must be wetted before use; the device requires the reservoir to be kept full.
- The psychrometric coefficient depends on the rate of ventilation and air temperature.
- The results of measurements depend on the atmospheric pressure.
- The psychrometer fan takes the air only along a few centimeters, which makes it impossible to determine accurately the humidity in certain areas.
- The necessity to use psychrometric tables.
- The spring-wound fan provides 8 minutes of aspiration per winding.

**Example**    An Assmann psychrometer gives an air temperature $T_a = 27°C$ and a wet-bulb temperature $T_w = 18°C$. Find the vapor pressure $e$, saturation vapor pressure $E$, relative humidity $r$, vapor deficit $d$, and dew point $T_d$. Atmospheric pressure $p_A = 1.01 \times 10^5$ Pa.

**Solution**    Table 2 (see Appendix) gives the saturation vapor pressure $E_1$, which corresponds to the temperature of wet thermometer $T_w = 18°C$:

$$E_1 = 2063 \text{ Pa.}$$

The saturation vapor pressure $E$, which corresponds to the temperature of dry thermometer $T_a = 27°C$:

$$E = 3565 \text{ Pa.}$$

Using the psychrometer Equation 8.1 gives the water vapor pressure:

$$e = E_1 - A(T_a - T_w)p_A$$

$$= 2063 \text{ Pa} - 6.62 \times 10^{-4} \text{ K}^{-1} (27°C - 18°C) \cdot 1.01 \times 10^5 \text{ Pa} = 1461 \text{ Pa}.$$

The relative humidity is

$$r = \frac{e}{E} = \frac{1461}{3565} = 0.41.$$

The vapor deficit is determined as

$$d = E - e = 3565 - 1461 = 2104 \text{ Pa}.$$

The dew point temperature $T_d$ can be determined from the following expression (Buck, 1981):

$$T_d = \frac{c \ln(e/a)}{b - \ln(e/a)},$$

where $a = 0.611$ kPa; $b = 17.502$; and $c = 240.97°C$. Using this expression the dew pint temperature is obtained as

$$T_d = \frac{240.97(\ln e - \ln 0.611)}{17.502 - \ln e + \ln 0.611} = \frac{240.97(\ln e + 0.493)}{17.009 - \ln e}$$

$$= \frac{240.97 \cdot \ln 1.461 + 118.80}{17.009 - \ln 1.461} = \frac{240.97 \cdot 0.379 + 118.80}{17.009 - 0.379} = 12.64°C.$$

**Control Exercise**   Find vapor pressure, saturated vapor pressure, and relative humidity if air temperature is 24°C and wet-bulb temperature is 18°C.

## 8.3  HAIR HYGROMETER

The operative principle of a *hair hygrometer* is the ability of fat-free hair to change its length in response to changing humidity. Water vapor can condense in the capillary pores of human hair and increased humidity reduces the curvature of the meniscus of the water in the pores causing the hair to extend. The length of hair elongation is proportional to the logarithm of the relative humidity. A number of hairs are used in the *hygrograph*, a device used for the continuous recording of relative humidity (Figure 8.2).

Other materials may be used as sensors in a hygrometer, for example, nylon, cotton, or the intestinal membrane of cows or pigs.

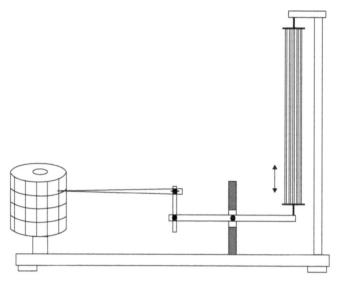

**FIGURE 8.2**  Diagram of a hair hygrograph.

*The advantages of a hair hygrometer:*

- It is simple in construction and cheap.
- The measurements are independent of the temperature; a temperature change from −30°C to 40°C alters hair elongation by only 1–3%.
- Inexpensive.
- Accuracy: ±3% at <50% RH and ±5% at >50% RH.

*The disadvantages of a hair hygrometer:*

- The time response increases with temperature. The decreasing temperature leads to a significant increase in response time.
- Another deficiency is caused by hysteresis; if a hair hygrometer is kept for several days in a dry place, humidity measurements can differ significantly (i.e., up to 15%) from those obtained by hygrometers kept in a damp room. Therefore, this type of hygrometer requires calibration.
- Errors are induced by dust, pollutants, rain, or fog.
- Hair must be cleaned and replaced often.
- Each hair has different expansion coefficients in bundle.
- Changes in zero drift can be significant.

## 8.4  CAPACITIVE HYGROMETER

The operative principle of a capacitive hygrometer is the humidity-dependent change in the capacity of a condenser. The condenser is formed by a hygroscopic polymer

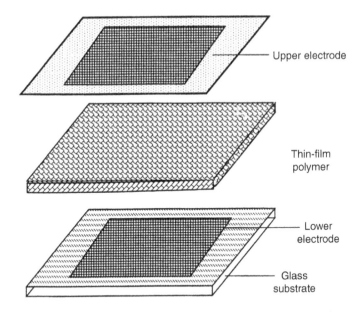

**FIGURE 8.3**    Diagram showing the principle of operation of a capacitive hygrometer.

film upon which are placed porous metallic electrodes (Figure 8.3). The polymer film absorbs or desorbs water molecules and the change in its volume results in a corresponding change in the distance between the electrodes and thereby the capacity of the condenser ($C = \varepsilon\varepsilon_0 A/d$). The change in capacity is proportional to the change in the relative humidity. The sensor is positioned on a square (6 mm × 6 mm) base of thin glass.

*The advantages of a capacitive hygrometer:*

- It has small size.
- It does not have any moving parts.
- There is the minimal effect of external temperature and water on the accuracy of measurements and the linearity of the response.
- Capacitive hygrometers are commonly installed in automated weather stations.

*The disadvantages of a capacitive hygrometer:*

- Its electronics is complex.
- It is affected by pollution.

## 8.5    CONDENSATION HYGROMETER

This method of determining the absolute humidity is based on measuring the dew point. If the surface of a metallic mirror (e.g., silver or copper which have good thermal

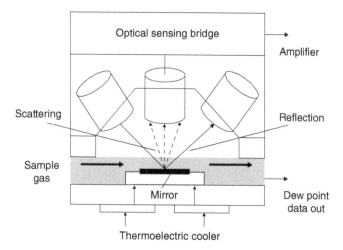

**FIGURE 8.4**    Diagram of dew point hygrometer.

conductivity) is cooled, moisture condenses on its surface. Surface temperature at this point is very close to the temperature at which water on the mirror surface is in equilibrium with the water vapor pressure in the gaseous atmosphere above the surface which is the dew point.

The mirror is cooled by a semiconductor element utilizing the Peltier effect. The surface of the mirror is illuminated by a photodiode. The formation of dew is accompanied by light scattering and decreased reflected light that corresponds to a decline in the photodetector output (Figure 8.4).

*The advantages of a condensation hygrometer:*

- The device demonstrates high sensitivity and the ability to measure the absolute humidity over in a wide temperature range (−80 − +100°C) with an accuracy of about 1°C.
- It is reliable.
- The devices of this type can be used in extreme climatic conditions—in cold and tropical environments.
- Can be used in remote sensing.

*The disadvantages of a condensation hygrometer:*

- There is the distortion of measurements at low temperatures.
- It is necessary to control the quality of the mirror's surface.
- The device has complex design and electronics.
- It has high cost.
- This hygrometer requires air flow over the surface.

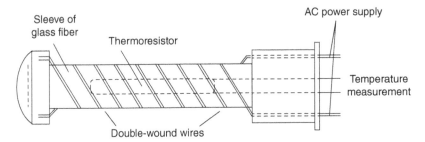

**FIGURE 8.5**   Schematic representation of the sensor of an electrolytic hygrometer.

## 8.6   ELECTROLYTIC HYGROMETER

The electrolytic hygrometer contains a tube coated with a thin layer of phosphorus pentoxide ($P_2O_5$) that fills the space between a spiral of two parallel conductors (Figure 8.5). If the gas passes through the tube, the moisture in the gas is absorbed by this layer. An electric current passing through the conductors breaks down water into hydrogen and oxygen via electrolysis. The current required for electrolysis is proportional to the volumetric concentration of water in the gas.

*The advantages of an electrolytic hygrometer:*

- It is available over a wide range of relative humidity (between 10% and 95%).
- It has high accuracy.

*The disadvantages of an electrolytic hygrometer:*

- It requires constant power supply.
- The ventilation rate affects the results of measurements.
- The device is sensitive to ambient pollution, salts, and water.
- The response time is rather long.

## 8.7   RADIATION ABSORPTION HYGROMETER (GAS ANALYZER)

Water vapor displays intense absorption bands in the infrared part of the spectrum (i.e., 1.10, 1.38, 1.87, 2.70, 6.30 μm). Estimation of the absorption at these bands makes it possible to measure the humidity of the air.

   A diagram of a two-beam gas analyzer is shown in Figure 8.6. Radiation from infrared sources passes through two cells, one filled with the reference (dry air) and the other with the sample (gas to be analyzed). If the intensity of the two beams is the same, the system indicates a state of equilibrium. If the gas concentration changes, an imbalance in the system occurs with the signal being amplified and recorded.

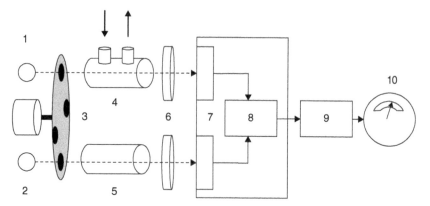

**FIGURE 8.6**  Two-beam diagram of the gas analyzer. 1,2, the sources of infrared radiation; 3, modulator; 4, cuvette with gas that is analyzed; 5, reference cuvette; 6, optical filters; 7, detectors; 8, system of registration; 9, amplifier; 10, indicator.

*The advantages of a radiation absorption hygrometer:*

- It can be used for remote sensing.
- This device provides measuring humidity profiles in flux studies.
- A two-beam system makes it possible to exclude the changes in the intensity of sources of radiation.
- These analyzers are portable, reliable, and robust.

*The disadvantages of a radiation absorption hygrometer:*

- It needs a very accurate thermostat to keep the detector at a temperature above that of the surroundings.
- The hygrometers of this type require daily calibration.
- The analyzers of this type are expensive.

## 8.8    AN OPEN-PATH SYSTEM FOR MEASURING HUMIDITY

A hygrometer with an open light circuit can be used to avoid the influence of pollution on the measurement results. Such open systems contain a source of modulated optical radiation that allows neglecting the influence of ambient light. The hygrometer also contains a disk with interference filters, whose bandwidth corresponds to wavelengths of 2.7 μm (water absorption) and 1.61 μm (where absorption of water is absent). A diagram showing the principle of operation of such a hygrometer is presented in Figure 8.7.

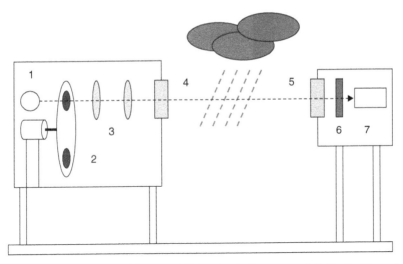

**FIGURE 8.7**  Hygrometer with opened light circuit. 1, source of light; 2, filter wheel; 3, collimator; 4,5, polyethylene windows; 6, interference filter; 7, thermostated detector.

*The advantages of an open-path system:*

- Such a system can be used remotely in the field conditions.
- It is possible to realize continuous humidity monitoring over long periods of time.

## 8.9   REMOTE SENSING HUMIDITY

*Satellites*. Humidity can be measured on a global scale using satellites. The basis of measurement is detecting the concentration of water vapor in the troposphere at altitudes of 4–12 km using a sensor sensitive to infrared radiation, which is absorbed by water vapor. Satellite systems are used to monitor climate and weather forecasting situations. For example, the Nimbus 7 temperature-humidity infrared radiometer (THIR) detected the 6.5- to 7.0-$\mu$m region (water vapor) and obtained the information on the moisture and cirrus cloud content of the upper troposphere and stratosphere. The ground resolution of the system was 20 km for the water vapor channel.

*Airborne Soil Moisture Measurement*. It is possible to use gamma radiation from radioactive materials that can be detected by low-flying aircrafts or helicopters. The gamma radiation originates from the potassium, uranium, and thorium radioisotopes in the soil. About 91% of the gamma radiation is emitted from the top 10 cm of the soil, 96% from the top 20 cm, and 99% from the top 30 cm.

The principle of operation of registration of airborne gamma radiation is based on the attenuation of natural terrestrial gamma radiation by the mass of water contained in the soil and snow. There is certain difference between the results of measuring natural terrestrial gamma radiation flux for wet and dry soils.

The system of registration of airborne gamma radiation consists of NaI(Tl) scintillation detectors, a pulse height analyzer, mini computer, temperature, pressure, and radar altitude sensors.

For example, the model Vaisala HUMICAP HMT100 of humidity sensor for airborne gamma radiation measurement of snow and soil moisture is characterized by accuracy ±1.7% of relative humidity.

### Extreme Situations

*The most humid cities are such cities in South and Southeast Asia as Kolkata, Chennai, and Cochin in India, Manila in the Philippines, Mogadishu in Somalia, Bangkok in Thailand, and the extremely humid Lahore in Pakistan.*

*Darwin is the most humid city in Australia; it experiences an extremely humid wet season from December to April. Kuala Lumpur in Malaysia and Singapore, a Southeast Asian city-state, have very high humidity all year round.*

# PRACTICAL EXERCISE 4

# MEASURING PARAMETERS OF HUMIDITY

## 1  OBJECTIVES

1. To measure parameters of humidity such as partial pressure ($e$), saturation vapor pressure ($E$), relative humidity ($r$), vapor deficit ($d$), dew point ($T_d$) using an Assmann psychrometer.
2. To understand the use of the psychrometric formula and tables.
3. To determine humidity parameters under varying atmospheric conditions.

## 2  MATERIALS SUPPLIED

1. Aspirative Assmann psychrometer (accuracy of each thermometer is $\pm 0.2°C$).
2. Barometer.
3. Psychrometric tables.

## 3  PRINCIPLE OF OPERATION

Assmann psychrometer consists of two mercury thermometers. The first one, the dry thermometer, provides the measurement of the temperature of the surrounding air $T_a$; the second one, the wet thermometer, contains a wet muslin that covers the thermometer bulb.

*Methods of Measuring Environmental Parameters*, First Edition. Yuriy Posudin.

The temperature of the wet thermometer is always lower than the temperature of the dry thermometer because water evaporates from the wet-bulb thermometer, removing high-energy molecules of water and cooling it down.

## 4  EXPERIMENTAL PROCEDURE

1. Wet the muslin that covers the bulb of the wet thermometer with distillated water;
2. Initiate the rotation of the mechanism of a spring-wound fan;
3. Watch the change of temperature of the wet thermometer and write the value of $T_w$;
4. Wait until the fan stops rotating and the temperature of the wet thermometer returns to the initial state;
5. Read the atmospheric pressure $p_A$ of the barometer;
6. Use psychrometric tables (see Appendix) and determine the saturation vapor pressure $E_1$ in mmHg or Pa, which corresponds to the temperature of wet thermometer $T_w$;
7. Determine the partial pressure $e$ using psychrometer Equation 8.1;
8. Use psychrometric tables (see Appendix) and determine the saturation vapor pressure $E$ in mmHg or Pa, which corresponds to the temperature of dry thermometer $T_a$;
9. Determine the relative humidity $r = \frac{e}{E} \cdot 100$;
10. Determine vapor deficit $d = E - e$;
11. Calculate the errors of measurements:

$$\Delta_{ce} = \Delta_{cE1} + A(\Delta_{cTDry} + \Delta_{cTWet})p_A + A(T_{Dry} - T_{Wet})\Delta_{cpp}; \quad (P4.1)$$

$$\Delta_{cr} = r(\varepsilon_e + \varepsilon_E), \quad (P4.2)$$

where index "$c$" corresponds to confidence interval of systematic error ($\Delta_c$);
12. Calculate the relative errors of the results of measurement of $e$ and $r$;
13. Calculate and compare the humidity parameters in a classroom before and after lecture; in the open air; and in the dining room.

## Constructive Test

1. You have just determined the parameters of humidity using an Assmann psychrometer.
2. Study the graphical relationships between air temperature, wet temperature, vapor pressure, saturation vapor pressure, relative humidity, and water deficit

which are presented in the scientific literature (see, e.g., Campbell and Norman, Environmental Biophysics, Springer, 1998).

3. Find the site http://www.met.rdg.ac.uk/~swshargi/MicroMetSoft.html where calculators of humidity parameters are proposed.

Compare your own results with those calculated by the graphical method and calculator. Analyze the advantages and limitations of these three methods.

## QUESTIONS AND PROBLEMS

1. Name the principal parameters of humidity.

2. What are the units of absolute and relative humidity?

3. Define deficit humidity and dew point.

4. Explain the psychrometric equation.

5. Make a comparative analysis of capacitive and condensation hygrometers.

6. Identify the advantages of an open hygrometer.

## REFERENCE

Posudin, Y. 2011. *Environmental Biophysics*. Fukuoka-Kiev, Ukraine. http://www.ekmair. ukma.kiev.ua/handle/123456789/951 (accessed March 3, 2013).

## FURTHER READING

Buck, A.L. 1981. New equations for computing vapour pressure and enhancement factor. *J. Appl. Meteorol.* 20:1527–1532.

Campbell, G.S. and J.M. Norman. 1998. *Environmental Biophysics,* 2nd edition. Springer, New York.

Campbell, C.B., Grover, B.L., and Campbell, M.D. 1971. Dew-point hygrometer with constant resistance humidity transducer. *J. Appl. Meteorol.* 10:146–151.

Catsky, J. 1970. Méthodes et techniques de mesure de la concentration en anhydride carbonique dans l'air. In: *Techniques d'Étude des Facteurs Physiques de la Biosphere.* Institut National de la Recherche Agronomique, Paris (France), I.N.R.A. Publ. 70-4, 181–199.

Guyot, G. 1998. *Physics of the Environment and Climate.* John Wiley & Sons, Chichester.

Harrison, D.R. and Dimeff, J. 1973. A diode quad-bridge circuit for use with capacitance transducers. *Rev. Scient. Instr.* 44:1463–1472.

Henry, J.G. and Heike, G.W. 1996. *Environmental Science and Engineering.* Prentice-Hall, Inc., New Jersey.

Hyson, P. and Hicks, B.B. 1975. A single-beam infrared hygrometer for evaporation measurement. *J. Appl. Meteorol.* 14:301–307.

Meteorological Office. 1981. *Handbook of Meteorological Instruments,* Volume 3 Measurement of Humidity, 2nd edition. Her Majesty's Stationery Office.

Monteith, J.L. and Unsworth, M. 1990. *Environmental Physics,* 2nd edition. Edward Arnold, London.

## ELECTRONIC REFERENCE

Hygrometer, http://en.wikipedia.org/wiki/Psychrometer#Psychrometers (accessed March 6, 2013).

# 9

# PRECIPITATION

## 9.1 DEFINITIONS

*Precipitation* is water in liquid or solid state, which falls from clouds or directly from the air onto the Earth's surface or objects present there. Precipitation is any product of the condensation of atmospheric water vapor that falls under gravity.

Types of precipitation include rain, drizzle, snow, snow grains, snow pellets, diamond dust, hail, and ice pellets.

Fog and mist are not precipitation; these droplets are suspensions, because the water vapor does not condense sufficiently to precipitate.

*Rainfall* is the quantity of water falling in a given area within a given period of time; it is estimated based on the depth of water that has fallen into a rain gauge.

## 9.2 MECHANISMS OF PRECIPITATION

Precipitation is a result of saturated atmospheric water vapor; the saturated water vapor condenses on small particles—aerosols or ionized atoms (condensation nuclei). This process depends on the atmospheric temperature: drops of water are formed at temperatures $T_a < 273.2$ K and particles of ice at temperatures $T_a < 233$ K.

Vertical lifting of air containing water vapor leads to cooling at a rate of about 1°C/100 m if condensation does not take place and 0.5°C/100 m if there is condensation. The typical temperature gradient in the troposphere is believed to be 0.65°C/100 m.

*Methods of Measuring Environmental Parameters*, First Edition. Yuriy Posudin.
© 2014 John Wiley & Sons, Inc. Published 2014 by John Wiley & Sons, Inc.

The transition of fog (cloud) to rain is determined by the velocity of the movement of the raindrops which is related to their size. Cloud droplets have a spherical shape and a size of less than 0.1 mm; raindrops are oblate in shape and range from 0.1 to 9 mm in size.

The presence of water vapor, condensation centers, and significant vertical lifting of the air leads to cooling of air up to the dew point and cloud formation.

## 9.3   PARAMETERS OF PRECIPITATION

*Total amount of precipitation* is expressed by the vertical depth (in millimeters) of water reaching the ground that would cover a horizontal area of the Earth's surface if there is no runoff, infiltration, or evaporation.

*Rate of precipitation (intensity)* is expressed in millimeters per hour and can be classified into different types of precipitation. For example, the intensity of cloudburst (average diameter $d$ of drop 2.85 mm) is 10.2 cm/h; heavy rain ($d = 2.05$ mm) is 1.52 cm/h; and light rain ($d = 1.24$ mm) is 1.02 cm/h (Lull, 1959).

*Duration of precipitation* is a parameter that characterizes the length of time the precipitation takes place. High intensity precipitation is likely to be of short duration while low intensity precipitation can occur over an extended duration.

## 9.4   ACID RAIN

*Acid precipitation* arises due to natural (volcanic gases, forest fires, ocean surface evaporation) and industrial (automobile exhausts, burning of industrial fuels) discharges of sulfur and nitrogen oxides into the atmosphere. Here, they are transformed into sulfate and nitrate particles that are mixed with water forming sulfuric and nitric acids and return to the Earth's surface via sedimentation and/or precipitation.

The formation of acid rain involves the following reactions:

$$SO_2 + OH \rightarrow HSO_3; \tag{9.1}$$

$$HSO_3 + O_2 \rightarrow HO_2 + SO_3; \tag{9.2}$$

$$SO_3 + H_2O \rightarrow H_2SO_4; \tag{9.3}$$

$$NO_2 + OH \rightarrow HNO_3. \tag{9.4}$$

*Dry deposition* of acids can occur in the absence of precipitation. The principal causes of dry deposition are gravitational sedimentation, interception, interaction of small particles with bigger obstacles, and the collision of particles due to diffusion and turbulence.

*Wet deposition* of acids takes place when they are removed from the atmosphere by precipitation. Falling rain droplets collide with aerosol particles during the interaction of small particles with bigger obstacles and the collision of particles due to diffusion and turbulence.

## 9.5   INTERCEPTION

The process by which flowing rain droplets falling on the leaves, branches, limbs and stems of plants, and a portion of the precipitation is retained and does not reach the ground is called *interception.* Intercepted precipitation evaporates back into the atmosphere. Rainfall that is not intercepted is called *throughfall.*

The process of canopy interception depends on the type of foliage (e.g., coniferous trees have greater interception than deciduous trees), growth form (trees, grasses, forbs), density of vegetation, plant structure, and meteorological factors (e.g., intensity, duration and frequency of precipitation, wind, or solar radiation).

## 9.6   GENERAL CHARACTERISTICS OF ISOTOPES

Two or more forms of the same element that contain equal numbers of protons but different numbers of neutrons in their nuclei and correspond to the same atomic number and position in the periodic table are called *isotopes.* Therefore, isotopes have different atomic masses and physical properties. Each chemical element has one or more isotopes.

The term "isotope" is derived from the Greek terms ίσος (iso-, "equal", "same") and τόπος ("place") since the same position in the periodic table is occupied by the different isotopes of the element.

*Stable isotopes* are those that are stable and do not undergo radioactive decay over time.

## 9.7   STABLE ISOTOPES OF WATER

Water ($H_2O$) has two elements: hydrogen $H$ and oxygen $O$. Hydrogen and oxygen have the following stable isotopes:

$^1H$: 99.9844%;
$^2D$: 0.0156%;
$^{16}O$: 99.763%;
$^{17}O$: 0.0375%;
$^{18}O$: 0.1995%.

It is clear that almost all water atoms consist of the lightest atoms, $^1H$ and $^{16}O$. However, it is possible to find water with one of the three atoms replaced by a heavier nuclide; for instance, $^{16}O$ replaced by $^{18}O$ and $^1H$ replaced by $^2D$. Therefore, water in different situations can consist of various combinations of $^1H$, $^2H$, $^{16}O$, and $^{18}O$ (i.e., $^1H^1H^{16}O$, $^1H^1H^{18}O$, $^1H^2H^{16}O$, $^1H^2H^{18}O$, $^2H^1H^{18}O$, $^2H^2H^{18}O$).

## 9.8    ISOTOPIC FRACTIONATION

Modern analytical equipment has demonstrated that various isotopes of any element behave differently in both physical processes (heavier isotopic molecules have a lower mobility and diffusion velocity, and higher binding energy) and chemical reactions (since the atoms of different isotopes are of different sizes and atomic weights). The separation of isotopes of an element during naturally occurring processes as a result of the mass differences between their nuclei is called *isotopic fractionation*.

Stable isotope abundances are expressed as the ratio of the two most abundant isotopes in the sample compared to the same ratio in an international standard, using the "delta" ($\delta$) notation. Because the differences in ratios between the sample and standard are very small, they are expressed as parts per thousand or "per mil" (‰) deviation from the standard. The $\delta$ notation can be expressed as

$$\delta_{sample} = [(R_{sample} - R_{std})/(R_{std})] \times 1000, \tag{9.5}$$

where, using water as the example, $R_{sample}$ is the ratio of $^{18}O/^{16}O$ or $^2H/^1H$ in the sample and $R_{std}$ is the ratio of the international standard for oxygen and hydrogen.

The standard is defined as 0‰. Let us note the international standards and their absolute isotope ratios for several environmentally important isotopes:

Measurement of hydrogen ratio $^2H/^1H$ requires Vienna Standard Mean Ocean Water (VSMOW) which has value $R = 0.00015575$; oxygen ratio $^{18}O/^{16}O$ can be measured with VSMOW which has value $R = 0.0020052$. Certain elements (e.g., oxygen, hydrogen) have more than one international standard.

Seawater has a $\delta^{18}O$ value of 0‰. Therefore, water with negative $\delta^{18}O$ or $\delta^2H$ is said to be *depleted* relative to seawater, while water with positive $\delta$ values is said to be *enriched*.

## 9.9    STABLE ISOTOPES IN PRECIPITATION PROCESSES

Changes in isotope ratios during the water cycle are related mainly to the processes of evaporation and condensation. Global weather, regional topography, and moisture distribution affect the isotope ratios. The nucleus of an atom is responsible for its physical properties. Lighter molecules of water have a higher vapor pressure; thus as water evaporates from the warm ocean surface, its vapor is enriched with the lighter isotope $^{16}O$. Plants also transfer the lighter isotope $^{16}O$ into the atmosphere. As moist air mass moves toward the north where it is cooled, precipitation is enriched with the heavier isotope $^{18}O$ while the remaining air consists of progressively more of the $^{16}O$ isotope (Figure 9.1).

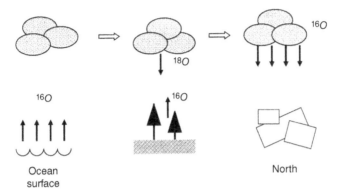

**FIGURE 9.1**   The movement of air masses with light isotopes.

Thus, the application of isotope techniques to ecohydrology makes it possible to determine the origins and ages of different water bodies, estimate the level of mixing, determine the location and proportion of water recharge, and can indicate the velocity of ground water flow.

## 9.10   APPLICATION OF STABLE ISOTOPES

The application of stable isotope analytical techniques in ecohydrology and climatology allows determining the origin and age of large masses of water, assessing their level of mixing, determining the location and proportion of water recharge, and indicating the velocity of ground water flow. Stable isotope analysis of atmospheric water vapor makes it possible to estimate atmospheric circulation and global weather patterns, and the transport of moisture (Ehleringer, 2002; Mook, 2006). Determining the stable isotope ratios of hydrogen and oxygen allows studying the distribution of stable isotopes and comparing different water sources and their movement into river and stream systems. Stable isotopes also allow investigating leaf evapotranspiration, metabolic processes, and water source competition among plant communities (Djomo et al., 2011; Reichstein et al., 2007; Gottlicher et al., 2006; Knohl and Buchmann, 2005; Knohl, 2003; Panferov et al., 2009; Helle and Panferov, 2004).

## 9.11   EFFECT OF PRECIPITATION ON LIVING ORGANISMS

Hail is a form of solid precipitation which consists of balls or irregular lumps of ice (hailstones). Hail is the product of condensation of water vapor in the thunderclouds, where the speed of air flow exceeds 10 m/s, and the temperature is −20°C to −30°C. Hail can cause serious damage to agricultural crops, people and animals, buildings, and transport. The largest hailstorm occurred in Bangladesh in 1986: hailstones weighed up to 1.02 kg (2.25 lb); the hailstorm killed 92 at Gopalganj.

Stormwater rains and floods provoke the damage of agricultural crops, soil erosion, flooding the farmlands and surrounding areas, salting, and removal of large quantities of land from agricultural use.

Acid rains can provoke serious human diseases such as asthma and allergy. Acid deposition can damage trees, forests, and crops. Acid runoff can destroy many forms of ecosystems, killing living organisms in soil and aquatic medium.

### Extreme Situations

*Highest 24-hour precipitation—1869.9 mm in Cilaos, La Reunion Island, March 15, 1952;*

*Highest precipitation in 1 month—9300 mm in Cherrapunji, India, July 1861;*

*Highest precipitation in 1 year—25,461.2 mm in Cherrapunji, India, August 1860–July 1861;*

*Highest average annual precipitation—possibly 11,684 mm, Mt. Waialeale, Kauai, Hawaii;*

*Lowest average annual precipitation—just over 0.0 mm, Arica, Atacama Desert, Chile;*

*Highest average number of days with thunder—322 days, Bogor, Indonesia;*

## 9.12   SNOW

*Snow* is a precipitation in the form of ice crystals that are formed directly from the water vapor of the air at a temperature of less than 0°C (32°F).

Falling of snow is called *snowfall.* It is accompanied by a formation of snowflakes (aggregation of ice crystals) that have a variety of sizes and shapes.

### 9.12.1   Parameters of Snow

There are three main parameters of snow:

**24-Hour Snowfall.** This is the amount of newly fallen snow in the past 24 hours. It is reported to the nearest 0.1 inch. It is not the total depth of snow on the ground, but the additional accumulation in the past 24 hours. If the ground was bare on the previous day's observation, then of course, it would be the total snow on the ground.

**Example**   Three separate snowsqualls affect your station during your 24-hour reporting day, say 3.0, 2.2, and 1.5 inches. The snow from each event melts off before the next accumulation and no snow is on the ground at your scheduled time of observation. The total snowfall for that reporting 24-hour day is the sum of the three separate snow squalls, 6.7 inches, even though the snow depth on your board at observation time was zero.

*Water Equivalent of Snow*. This is the water content obtained from melting snow collected in the gauge since the last observation. This measurement is taken once-a-day at your specified time of observation. Snow water equivalent is reported to the nearest 0.1 inch.

**Snow Depth.** This is the total depth of the old and new snow on the ground. This measurement is realized once-a-day at the scheduled time of observation with a measuring stick. It is reported to the nearest whole inch: for example, 3.4 inches is reported as 3 inches and 1.5 inches is reported as 2 inches.

### 9.12.2   Effect of Snow on Living Organisms

*Effect on Human Health.*   As soon as fresh snow reflects about 90% of incident ultraviolet radiation, it can provoke the so-called snow blindness (also known as ultraviolet keratitis, photokeratitis, or niphablepsia). This disease occurs in polar regions and at high altitudes where ultraviolet radiation is higher.

*Effect on Animals.*   Snow cover can protect animals from extreme cold. The winter temperature beneath the snow cover is significantly lesser than in the air above snow. For example, if the air temperature is −10°C, the temperature in the burrows that are created by polar animals (lemmings) will be about 10–20°C. The same thermal situation is observed in the snow holes of ptarmigans.

#### *Snowiest Places on Earth*

*There is an opinion that Mt. Baker in Washington State averages over 600″ a year and holds the world record snowfall in a season: over 1000 inches!*

*The Coast Mountains of western North America, stretching from Southern California to Alaska, produce the greatest snowfalls on the planet.*

#### *Extreme Situations*

*Greatest snowfall in one season—3110.2 cm, Paradise, Mt Rainer, Washington State, USA, 1971–1972;*

*Greatest snowfall in 24 hours—508.8 km/hr, Oklahoma City, Oklahoma, USA, May 3, 1999.*

### 9.13   FOG

A cloud in contact with the ground or water is called *fog*. If you can see the cloud of water droplets less than 1 km (0.62 mi), it is a fog; if you can see this cloud between 1 and 2 km (0.62–1.2 mi), it is a mist.

Fogs are important sources of moisture and nutrients required for forest ecosystems. Investigating the properties of fog gives useful information about acid deposition, water balance, and canopy–fog interaction.

### 9.13.1    Parameters of Fog

The density of fog is about 0.05 kg/m$^3$ and the diameter of droplets is 1–60 μm. The majority of the drops have a diameter of 10–30 μm at positive air temperatures and of 4–10 μm at negative temperatures.

Fog appears if the air temperature becomes nearer to the dew point (generally, when the difference between air temperature and dew point is less than 2.5°C (4°F)).

### 9.13.2    Effect of Fog on Living Organisms

People have learned to use the fog as a source of water for domestic and commercial uses. Fog can be used for collection of water in regions where rain cannot cover water needs throughout the year. The technique of fog collection provides the application of effective condensers (*fog collectors*) that contain a large (12 m × 4 m) piece of canvas that is placed between two poles and supplied with a long trough underneath. Such fog collection systems were approbated on Mt Sutton in Quebec, Canada, on El Tofo Mountain in northern Chile.

The efficiency of fog collector can reach about 15,000 L of water per day that can be used for household needs of the population inhabiting the surrounding areas.

Redwood forests in California receive approximately 30–40% of their moisture from coastal fog. A change in climate patterns could result in relative drought in these areas.

Some animals, including insects, depend on wet fog as a principal source of water, particularly in otherwise desert climates, as along many African coastal areas. Some coastal communities use fog nets to extract moisture from the atmosphere where groundwater pumping and rainwater collection are insufficient.

A beetle (*Stenocara gracilipes*) is an inhabitant of the Namib desert of southern Africa which is of the driest places in the world, receiving only 1.4 cm (0.5 in) of rain per year. But each night and morning fog from the Atlantic Ocean appears in this desert. The Stenocara beetle uses the specific configuration of hydrophilic bumps and hydrophobic troughs between the bumps on its shell, which makes it possible to collect moisture and channel water directly into its mouth.

A spider *Uloborus walckenaerius* can effectively collect water from the air. Its fibers are able to change its structure when they come into contact with water: the fibers form hydrophilic spindle knots while the joints in between the knots remain smooth. The condensing water droplets slide along the smooth surfaces and merge into bigger drops at the knots.

#### The Foggiest Places in the World

*The Grand Banks off the island of Newfoundland, Argentina, and Point Reyes, California are the foggiest places in the world, each with over 200 foggy days per year. Dense fog can be often found in the lower part of the Po Valley and the Arno and Tiber valleys in Italy or the Ebro Valley in north-eastern Spain, in the Seeland area (Switzerland), in coastal Chile, coastal Namibia, and the island Severnaya Zemlya.*

# REFERENCES

Djomo, A.N., Knohl, A., and Gravenhorst, G. 2011. Estimations of total ecosystem carbon pools distribution and carbon biomass current annual increment of a moist tropical forest. *Forest Ecol Manag.* 261:1448–1459.

Ehleringer, J.R. and Cerling, T.E. 2002. Stable Isotopes. Vol. 2. The Earth system: biological and ecological dimensions of global environmental change. In: *Encyclopedia of Global Environmental Change* (eds H.A. Mooney and J.G. Canadell). John Wiley & Sons, Ltd, Chichester, pp. 544–550.

Gottlicher, S., Knohl, A., Wanek, W., Buchmann, N., and Richter, A. 2006. Short-term changes in carbon isotope composition of soluble carbohydrates and starch: from canopy leaves to the root system. *Rapid Commun. Mass Spectrom.* 20:653–660.

Helle, G. and Panferov, O. 2004. Tree rings, isotopes, climate and environment–TRICE. *PAGES Newsletter*, 122:22–23.

Knohl, A. 2003. *Carbon dioxide exchange and isotopic signature ($^{13}C$) of an unmanaged 250-year-old deciduous forest*. Berichte des Forschungszentrum Waldoekosysteme der Universitaet Goettingen, Reihe A, Band 188, Goettingen, PhD thesis.

Knohl, A. and Buchmann, N. 2005. Partitioning the net $CO_2$ flux of a deciduous forest into respiration and assimilation using stable carbon isotopes. *Global Biogeochem. Cy.* 19:Art. No. GB4008.

Lull, H.W. 1959. *Soil Compaction on Forest and Range Lands*. U.S. Dept. of Agriculture, Forestry Service, Misc. Publication No.768.

Mook, W.G. 2006. *Introduction to Isotope Hydrology. Stable and Radioactive Isotopes of Hydrogen, Oxygen and Carbon*. Taylor & Francis, London.

Panferov, O., Ibrom, I., Kreilein, H., Oltchev, A., Rauf, A., June, T., Gravenhorst, G., and Knohl, A. 2009. Between deforestation and climate impact: the Bariri Flux tower site in the primary montane rainforest of Central Sulawesi, Indonesia. *Flux Lett.* 2/3: 17–19.

Reichstein, M., Papale, D.D., Valentini, R., Aubinet, M., Bernhofer, C., Knohl, A., Laurila, T. Lindroth, A., Moors, E., Pilegaard, K., and Seufert, G. 2007. Determinants of terrestrial ecosystem carbon balance inferred from European eddy covariance flux sites. *Geophys. Res. Lett.* 34:L01402.

# 10

# MEASUREMENT OF PRECIPITATION

## 10.1 MEASUREMENT OF PRECIPITATION PARAMETERS

A *rain gauge* is an instrument used by meteorologists and hydrologists to gather and measure the amount of liquid precipitation over a certain period of time (e.g., rainfall in millimeters per 24 hours). There are a number of types of rain gauges that can be used to measure the amount of rainfall.

A *disdrometer* is an instrument used to measure the drop-size distribution and velocity of falling raindrops.

### 10.1.1 Standard Rain Gauge

The standard National Oceanic and Atmospheric Administration (NOAA) rain gauge consists of a fiberglass polyester collector, funnel, and bottle (Figure 10.1). A cylindrical collector has an area of 200 cm$^2$ and height of 50 cm. The rain enters through the funnel into a graduated bottle. The procedure of measurement provides the calculation of the height (in millimeters) of the water in the bottle. The system is fixed to a concrete block.

*The advantage of a standard rain gauge:*

- Simple in operation.

*The disadvantage of a standard rain gauge:*

- Operator intervention.

*Methods of Measuring Environmental Parameters*, First Edition. Yuriy Posudin.
© 2014 John Wiley & Sons, Inc. Published 2014 by John Wiley & Sons, Inc.

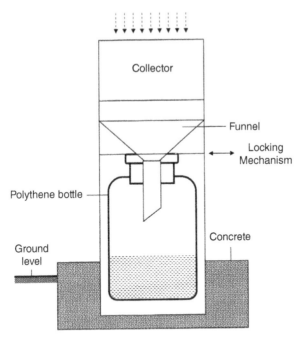

**FIGURE 10.1**  The standard National Oceanic and Atmospheric Administration (NOAA) rain gauge.

## 10.1.2   Tipping Bucket Rain Gauge

This device is used in automated weather observation stations. It consists of two buckets that are mounted on a fulcrum. Each bucket holds an exact amount of precipitation. One of them is positioned beneath the rain collector that funnels the precipitation into the bucket (Figure 10.2). The whole structure can be rotated around the horizontal axis. As one bucket (A) is filled with water during rain, it becomes overbalanced and tips down, emptying itself. The second bucket (B) begins to be filled with water.

A small magnet (C) serves as the contact that closes an electric circuit (D). The quantity of contacts corresponds to the total amount of precipitation while the time interval between two contacts represents the intensity of rain. If a wireless rain gauge is used, the information is transmitted via radio signals.

*The advantages of a tipping bucket rain gauge:*

- This type of rain gauge is simple and easy to operate and maintain.
- The minimal expected operational lifetime is 15 years without loss of functioning.
- Accuracy is 2–5%.

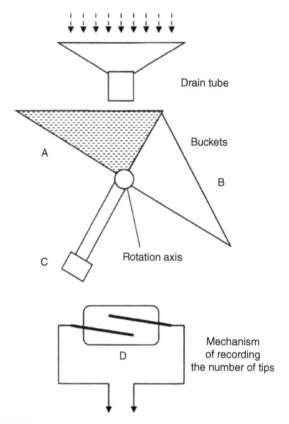

**FIGURE 10.2**    Principle of operation of the tipping-bucket rain gauge.

*The disadvantages of a tipping bucket rain gauge:*

- This type of rain gauge contains moving parts.
- It can withstand attacks by fungi, insects, and rodents; therefore all openings of the rain gauge except the collector are covered with stainless steel.
- It shall have a smooth and permanent surface finish to minimize evaporation loss.
- Tipping buckets have the tendency to underestimate the amount of rainfall, particularly in snowfall and heavy rainfall events.

### 10.1.3    Siphon Rain Gauge

A rain gauge of this type consists of a cylindrical chamber equipped with a float that is coupled to a pen recorder (Figure 10.3). This gauge also contains a siphon—a tube

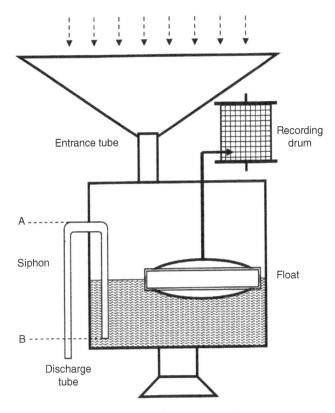

**FIGURE 10.3**   A siphon (float) rain gauge.

in an inverted-U shape which causes a liquid to flow uphill, above the surface of the reservoir.

Rain enters the gauge and as the rainwater increases in the chamber, the float rises and a pen mounted on the float draws on a graph attached to a drum. As soon as the float reaches the top of the chamber (which corresponds to the upper position of the siphon *A*), all the water in the chamber below the float empties and the float returns to its original position. When the water reaches the lower position *B*, the siphon stops its action.

Recording the rainfall can continue from 24 hours to 7 days.

*The advantages of a siphon rain gauge:*

- It is more reliable and accurate compared to the common rain gauge in use today (accuracy $\pm 1\%$ to 250 mm/hour and $\pm 3\%$ to 500 mm/hour).
- It does not have a tipping arm that moves and collects dirt.
- This type of rain gauge can be used for automated measurements.
- All materials are corrosion resistant.

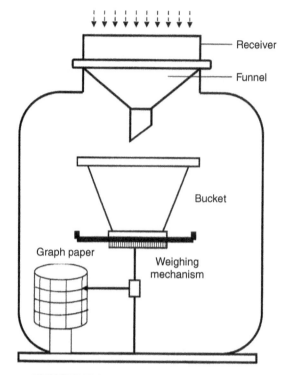

**FIGURE 10.4**    A weighing bucket rain gauge.

*The disadvantages of a siphon rain gauge:*

- Non-uniform siphoning process during the loss of water.
- The rain can be collected in an auxiliary reservoir.
- This rain gauge requires calibration.

### 10.1.4    Weighing Bucket Gauge

This device allows continuous monitoring of rainfall. Water is collected in a container that is connected to a weighing mechanism using a spring. A pen touching a drum with a graph paper records the weight (Figure 10.4). As the weight of the container increases with increasing rainfall, the pen indicates the accumulation of precipitation.

*The advantages of a weighing bucket gauge:*

- This type of rain gauge has no moving parts.
- This gauge is more accurate in comparison with the tipping bucket rain gauge.
- It is capable to measure the amount or level of chemicals present in the atmosphere.

- It can be used in automated measurements.
- It can distinguish other forms of precipitation such as rain, hail, and snow.

*The disadvantages of a weighing bucket gauge:*

- This type of rain gauge is expensive and requires more maintenance than tipping bucket gauges.
- It is complex.
- It is subjected to debris (dirt, leaves) and wind.

### 10.1.5   Optical Rain Gauge

The typical optical rain gauge (model RG-11, Hydreon Corporation) consists of the source of infrared (IR) radiation, spherical transparent surface, and detector. When drops hit the outside surface, it allows some of the beams to escape. The sensor detects the change in beam intensity, and determines the size of the rain drop that caused the change (Figure 10.5).

*The advantages of an optical rain gauge:*

- High accuracy.
- Compactness.

*The disadvantages of an optical rain gauge:*

- It is complex.

### 10.1.6   Laser Precipitation Monitor

The *laser precipitation monitor (LPM)* can measure the amount, intensity, and type of precipitation with great accuracy.

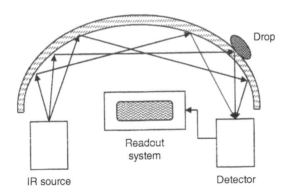

**FIGURE 10.5**   Principle of operation of optical rain gauge.

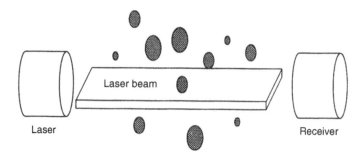

**FIGURE 10.6**    Laser precipitation monitor.

The LPM (Bristol Industrial and Research Associates Limited, Bristol, UK) estimates such precipitation parameters as the intensity, volume (water equivalent), and the precipitation spectrum (diameter and velocity); it provides a drop-size distribution with 400 classes of raindrops (22 diameters × 20 velocity). The LPM measures particles down to 0.16 mm diameter.

The LPM detects and discriminates the different types of precipitation as drizzle, rain, hail, snow, snow grains, graupel, and ice pellets.

The device consists of a source of optical radiation (laser diode, wavelength 785 nm, optical power 0.5 mW) and a signal processor (Figure 10.6). The laser produces a light beam (220 × 20 × 0.75 mm); when a precipitation drop passes through the beam with measuring area 45 cm² (7 inch²) the receiving signal is reduced. The vertical velocity of the drop is determined from the signal duration by using the measured drop diameter and the known thickness of the beam.

The range of precipitation parameters is

| | |
|---|---|
| Particle size | 0.16–8 mm |
| Particle velocity | 0.2–20 m/s |
| Minimum intensity | 0.005 mm/h drizzle |
| Maximum intensity | 250 mm/h |

*The advantages of a laser precipitation monitor:*

- High accuracy and sensitivity.
- The LPM is capable of measuring particles down to 0.16 mm diameter.
- Compactness.
- Extended heaters for use in the mountains.
- Remote support.

### 10.1.7    Acoustic Rain Gauge

The *acoustic rain gauge* (acoustic disdrometer) is a device designed to estimate rainfall parameters over water reservoirs (oceans, lakes). The principal component of

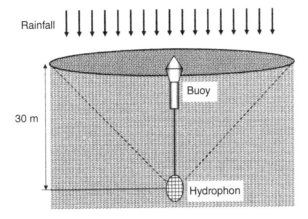

**FIGURE 10.7**    Acoustic rain gauge.

the acoustic rain gauge is a *hydrophone* (from Greek "hydro" meaning "water" and "phone" meaning "sound")—a microphone that is used under water for recording or listening to underwater sounds. This hydrophone contains a piezoelectric transducer that converts a sound signal (a pressure wave) into an electrical signal.

Each sound that is produced during rainfall on the water surface is unique. Analysis of the underwater sound which depends on the drop-size distribution makes it possible to estimate the character of the rainfall. The rainfall rate can be estimated through the analysis of the selected moments of the drop-size distribution. The area of the ocean surface above the hydrophone where the sound of rain is recorded can be increased if the hydrophone is placed deeper (about 30 m) in the ocean (Figure 10.7).

The analysis of spectral properties and intensity of sounds make it possible to distinguish rainfall and breaking waves: increasing wind speed raises sound levels across the range 500 Hz to 50 kHz; drizzle is characterized by a unique peak at frequencies from 13 to 25 kHz; rainfall produces relatively more sound between 5 and 15 kHz; heavy rain produces a flatter spectrum. The intensity of sound produced by heavy rain is significantly higher compared to the sound intensity produced by breaking waves.

*The advantage of an acoustic rain gauge:*

- It can estimate rainfall parameters over water reservoirs.

## 10.2    MEASUREMENT OF ACID RAIN POLLUTION

The principal methods of monitoring acid rain are measuring the pH and conductivity of collected rain samples and the use of ion-exchange chromatography for chemical analysis.

The collector for chemical analysis is used for the analysis of pH, ions, ammonium, aluminum, chloride, and stable isotope samples. The principal feature of such a collector is the prevention of water evaporation which is inhibited by using a narrow tube between the funnel and plastic reservoir can.

### 10.2.1  pH-metry

The operative principle of a pH-meter is based on measuring the concentration of hydrogen ions which determines the level of acidity. These ions are formed by the reaction of acids dissolved in water, with water. Acidity scales range from 0 (maximum acidity) to 7 (neutral situation) to 14 (maximum alkalinity). Pure rain has a pH of 5.6; a pH range between 0 and 5 indicates acid rain. Because pH is affected by chemicals in the water, pH is an important parameter of the water quality. The pH determines the level of nutrients and heavy metals that can be utilized by aquatic organisms.

### 10.2.2  Conductivity

The electrical conductance of a solution is a physical value that estimates the ability of a solution to conduct a current. The current is produced by the movement of ions through the solution. So, the electrical conductance is proportional to the ion concentration and therefore to the pH of the solution. The units of electrical conductivity are deciSiemens per meter (dS/m) or microSiemens per centimeter ($\mu$S/cm). The scale of electrical conductance begins with absolutely pure water (0.055 $\mu$S/cm) and extends through distilled water (0.5 $\mu$S/cm) and demineralized water (5 $\mu$S/cm) to water from streams (10–500 $\mu$S/cm), and concentrated acids and bases ($10^5$ $\mu$S/cm).

### 10.2.3  Ion-Exchange Chromatography

*Chromatography* (from Greek $\chi\rho\acute{\omega}\mu\alpha$: *chroma*, color and $\gamma\rho\alpha\varphi\varepsilon\iota\nu$: "grafein" to write) is a method of separating mixtures of molecules for subsequent identification of their individual components. It is based on differences in the partitioning behavior of analytes between the mobile phase and the stationary phase of the column. The mechanisms of interaction between the components of the mixture and the stationary phase can be due to differences in adsorption or solubility, charge interactions, van der Waal's forces, size separation, and differential affinity.

*Ion-exchange chromatography* (or *ion chromatography*) is a process that separates ions and polar molecules based on their charge properties and coulombic (ionic) interactions. The solution to be chromatographed is usually called a *sample*, and the individually separated components are called *analytes*.

The stationary phase surface has ionic functional groups that interact with the analyte ions of the opposite charge. Ion chromatography is the only technique that can quantitatively analyze anions at the parts per billion (ppb) level.

A typical ion-exchange chromatograph (e.g., the 761 Compact IC System) consists of a valve for sample injection (injection valve); a high-pressure pump; a pulsation dampener for protecting the column against damages even with low level pressure variations; a column chamber with insulation sufficient for thermally stable conditions and shielding of the system against electromagnetic interference; an ion-exchange column; a conductivity detector with precise temperature stability; and eluent and sample containers.

A sample is introduced into a sample loop of known volume. A buffered aqueous solution (mobile phase) carries the sample from the loop onto a column that contains some form of stationary phase material.

## 10.3   ISOTOPES IN PRECIPITATION

### 10.3.1   Isotope Ratio Mass Spectrometry

*Isotope ratio mass spectrometry (IRMS)* is a special field of mass spectrometry that provides a precise measurement of mixtures of stable isotopes and allows estimating the relative abundance of each of the isotopes.

An IRMS gives a precise analysis of the stable isotopic compositions of, for example, $\delta^{18}O$ and $\delta D$. The sample is converted to a gas and the pressures of sample and reference gases are adjusted by compressing or expanding the bellows until both pressures are equal. The dual inlet system of the gas isotope ratio spectrometer is equipped with changeover valves for rapid switching between the sample and reference gas. The valves are made of Teflon or gold to eliminate any leakage of the gases and contamination of one gas by the other.

Each gas passes through a capillary column (about 1 m of length), which has a crimp located at the end of the mass spectrometer, and enters the source region. Such a system prevents isotopic fractionation as the gas enters the high vacuum of the source region. Here the gas is ionized using a heating filament (e.g., tungsten, rhenium, thoriated indium) at a high temperature and the gas is bombarded with electrons, producing positively charged ions. The ions are accelerated by an electric field that is created by the application of a high voltage potential through a series of collimating electric lenses into an electron beam that passes through a strong magnetic field in the flight tube.

If an ion of mass $m$ and charge $z$ is accelerated at potential $U$ and injected into a uniform magnetic field $B$, then the ion experiences a force and moves in a circular orbit of radius $R$. The motion is described by the *mass spectrometer equation*:

$$m/z = B^2 R^2 / 2U. \tag{10.1}$$

For singly charged ions, the radius is determined by the choice of magnetic and electric fields. The combination of electric and magnetic fields selects ions of particular mass and forms a mass filter. In such a way, ions are deflected in a circular trajectory with the light ions being deflected more strongly than the heavy ones. Thus the spatial separation of different isotopes of the same molecule is achieved. The voltage of the separated ions is measured in individual Faraday collectors and the relative voltage of the sample to the standard is related to the isotopic composition of the sample gas.

The PRISM, typical analytical and isotope ratio analyzer, consists of an ion source, a dual inject system, a differential pumping option, electronic units, a high vacuum pumping system, an electrical wiring assembly, a main switch, a circuit breaker panel, and utility input and output panels.

The advantages of IRMS is high precision of measurement (e.g., stable IRMS VG PRISM provides high precision $\delta^{18}O$ and $\delta^{13}C$ analysis of silicates and carbonates with precision better than 0.02‰ and automated $\delta^{18}O$ measurement of water samples (24 samples a day) with precision better than 0.05‰).

The primary disadvantage of these spectrometers is their extremely high price.

### 10.3.2   Diode Laser: Principle of Operation

Semiconductor materials are characterized by certain spectral bands: the *valence band* that is completely filled with electrons and the *conduction band* that is partially filled with electrons (free or conduction electrons). These bands are separated by a *band gap*. Properties of solids depend on the filling zone electrons. The value of the band gap in semiconductors is 1–2 eV. Electrons can jump from the valence band to the conduction band due to the thermal motion of atoms. The hole is the absence of an electron in a particular place in an atom and occurs during the transition of a free electron from the valent band to the conduction band. Holes and electrons are the two types of charge carriers.

Emission of a photon occurs during the transition of an electron from the conduction band to the valence band. The emission wavelength is determined by the width of the band gap. The transition is accompanied by the recombination of an electron with a hole; conduction electrons and holes are annihilated and the emission of photons is stopped.

A *laser diode* is a semiconductor device that consists of two semiconductors of p-type and n-type.

A *p–n junction* (n-negative and p-positive) is a region of space at the junction of the two semiconductors, where the transition from one type of conduction to the other type takes place.

The dependence of free electrons depends strongly on the applied electric field. Figure 10.8a demonstrates the relative filling of the energy bands by electrons with no voltage applied to the diode. If a voltage is applied to the p–n junction, its value decreases (Figure 10.8b).

The concentration of electrons in the conduction band near the junction on the n-side and the concentration of holes in the valence band near the junction on the p-side is increased under these conditions.

When the conduction band electron moves into a vacant valence band, the transition is accompanied by the emission of a photon. The energy of the photon is equal to the energy gap.

Gallium arsenide, indium phosphide, gallium antimonide, and gallium nitride are the most typical semiconductor materials that are used in diode lasers. The frequency

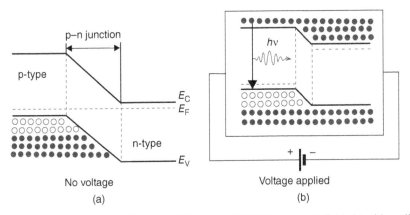

**FIGURE 10.8**    Energy-level diagram of diode laser. (a) Without electric field; (b) with applied electric field.

of laser radiation can be tuned by either changing its temperature or injection current density. The frequency range of some diode lasers depends on the semiconductor material (Werle et al., 2002): typical emission wavelength is 0.78 and 0.83 $\mu$m for GaAs/AlGaAs; 1.3–1.55 $\mu$m for InGaAsP/InP; 1.6–2.2 $\mu$m for InGaAsP/InAsP; 2–2.4 $\mu$m for antimonide lasers (AlGaAsSb, InGaAsSb, InGaAsSb, InAsSbP); 3–30 $\mu$m for lead-salt lasers.

*The advantages of a diode laser:*

- High resolution.
- Compactness.
- Low cost.
- Requires low power supply.

*The disadvantage of a diode laser:*

- High divergence of laser radiation.

### 10.3.3  Tunable Diode Laser Absorption Spectroscopy

*Tunable diode laser absorption spectroscopy (TDLAS)* is an analytical technique that uses tunable diode lasers to estimate the concentration of a gas sample using absorption spectroscopy and measuring the absorption of radiation which is a function of the frequency or wavelength.

A tunable diode laser absorption spectrometer consists of tunable diode lasers and a laser absorption spectrometer. It can be used to measure the concentration of water vapor and estimate its principal molecules that differ only in their isotopic composition (isotopologues: $H_2O$, $HDO$, $H_2{}^{18}O$, $H_2{}^{17}O$, $H^{16}OH$, $H^{17}OH$) in natural samples.

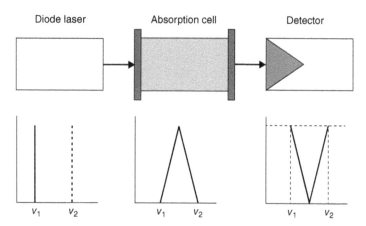

**FIGURE 10.9**    Tunable diode laser absorption spectroscopy (TDLAS).

The principle of TDLAS operation is repeatedly scanning laser emission across the absorption spectral line of a molecule of interest. This system includes the diode laser as a source of radiation, multi-pass-absorption cell with sample gas and reference cell, telescope system, beam splitters, and sample and reference detectors.

The principle of operation of TDLAS is presented in Figure 10.9.

TDLAS can be effectively utilized for stable isotope analysis. The sensitivity of the analyzer depends on the absorption strength of the line chosen and on the absorption path length.

The technique provides information about sources, sinks, and transfer of atmospheric gases and makes it possible to measure the isotopic composition of water vapor with very high precision (in the order of ppb). It is also possible to measure the temperature, pressure, velocity, and mass flux of the gas under observation (Arroyo and Hanson, 1993; Linnerud et al., 1998; Gagliardi et al., 2001; Kammer et al., 2011; Aemisegger et al., 2012; Sturm and Knohl, 2010; Werle et al., 2008).

However, the instrumentation is expensive and cumbersome, and it requires a constant power supply and liquid nitrogen for cooling.

### 10.3.4    Modulated Techniques

When a diode laser is used in laser absorption spectroscopy, a non-desired fringe structure appears due to the numerous reflections from the optical elements inside the optical resonator of the laser. These fringes are superimposed on the signal of interest producing spectral noise. A sample modulation technique is used to avoid this noise (Werle et al., 2005).

### 10.3.5    Cavity Ring-Down Spectroscopy

*Cavity ring-down spectroscopy (CRDS)* is a highly sensitive method for quantitative diagnostics of gaseous samples.

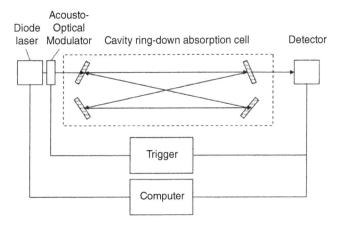

**FIGURE 10.10**  A typical cavity ring-down spectrometer.

A typical cavity ring-down spectrometer consists of a tunable diode laser and an optical ring resonator (Figure 10.10) where light can circulate in two different directions.

Laser radiation in the cavity produces a standing wave that occurs during a phenomenon known as resonance. The tendency of a system is to oscillate with a greater amplitude at certain frequencies than at others. When the laser radiation is in resonance with a cavity mode (distribution of light in the cavity), the intensity in the cavity increases due to constructive interference.

If the laser is switched off, the radiation intensity exponentially decays. If a material that absorbs light is placed in the cavity, the decay of radiation occurs faster.

A ring-down spectrometer measures the decay of the radiation to $1/e$ where the number $e$ is a mathematical constant (about 2.71828) that is the base of the natural logarithm.

The decay constant (or the ring-down time) $\tau$ depends on the loss mechanisms within the cavity. The ring-down time can be used to calculate the concentration of the absorbing substance in the gas sample that fills the cavity.

The intensity of light within the cavity can be determined as an exponential function of time:

$$A = (1 - R)\left(\frac{\tau_0 - \tau}{\tau}\right),\qquad (10.2)$$

where $A$ is the absorption; $R$ is the mirror reflectivity; $\tau_0$ is the decay constant for an empty cavity; and $\tau$ is the decay constant for the cavity with the sample.

Measuring the decay rate is more sensitive than measuring the absolute absorbance since this technique makes it possible to avoid the effect of laser fluctuations. The sensitivity of CRDS can reach the parts per trillion level.

## 10.4 REMOTE SENSING OF PRECIPITATION

### 10.4.1 Types of Remote Sensing Techniques

Remote sensing of precipitation may be defined as measurement and estimation of the precipitation parameters in a volume which is distant from the sensor. This technology allows determining the type, range, location, motion, and intensity of precipitation using radars and satellites. There are two principal types of remote sensing technique: passive remote sensing and active remote sensing.

*Passive remote sensing of precipitation* relies on the reception of naturally emitted electromagnetic energy by sensors. The sources of energy that are naturally available, such as the surface emission of the Earth, cosmic background, and rain emission are used as the sources of electromagnetic radiation.

*Active remote sensing of precipitation* provides the emission of electromagnetic energy and detection of those part of radiation that is reflected or scattered from the target (rainfall, clouds, etc.). The source and sensor of electromagnetic radiation are established at the board of satellite or air carrier.

### 10.4.2 Radars

Modern radar systems include Doppler radars that are able to estimate the velocity of movement of rain droplets (see Section 4.5.5) with high sensitivity and resolution. For example, the Nexrad (Next-Generation Radar) WSR-88D demonstrates spatial resolution about 4 km, and temporal resolution at 6–10 minutes. There are 160 Doppler radars in the United States which are used to measure raindrop velocity, predict tornadoes, and estimate cold fronts and thunderstorm gust fronts.

One modification of the weather radar is the use of *polarimetric radar* that transmits radio wave pulses with both horizontal and vertical orientations. This type of radar makes it possible to determine the horizontal and vertical dimensions of clouds and precipitation particles (Figure 10.11).

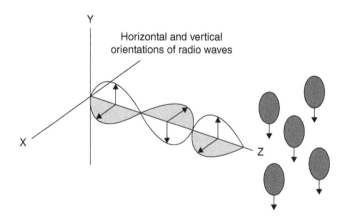

**FIGURE 10.11** Principle of operation of polarimetric radar.

*The advantages of a radar remote sensing:*

- Excellent space and time resolution.
- Observations in real time.

*The disadvantages of a radar remote sensing:*

- Little coverage over oceans or remote regions.
- It requires the signal calibration.
- Corrections required for beam filling, bright band, anomalous propagation, attenuation, etc.
- Expensive to operate.

### 10.4.3   Satellites

The satellite systems allow studying tropical and subtropical rainfall, predicting flooding, and helping scientists to understand global climate change. Satellites are used for precipitation analysis and forecasting. Some are geostationary (geosynchronous) satellites that have circular orbits over the equator (e.g., GOES-8, GOES-10, GMS-5, Metsat-6, Metsat-7); others are low orbital satellites (e.g., TRMM, NOAA-15, -16, -17, DMSP F13, F14, F15).

*Global precipitation measurement (GPM) mission* as an example of passive remote sensing technique is an international satellite mission (NASA and the Japanese Aerospace Exploration Agency) to provide next-generation observations of rain and snow. The GPM mission makes it possible to understand the Earth's water and energy cycles, extreme events that cause natural disasters, and obtain the useful precipitation information from space.

The GPM instruments include dual-frequency precipitation radar (DPR) and a multi-channel GPM microwave imager (GMI). The system of these instruments provides three-dimensional measurements of precipitation structure over 78- and 152-mile (125 and 245 km) swaths, to obtain new information on particle drop-size distributions over moderate precipitation intensities, to measure cloud and aerosol parameters, and to identify heavy, moderate, and light precipitation using the polarization difference as an indicator of the optical thickness and water content.

The *Tropical Rainfall Measuring Mission (TRMM)* satellite is an example of active remote sensing of precipitation. TRMM is a joint project between the United States (under the leadership of NASA's Goddard Space Flight Center) and Japan (under the leadership of the National Space Development Agency). The satellite was launched from Tanegashima, Japan, in 1997 to monitor rainfall in the tropics. It was placed in low earth orbit with a precipitation radar (PR), a passive TRMM microwave imager (TMI), a visible-infrared radiometer (VIRS), a lightning sensor, and a cloud sensor.

The TRMM satellite provides an analysis of the inter-annual variability in global rainfall and radar images of rain from space. The TRMM system measures the size and number of rain/snow drops at multiple vertical layers in the cloud, provides three-dimensional maps of storm structures, obtains invaluable information on the intensity and distribution of the rain, on the rain type, on the storm depth, and on the height at which the snow melts into rain.

The TRMM Precipitation Radar has a horizontal resolution at the ground of about 3.1 miles (5 km) and a swath width of 154 miles (247 km); it provides the measurement of vertical profiles of the rain and snow from the surface up to a height of about 12 miles (20 km), the detection of fairly light rain rates down to about 0.027 inches (0.7 mm) per hour.

*The advantages of a geostationary satellite remote sensing:*

- Good space and time resolution.
- Observations in near real time.
- Samples oceans and remote regions.
- Consistent measurement system.

*The disadvantage of a geostationary satellite remote sensing:*

- Measures cloud-top properties instead of rain.

### 10.4.4   Estimation and Analysis of Precipitation Parameters

*GOES Precipitation Index (GPI)* is a system of quantitative estimation of precipitation that is based on the recording of IR radiation (11–13 μm) emitted by terrestrial surfaces. As soon as clouds are opaque for IR radiation and high level cloudiness is well correlated with precipitation, it is possible to use GPI to estimate precipitation over the tropics and warm-season extratropics where the GPI yields useful results.

The GPI is a precipitation estimation algorithm. The estimation procedure includes the measurement of (Arkin and Meisner, 1987):

$$\text{GPI (mm)} = F \times R \times T, \tag{10.3}$$

where $F$ is the fractional coverage of IR pixels less than 235 K over a reasonably large domain (50 km × 50 km and larger); $R$ is the precipitation rate (3 mm/hour); $T$ is the number of hours over which the fractional coverage $F$ was compiled.

The main objective of the project *Global Precipitation Climatology Project (GPCP)* is to develop a more complete understanding of the spatial and temporal patterns of global precipitation over the global oceans and over land. It is based on the analysis and elaboration of data from over 6000 rain gauge stations, and satellite geostationary and low-orbit IR, passive microwave, and sounding observations. It makes it possible to estimate the monthly rainfall on a 2.5-degree global grid from 1979 to the present.

The operational *PERSIANN (Precipitation Estimation from Remotely Sensed Information using Artificial Neural Networks)* system is based on computed estimation of rainfall rate at each $0.25° \times 0.25°$ pixel of the IR brightness temperature image provided by geostationary satellites. Global rainfall product covers 50°S–50°N.

## 10.5   SNOW MEASUREMENT

### 10.5.1   Measurement of Snowfall

Snowfall can be estimated as the maximum depth of snow on a *snowboard* (typically a piece of plywood painted white). Daily total snowfall provides four 6-hour snowfall measurements.

A snowboard is a meteorological tool which can be used for measuring snow accumulation. It is prepared from a flat piece of plywood painted a white color, around 41–61 cm (16–24 inches) in length and width and around 13–19 mm (0.5 to 0.75 inches) thick.

### Constructive Tests

1. Why should the snowboard be painted white?
2. Why should the snowboard be made of plywood?

### 10.5.2   Snow Gauge

This device consists of a copper container and the funnel-shaped gauge itself. The size of the container is 51.5 cm (201/4 inches). The whole structure is mounted on a pipe. Then the snow from container is melted and the amount of water is measured.

### 10.5.3   Ultrasonic Snow Depth Sensor

The principle of operation of this sensor is based on measuring the time required for an ultrasonic pulse to travel to and from a target surface. Distances from the ground surface and snow cover are measured by emitting a 40 KHz burst and then recording for the return echo (Figure 10.12).

*The advantages of an ultrasonic snow depth sensor:*

- There are no moving parts.
- High accuracy and resolution.
- Wide temperature range.
- Operating reliability.
- Low energy consumption.

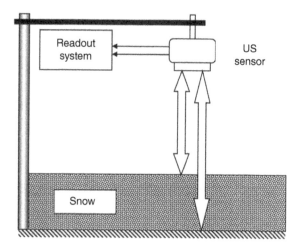

**FIGURE 10.12**  Ultrasonic snow depth sensor.

For example, the 260–700 Ultrasonic Snow Depth Sensor is characterized by the following parameters:

Range: 0.5–10 m (1.6 to 32.8 feet);
Beam width: 22°;
Accuracy: ±1 cm or 0.4% distance to target;
Resolution: 3 mm (0.12 inches);
Temperature range: −40°C to +70°C (−40°F to 158°F).

### 10.5.4  Laser Snow Depth Sensor

The principle of operation of laser snow depth sensor is based on the emission of short laser pulses that travel to the target and is reflected back to the receiver. The depth of snow is determined by measuring the time interval between the received signals reflected from the ground and snow surfaces.

For example, the Laser Snow Depth Sensor USH-8 (Hoskin Scientific, Canada) demonstrates the following characteristics:

- Measuring range: 0–15 m.
- Uncertainty better 5 mm.
- Operating temperature: −40°C to 50°C.

### 10.5.5  Remote Sensing of Snow Cover

*Gamma Radiation.* This technology is described in Section 8.9. The estimation of snow cover is realized through the measurement of its depth by air-carrier before

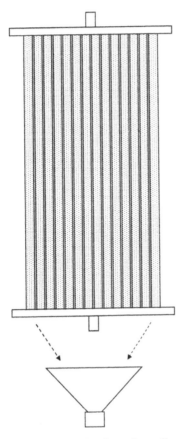

**FIGURE 10.13**   Passive string collector.

and after snowfall. The measurement of the attenuation of gamma radiation by snow layer provides the information concerning the average snow water equivalent (Carroll et al., 1995). This technique requires the correction for the background soil moisture.

*Visible/Near Infrared Technique*. Satellite sensors in the visible and near-IR wavelengths provide information on the spatial distribution of snow-covered areas, rates of snow-cover depletion, and surface albedo.

The satellite technology is based on using Landsat Thematic Mapper (TM), the Landsat Multispectral Scanning System (MSS), Advanced Very High Resolution Radiometer (AVHRR), the French SPOT satellite (Satellite Pour l'Observation de la Terre), and the Moderate-Resolution Imaging Spectroradiometer (MODIS).

These systems are widely used for mapping snow cover with high spatial resolution (80 m for The Landsat MSS, 30 m for the Landsat TM, 10–20 m for the SPOT satellite).

*Thermal Infrared*. This technology is based on the measuring of the snowpack temperature, which depends on the spectral emissivity of the snow, liquid water content, and the snow grain size. Thermal IR data can be used for the identification of snow cover boundaries.

***Microwave Remote Sensing***. The main idea of this technology is the estimation of such parameters as snow cover area, snow depth, and the content of liquid water in the snowpack which depend on the process of snow melting (Kunzi et al., 1982). The aerial sea-ice climatological measurements in the Arctic and Antarctic regions were performed by the Seasat and Nimbus 7 satellites which were equipped with the scanning multichannel microwave radiometer (SMMR).

Active microwave systems demonstrate good spatial resolution, while passive systems are simpler in data interpretation. The limitation of microwave remote sensing systems is high cost.

## 10.6   FOG-WATER MEASUREMENT

The water in fog is collected usually using a passive string collector. It is a cylinder (21 × 45 cm) with a total collection surface of 945 cm$^2$. The fog droplets are collected by 460 vertical Teflon strings arranged cylindrically (Figure 10.13). The droplets combine to form larger drops that run down the strings and are accumulated for analysis.

## REFERENCES

Aemisegger, F., Sturm, P., Graf, P., Sodemann, H., Pfahl, S., Knohl, A., and Wernli, H. 2012. Measuring variations of $\delta18O$ and $\delta2H$ in atmospheric water vapour using two commercial laser-based spectrometers: an instrument characterisation study. *Atmos. Meas. Tech.* 5:1491–1511. doi:10.5194/amt-5-1491-2012

Arkin, P.A. and Meisner, B.N. 1987. The relationship between large-scale convective rainfall and cold cloud over the western hemisphere during 1982–84. *Mon. Weather Rev.* 115:51–74.

Arroyo, M.P. and Hanson, R.K. 1993. Absorption measurements of water-vapour concentration, temperature, and line-shape parameters using tunable InGaAsP diode laser. *Appl. Opt.* 32:6104–6116.

Carroll, S., Day, G., Cressie, N., and Carroll, N. 1995. Spatial modeling of snow water equivalent using airborne and ground-based snow data. *Environmetrics*, 6:127–139.

Gagliardi, G., Restieri, R., De Biasio, G., De Natale, P., Cotrufo, F., and Gianfrani, L. 2001. Quantitative diode laser absorption spectroscopy near 2 $\mu$m with high precision measurements of $CO_2$ concentration. *Rev. Sci. Instr.* 72:4228–4233.

Kammer, A., Tuzson, B., Emmenegger, L., Knohl, A., Mohn, J., and Hagedorn, F. 2011. Application of a quantum cascade laser-based spectrometer in a closed chamber system for real-time delta C-13 and delta O-18 measurements of soil-respired $CO_2$. *Agr. Forest Meteorol.*, 151(1):39–48.

Kunzi, K.F., Patil, S., and Rott, H. 1982. Snow-cover parameters retrieved from nimbus-7 scanning multichannel microwave radiometer (SMMR) data. *IEEE Trans. Geosci. Remote Sens.* GE-20(4):452–467.

Linnerud I., Kaspersen, P., and Jaeger, T. 1998. Gas monitoring in the process industry using diode laser spectroscopy. *Appl. Phys. B*, 67:297–305.

Sturm, P. and Knohl, A. 2010. Water vapour $D^2H$ and $D^{18}O$ measurements using off-axis integrated cavity output spectroscopy. *Atmos. Meas. Tech.* 3:67–77.

Werle, P., Slemra, F., Maurera, R., Kormann, R., Mucke, R., and Janke, B. 2002. Near- and mid-infrared laser-optical sensors for gas analysis. *Opt. Lasers Eng.* 37:101–114.

Werle, P., Dyroff, C., Zahn, A., Mazzinghi, P., and D'Amato, F. 2005. A new concept for sensitive *in situ* stable isotope ratio infrared spectroscopy based on sample modulation. *Isot. Environ. Health Stud.* 41(4):323–333.

Werle, P., D'Amato, F., and Viciani, S. 2008. Tunable diode-laser spectroscopy: principles, performance, perspectives. In: *Lasers in Chemistry. Optical Probes and Reaction Starters* (ed. M. Lackner). Wiley-VCH, Weinheim, pp. 255–275.

# PRACTICAL EXERCISE 5

# VELOCITY OF A FALLING RAINDROP

## 1  BALANCE OF FORCES

Let us consider a single raindrop falling at velocity $v$. This velocity can be determined by making a force balance on the drop:

$$F = F_g - F_B - F_D \qquad (P5.1)$$

where $F$ is the resultant force; $F_g$ is the gravitational force; $F_B$ is the buoyant force; and $F_D$ is the friction or drag force. The resultant force $F = 0$ if terminal velocity is reached.

The forces from Equation P5.1 can be expressed as follows:

$$F = M\frac{dv}{dt}. \qquad (P5.2)$$

$$F_g = Mg; \qquad (P5.3)$$

$$F_B = \frac{\rho_{air}}{\rho_{drop}}Mg; \qquad (P5.4)$$

$$F_D = \frac{C_D A_{drop}\rho_{drop}v^2}{2}, \qquad (P5.5)$$

where $C_D$ is the coefficient of drag (or friction); $A_{drop}$ is the cross-sectional area of the drop at right angles to direction of falling.

*Methods of Measuring Environmental Parameters*, First Edition. Yuriy Posudin.
© 2014 John Wiley & Sons, Inc. Published 2014 by John Wiley & Sons, Inc.

As soon as $\rho_{air} = 1.205\ kg/m^3$; $\rho_{drop} = 998.2\ kg/m^3$ at 20°C, the ratio $\frac{\rho_{air}}{\rho_{drop}} \approx 10^{-3}$.

So, it is possible to neglect buoyant force ($F_B = 0$).

The equation of balance can be written as:

$$Mg - \frac{C_D A_{drop} \rho v^2}{2} = 0. \qquad (P5.6)$$

It is possible to determine the terminal velocity of the falling raindrop:

$$\rho_{drop} \left( \frac{4}{3} \pi r^3 \right) g = \frac{C_D \pi r^2 \rho v^2}{2}. \qquad (P5.7)$$

Solving for $v$, we get:

$$v = \sqrt{\frac{8 r \rho_{drop} g}{3 \rho_{air} C_D}}. \qquad (P5.8)$$

## 2   THE SIZE AND SHAPE OF RAINDROPS

Small drops (0.14–0.50 mm) are spherical. Larger drops (0.50–1.4 mm) become flattened. Very large drops (about 2 mm) demonstrates the concavity of the flattened base; they are shaped like parachutes (Figure P5.1). At 5–7 mm the force of the air through which it is falling causes the drop to break up (Horstmeyer, 2008).

Typically, the drops that have diameter less than 100 $\mu$m belong to clouds; raindrops are 1–2 mm in diameter.

## 3   THE DRAG COEFFICIENT

The *drag coefficient* $C_D$ is a dimensionless quantity that is used to quantify the drag or resistance of a drop in an air environment. This coefficient depends on the velocity of the drop relative to the air, drop size and shape, fluid density and fluid viscosity. The drag coefficient is a function of Reynolds number.

## 4   THE REYNOLDS NUMBER

The Reynolds number characterizes different flow regimes, such as laminar or turbulent flow.

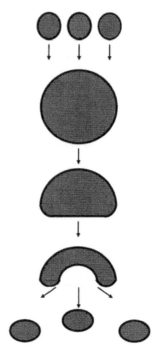

**FIGURE P5.1**   The size and shape of falling raindrops.

*Laminar flow* occurs when a fluid flows in parallel layers, with no disruption between the layers; it is characterized by low Reynolds numbers, where viscous forces are dominant.

*Turbulent flow* is a flow regime that is characterized by chaotic eddies, vortices, and other flow instabilities; it occurs at high Reynolds numbers and is dominated by inertial forces.

The Reynolds number is found from the equation:

$$\text{Re} = \frac{\rho v_0 d}{\mu_0} \tag{P5.9}$$

with $\rho$ as the density of the fluid, $\mu_0$ as the viscosity of the same fluid, and $v$ which is velocity. As one may see, the Reynolds number is dependent on the velocity at which the object is traveling.

For low Reynolds numbers ($\text{Re} \ll 1$) associated with a sphere, the fluid flows around the sphere without separating. Therefore, the drag coefficient, which is found empirically, is inversely proportional to the Reynolds number (friction drag predominates) and can be defined as (Henry and Heike, 1996)

$$C_D = 24/\text{Re.} \tag{P5.10}$$

The dependence of the drag coefficient on the Reynolds number in the intermediate range $2 < R < 500$ (both friction and form drag are important) is described as follows:

$$C_D = 18.5/\text{Re}^{0.6}. \tag{P5.11}$$

The turbulent regime of flow (form drag predominates) is described as:

$$C_D = 0.44/\text{Re.} \tag{P5.12}$$

We shall be interested with the spherical precipitating drops with $10 < R < 1000$ and $0.1$ mm $< r < 1$ mm.

Gunn and Kinzer (1949) presented a table with terminal velocity data as a function of drop diameter, measured in the laboratory. These data can be fitted with the following function:

$$v = 9.40[1 - \exp(-1.57 \cdot 10^3 \, d^{1.15})]. \tag{P5.13}$$

**Example**   Find the terminal velocity of rain drop, if the drop diameter $d = 0.2$ mm.

*Solution*   Using Equation P5.13, we get:

1. $d^{1.15} = (2 \times 10^{-4})^{1.15} = 5.57 \times 10^{-5}$;
2. $-1.57 \times 10^3 d^{1.15} = -1.57 \times 10^3 \times 5.57 \times 10^{-5} = -8.7449 \times 10^{-2}$;
3. $\exp(-1.57 \times 10^3 d^{1.15}) = \exp(-8.7449 \times 10^{-2}) = 0.91627$;
4. $1 - \exp(-1.57 \times 10^3 d^{1.15}) = 1 - 0.91627 = 0.0837$;
5. $v = 9.40 [1 - \exp(-1.57 \times 10^3 d^{1.15})] = 9.40 \times 0.0837 = 0.78678$ m/s.

**Example**   Calculate the terminal velocity of a rain drop for which radius $r = 0.05$ mm; the Reynolds number $Re = 10^2$; the coefficient of drag $C_D = 1.0$.

*Solution*   We can find from Equation P5.8 that:

$$v = \sqrt{\frac{8 r \rho_{\text{drop}} g}{3 \rho_{\text{air}} C_D}} = \sqrt{\frac{8 \times 5 \times 10^{-5} \times 998.2 \times 9.8}{3 \times 1.205 \times 1.0}} = 1.04 \text{ m/s.}$$

**Control Exercise**   Calculate the terminal velocity of a rain drop for which radius $r = 1$ mm; the Reynolds number $Re = 10^3$; the coefficient of drag $C_D = 0.4$.

**Constructive Test**   What speed is called the terminal velocity?

**Constructive Tests**   Plot the dependence of the terminal velocity of a water drop on its diameter. What is the typical cloud droplet size and the typical rain drop size?

## QUESTIONS AND PROBLEMS

1. Describe the main types and forms of precipitation.

2. What processes are accompanied by the appearance of rain?

3. Explain the mechanisms of cloud formation.

4. Compare different types of rain gauges.

5. Explain the principle of operation of polarimetric Doppler radar.

6. Describe the satellite Tropical Rainfall Measuring Mission (TRMM).

7. Explain the principle of operation of a diode laser.

8. Explain the principle of operation of a cavity ring-down spectrometer.

## REFERENCES

Gunn, R. and Kinzer, G.D. 1949. The terminal velocity of fall for water droplets in stagnant air. *J. Atmos. Sci.*, 6(4):243–248.

Henry, J.G. and Heike, G.W. 1996. *Environmental Science and Engineering*. Prentice-Hall, Inc., New Jersey.

## FURTHER READING

Andersen, T. 1982. Operational snow mapping by satellites. In: *Hydrological Aspects of Alpine and High Mountain Areas* (ed. J.W. Glen). International Association of Hydrological Sciences, Wallingford, UK, pp. 149–154.

Boyd, F.E., Joseph, W.W., and Scime, E.E. 2001. Dynamics of falling raindrops. *Eur. J. Phys.* 22:113–118.

Craig, H. and Gordon, L. 1965. Deuterium and oxygen-18 variations in the ocean and the marine atmosphere. P. 277. In: *Symposium on marine geochemistry*. Graduate school of Oceanograophy, Univ. Rhode Island, Occ Publ. No 3.

Dyroff, C., Fütterer, D., and Zahn, A. 2010. Compact diode-laser spectrometer ISOWAT for highly sensitive airborne measurements of water-isotope ratios. *Appl Phys B*, 98:537–548.

Gianfrani, L., Gagliardi, G., van Burgel, M., and Kerstel, E. 2003. Isotope analysis of water by means of near infrared dual-wavelength diode laser spectroscopy. *Opt. Express* 11(13):1566–1576.

Hoefs, J. 2004. *Stable Isotope Geochemistry*. Springer-Verlag, Berlin.

Khrgian, A.K. 1969. *Fizika Atmosfery*. Leningrad. Section 74.

Kume, A., Bekku, Yu.S., Hanba, Yu.T., and Kanda, H. 2003. High arctic glacier foreland. *Arct. Antarct. Alp. Res.* 35(3):377–383.

Martinec, J. and Rango, A. 1986. Parameter values for snowmelt runoff modelling. *J. Hydrol.* 84:197–219.

Nystuen, J.A., Proni, J.R., Black, P.G., and Wilkerson, J.C. 1996. A comparison of automatic rain gauges. *J. Atmos. Oceanic. Technol.* 13:62–73.

Parker, A.R. and Lawrence, C.R. 2001. Water capture by a desert beetle. *Nature,* 414(6859):33–34.

Peters, L.I. and Yakir, D. 2010. A rapid method for the sampling of atmospheric water vapour for isotopic analysis. *Rapid Commun. Mass Spectrom.* 24:103–108.

Rango, A., Salomonson, V.V., and Foster, J.L. 1977. Seasonal streamflow eshmation in the Himalayan region employing meteorological satellite snow cover observadons. *Water Resour. Res.* 14:359–373.

Seinfeld, J.H. and Pandis, S.N. 2006. *Atmospheric Chemistry and Physics - From Air Pollution to Climate Change*, 2nd edition. John Wiley & Sons, Inc.

Strangeways, I. 2006. *Precipitation: Theory, Measurement and Distribution*. Cambridge University Press.

Viessman, W. and Lewis, G.L. 2003. *Introduction to Hydrology*, 5th edition. Harper Collins College Publishers.

Yongmei, Zheng, Bai, H., Huang, Z., Tian, X., Nie, F.-Q., Zhao, Y., Zhail, J., and Jiang, L. 2010. Directional water collection on wetted spider silk. *Nature,* 463:640–643.

Zverev, A.S. 1977. Tumany i ikh predskazanie. Leningrad, Gidrometeoizdat. (Russian).

## ELECTRONIC REFERENCES

Acoustic rain gauge (ARG), http://www.dosits.org/technology/currentstemperature/acousticraingauge/ (accessed April 1, 2013).

Acoustic rain gauges, http://www.noc.soton.ac.uk/JRD/SAT/pers/gdq_others/Quartly2012_ARG/ARG.html (accessed April 1, 2013).

Balmer, D. Separation of boundary layers, http://www.see.ed.ac.uk/~johnc/teaching/fluidmechanics4/2003-04/fluids14/separation.html (accessed April 5, 2013).

Carbone, R.E. Active remote sensing, an overview, http://www.asp.ucar.edu/colloquium/1992/notes/part2/chapt1.html (accessed April 5, 2013).

Global precipitation climatology project (GPCP), http://www.gewex.org/gpcp.html (accessed April 6, 2013).

GOES precipitation index, http://www.cpc.ncep.noaa.gov/products/global_precip/html/wpage.gpi.shtml (accessed April 6, 2013).

Horstmeyer, Steven L. 2008 Clouds do not float, so .... if clouds are not lighter than air, how do they stay up? http://www.shorstmeyer.com/wxfaqs/float/float.html (accessed April 6, 2013).

Horstmeyer, Steven L. 2008. Deformation of falling rain drops, http://www.shorstmeyer.com/wxfaqs/float/dropdeform.html (accessed April 6, 2013).

Laser precipitation monitor (Distrometer). State of the art technology with best cost-performance ratio, http://www.thiesclima.com/disdrometer.html (accessed April 10, 2013).

Lasers: the new snow measuring tool, http://www.nbcnews.com/id/46434177/ns/technology_and_science-science/t/lasers-new-snow-measuring-tool/ (accessed April 12, 2013).

List of snowiest places on Earth? http://wiki.answers.com/Q/List_of_snowiest_places_on_Earth (accessed April 14, 2013).

Optical properties of snow, http://www.civil.utah.edu/ícv5450/Remote/AVIRIS/optics.html (accessed April 15, 2013).

Posudin, Y. 2011. Environmental Biophysics. Fukuoka-Kiev, http://www.ekmair.ukma.kiev.ua/handle/123456789/951 (accessed April 17, 2013).

Scholl, M. Precipitation isotope collector designs, http://water.usgs.gov/nrp/proj.bib/hawaii/precip_methods.htm (accessed April 18, 2013).

Snow. From Wikipedia, the free encyclopedia, http://en.wikipedia.org/wiki/Snow (accessed April 20, 2013).

Snow gauge. From Wikipedia, the free encyclopedia, http://en.wikipedia.org/wiki/Snow_gauge (accessed April 24, 2013).

Snow measurement guidelines, http://www.weather.gov/gsp/snow (accessed April 29, 2013).

Snow measurement reminders, http://www.crh.noaa.gov/gid/?n=snowmeasurement (accessed April 29, 2013).

The PERSIANN System, http://www.chrs.web.uci.edu/research/satellite_precipitation/activities00.html (accessed April 29, 2013).

260–700 Ultrasonic Snow Depth Sensor, http://www.novalynx.com/260-700.html (accessed April 29, 2013).

Unterman, N.A. Fog particle size, http://www.dri.edu/People/pat/webWPA/pdfdocs/dropletRet.pdf (accessed April 30, 2013).

World Weather Records, http://www.dandantheweatherman.com/Pikanto/Worldrec.htm (accessed April 30, 2013).

# 11

# SOLAR RADIATION

## 11.1  SI RADIOMETRY AND PHOTOMETRY UNITS

There are two parallel systems of quantities known as radiometric and photometric quantities. Each quantity in radiometric system has an analogous quantity in photometric system.

If the source of optical radiation is characterized by flux $dQ/dt$, where $Q$ is the energy of the source, the corresponding quantities in both systems are

Radiant flux $\Phi_e$(W);
Luminous Flux $\Phi_v$(lumen = cd · sr).

Intensity of optical radiation or power per unit direction $dQ/dtd\omega$ corresponds to the following pairs:

Radiant Intensity $I_e$(W/sr);
Luminous intensity $I_v$(candela = lm/sr).

Flux density (power per unit area) $dQ/dA$ corresponds to the following pairs:

Irradiance $E_e$(W/m$^2$);
Illuminance $E_v$(lux = lm/m$^2$);

*Methods of Measuring Environmental Parameters*, First Edition. Yuriy Posudin.
© 2014 John Wiley & Sons, Inc. Published 2014 by John Wiley & Sons, Inc.

Radiant emittance $M_e$(W/m$^2$);
Luminous emittance $M_v$(lux = lm/m$^2$);

Radiance $L_e$(W/sr · m$^2$);
Luminance $L_v$(cd/m$^2$).

## 11.2    THE PHOTOSYNTHETIC PHOTON FLUX DENSITY

Photosynthetically active radiation (PAR) is related to the spectral range of solar radiation from 400 to 700 nm that terrestrial, aquatic plants, and algae are able to use during photosynthesis.

PAR is normally quantified as $\mu$mol·photons/m$^2$/s, which is a measure of the *photosynthetic photon flux density (PPFD)*, or Einstein/m$^2$/s or $\mu$moles/m$^2$/s. One Einstein = 1 mole of photons = $6.022 \times 10^{23}$ photons, hence, 1 $\mu$Einstein = $6.022 \times 10^{17}$ photons.

## 11.3    PARAMETERS OF SUN

*Solar radiation* is the electromagnetic radiation and particles emitted by the Sun; it covers a wavelength range in the visible and near-visible (ultraviolet (UV) and near-infrared (NIR)) parts of spectrum.

The principal characteristics of the Sun are

- mean distance from Earth $149.6 \times 10^6$ km;
- mean diameter $1.392 \times 10^6$ km;
- mass $1.988 \times 10^{30}$ kg;
- density 1.408 g/cm$^3$;
- surface temperature 5785 K;
- temperature of corona 5 MK;
- core temperature $\sim$13.6 MK.

The main parameters of solar radiation are intensity, periodicity of solar activity, and spectral composition.

## 11.4    INTENSITY OF THE SUN

The amount of incoming solar electromagnetic radiation per unit area that reaches our atmosphere is called the *solar constant*. This value was measured using the satellite system such as the *Nimbus 6, 7,* and *Solar Maximum Mission (SMM)* spacecraft and was estimated to be 1373 W/m$^2$.

The total solar energy for a sphere of radius $d = 1496 \times 10^{11}$ m (distance between the Sun and Earth) is

$$E = I \times 4\pi d^2 = 1373 \text{ W/m}^2 \times 4 \times 3.14 \times (1.496 \times 10^{11})^2 \text{m}^2 = 3.88 \times 10^{26} \text{ W.}$$

The Sun is an absolute black body; the rate at which it emits radiant energy is proportional to the fourth power of its absolute temperature. This is known as the *Stefan–Boltzmann law*:

$$\sigma T^4 = 3.88 \times 10^{26} \text{ W}/4\pi R_S^2,$$

where $\sigma$ is the Stefan–Boltzmann constant ($5.67 \times 10^{-8}$ W/m$^2 \cdot$K$^4$).

From here, the absolute temperature of the Sun $T = 5770$ K.
The radius of the Earth is $R_E = 6.37 \times 10^6$ m; $R_E^2 = 4.06 \times 10^{13}$ m$^2$;
Cross-sectional area of the Earth is $\pi R_E^2 = 3.14 \times 4.06 \times 10^{13}$ m$^2 = 12.75 \times 10^{13}$ m$^2$.
The area of the entire Earth's surface is $4\pi R_E^2 = 51 \times 10^{13}$ m$^2$.

The power of solar radiation reaching the Earth's surface (consider that the Earth rotates and the surface that is irradiated by the Sun is four times as great as a disk) is

$$P = 1373 \text{ W/m}^2 \times \pi R_E^2 = 1373 \times 12.75 \times 10^{13} \text{m}^2 = 1.74 \times 10^{17} \text{ W.}$$

Thus, the intensity of sunlight that falls on the Earth's surface is

$$I = P/A = (1.74 \times 10^{17} \text{ W}/51 \times 10^{13} \text{ m}^2) = 342 \text{ W/m}^2.$$

The solar constant characterizes solar radiation that reaches the Earth's atmosphere. While entering the atmosphere, solar radiation can be absorbed or scattered by atmosphere particles (aerosols, water vapor) and molecules.

The part of solar radiation reaching the radiometer directly from the Sun is called *direct solar radiation*.

Radiation that has been scattered outward from the direct beam by atmospheric particles and molecules is called *diffuse solar radiation*.

The sum total of direct solar radiation and diffuse radiation that fall together on a horizontal surface is the *global solar radiation*.

Global solar radiation is described by the equation:

$$E_{G\downarrow} = E_S \cos\theta + E_{D\downarrow}, \tag{11.1}$$

where $E_{G\downarrow}$ is the global irradiance on a horizontal surface; $E_S$ is the direct beam irradiance on a surface perpendicular to the direct beam; $E_{D\downarrow}$ is the diffuse irradiance; and $\theta$ is the Sun's zenith angle (Figure 11.1).

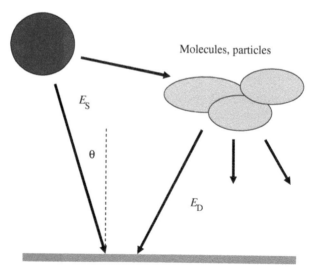

**FIGURE 11.1**   Direct and diffuse solar radiation.

Here *irradiance* is the power of electromagnetic radiation per unit area incident on a surface.

## 11.5   PERIODICITY OF SOLAR ACTIVITY

Cyclic changes of solar activity are known as *solar cycles*, which are characterized by cyclic increases and decreases in sunspots in cycles of approximately 11, 22, 87 (70–100), 210, and 2300 years.

## 11.6   SPECTRAL COMPOSITION OF SOLAR RADIATION

The entire frequency range of electromagnetic waves is called the *electromagnetic spectrum*. The solar spectrum occupies the range of 200–5000 nm and consists of several important bands—*ultraviolet* (UV) 0.20–0.40 $\mu$m, *visible* (VIS) 0.40–0.70 $\mu$m, *near-infrared* (NIR) 0.7–4.00 $\mu$m, and *infrared* (IR) 4.00–100.00 $\mu$m. The UV portion of the spectrum occupies 5%, visible 35%, and infrared (IR) 60% of the total solar radiation.

## 11.7   ATMOSPHERIC RADIATION

*Atmospheric radiation* is long-wave (IR) radiation emitted by or being propagated through the atmosphere.

Components of the atmosphere such as water vapor, carbon dioxide, and ozone are responsible for the absorption and emission of long-wave radiation.

The radiation spectrum of the atmosphere occupies the 5–100 $\mu$m region. The atmosphere can be modeled as a black body, the irradiance of which is proportional to the fourth power of the black body's thermodynamic temperature, $T$:

$$M_{A\uparrow} = \sigma T_A^4, \tag{11.2}$$

where $M_{A\uparrow}$ is the black-body irradiance (W/m$^2$) and $T_A$ is the absolute temperature (K).

Average values of atmospheric radiation are: $M_{A\uparrow} = 170$ W/m$^2$, clear sky, dry air, temperature $-39°$C; $M_{A\uparrow} = 310$ W/m$^2$, clear sky, moist air, temperature $-1°$C; $M_{A\uparrow} = 380$ W/m$^2$, the sky is half covered with thick clouds, temperature $+13°$C (Guyot, 1998).

## 11.8    TERRESTRIAL RADIATION

Long-wave electromagnetic radiation emitted by the Earth and its atmosphere is called *terrestrial radiation.*

The terrestrial surface absorbs radiation and behaves like a gray body at a temperature of 288 K. The spectral region this body occupies is the 4–50 $\mu$m range with a maximum at 10 $\mu$m.

The gray-body irradiance $M_{T\uparrow}$ is determined as

$$M_{T\uparrow} = \varepsilon \sigma T_T^4, \tag{11.3}$$

where $\varepsilon$ is the emissivity of the terrestrial surface, $\sigma$ is the Stefan–Boltzmann constant, and $T_T$ is the absolute temperature of the terrestrial surface.

Terrestrial radiation is almost completely absorbed by the atmosphere (primarily by water vapor, carbon dioxide, and ozone), except for specific atmospheric windows.

## 11.9    EFFECT OF SOLAR ULTRAVIOLET RADIATION ON LIVING ORGANISMS

Ultraviolet range can be divided into three parts: UV-A (400–315 nm), UV-B (315–280 nm), and UV-C (<280 nm). It is considered that UV-C radiation is the most hazardous for living organisms; UV-B may provoke specific but not always dangerous effects in living organisms; UV-A is safe radiation.

Solar ultraviolet radiation is characterized by a substantial impact on human health, terrestrial plants, aquatic ecosystems, and air quality. Ultraviolet radiation causes a number of effects on the human organism (positive and negative) such as cancerogenic effects, erythema, antirickets effect, conjunctivitis, and germicidal effects. Action spectra of these effects are often very close to each other or overlap. Therefore, UV radiation should be used with caution and understanding the nature and mechanisms of its action.

The primary organs of humans and animals which are exposed by natural UV-B radiation are eyes and skin. Thus, UV-B radiation induces the cataracts, erythema, ageing of the skin, photodermatoses, and skin cancer. The exposure of unprotected eyes to solar radiation (mainly UV-B and UV-C) provokes the photokeratoconjunctivitis (snow blindness). The positive effect of this radiation is the formation of vitamin D in the skin that is important for the maintenance of bone tissues. It is necessary to mention the effect of UV-B radiation on the immune system.

Solar UV-B radiation provides the effects on physiological and developmental processes of plants and algae, including the changes in plant morphology, phenology, biomass accumulation, inhibition of photosynthesis, and DNA damage.

Solar UV-B radiation demonstrates strong effects on inhabitants of aquatic ecosystems. It inhibits the motility of phytoplankton and spatial orientation, causes damage to early development stages of fish, shrimp, crab, amphibians, and impairs larval development. The photosynthesis of red, brown, and green algae is inhibited by UV-B radiation.

Energetic UV-B radiation can break the bonds of atmospheric gases such as ozone, nitrogen dioxide, formaldehyde, hydrogen peroxide, nitric acid producing highly reactive atomic, and molecular radical species (O, H, OH, $HO_2$) that are responsible for adverse effects on human health and air quality.

## 11.10   EFFECT OF SOLAR VISIBLE RADIATION ON LIVING ORGANISMS

Electromagnetic radiation in the range of wavelengths from about 400 to 700 nm is called *visible radiation*; a typical human eye is sensitive to this part of the spectrum.

Solar visible radiation induces certain photobiological responses in living organisms. The following are the types of photobiological responses (Konev and Volotovsky, 1974):

1. *Energy responses* during which light energy is transformed into chemical energy due to the synthesis of new organic molecules. *Photosynthesis* as an example of energy reaction is the process of transformation of solar energy into energy of chemical bonds of organic compounds by green plants and photosynthetic microorganisms. This is a complex process by which carbon dioxide and water are converted into carbohydrates and oxygen using energy from the Sun and chlorophyll.

2. *Information responses* are those reactions in which light acts as a control signal that induces a specialized mechanism of formation of photoproducts, and provides the information about the environment. The responses of this type include photomovement, photoperiodism, photomorphogenesis, and phototropism.

   *Photomovement*, in a wide meaning, is the movement of microorganisms (algae, protozoa, bacteria) or its alteration induced by light. The following are representatives of various divisions of cyanoprocariotic algae— Cyanophyta (*Oscillatoria geminata*, *Synechococcus aeruginosus*), eukaryotic algae—Euglenophyta (*Euglena gracilis, Astasia longa, Peranema trichopho-rum*), Dinophyta (*Peridinium gatunense*), Bacillariophyta (*Pinnularia strep-toraphe, Anomoeoneis sculpta*), Cryptophyta (*Cryptomonas* spp., *Chroomonas* spp.), Chlorophyta (*Dunaliella salina, D. viridis, Chlamydomonas reinhardtii, Chloromonas* sp., *Haematococcus pluvialis, Stephanosphaera* sp., *Gonium* sp., *Eudorina* sp., *Volvox* sp.), and spores and gametes of green (Chlorophyta), golden (Chrysophyta), yellow-green (Xanthophyta), red (Rhodophyta) and brown (Phaeophyta*)* algae are currently or have been used as models in these investigations.

*Photoperiodism* is the physiological or behavioral response of an organism or population of organisms to changes of duration in daily, seasonal, or early cycles of solar activity.

*Phototropism* is the change in the directional growth of the plants under the influence of unilateral illuminance.

*Photomorphogenesis* is the growth and development of plants that are stim-ulated by the light of different spectral composition and intensity.

3. *Destructive-modifying responses* are reactions associated with the damage of biological molecules by light; these reactions can provoke lethal or mutational effects. This type of reactions includes photosensitization, photoreactivation, responses of biological systems to ultraviolet radiation.

*Photosensitization* is the process of sensitization of living organism to simultane-ous action of light and certain chemicals that leads to the disturbance of its viability is called *photosensitization*.

This phenomenon is spread in nature widely. Some herbivorous animals eat plants that contain some chemicals such as species of Guttiferae (*Hypericum perforatum, H. crispum, H. pulchrum, H. leucoptycodes, H. maculatum*); *Fagopyrum* (*F. escu-lentum* or *Polygonum fagopyrum*); genus *Tribulus* (*T. terrestris* and *T. ubis*, fam-ily Zygophyllaceae); some species *Lippia* (*L. rehmanni* and *L. pretoriensis*, family Verbenaceae); grass *Panicum* (*P. laevifolium* and *P. coloratum*, family Gramineae). In Ukraine, such plants as *Trifolium* L. and *Sisymbrium altissimum* L. exhibit photosensitization.

Plant compounds or pigments are absorbed by the digestive system and provoke a direct effect on the non-pigmented parts of the skin such as near eyes, mouth, ears, and hoofs when they are exposed to sunlight. The final result is the necrosis, itch, scab formation, and infection of the organism. The animal can die in 8–10 hours.

## REFERENCES

Guyot, G. 1998. *Physics of the Environment and Climate*. John Wiley & Sons. Chichester.

Konev, S. and Volotovsky, I. 1974. *Fotobiologiya*. Minsk: Bel. Gos. Univ. (In Russian).

# 12

# MEASUREMENT OF SOLAR RADIATION

## 12.1 CLASSIFICATION OF RADIOMETERS

There are three main classes of instruments used to measure solar radiation:

A *radiometer* is used for measuring the radiant flux or power of electromagnetic radiation. Here we shall discuss such radiometers as pyrheliometer, pyranometer, pyrgeometer, albedometer, and net pyrradiometer which are used in measuring the environmental radiation. Radiometers such as thermocouples, thermopiles, and thermistors respond to radiant energy. These devices measure *irradiance*—the power of electromagnetic radiation per unit area incident on a surface. The SI unit for this quantity is $W/m^2$.

A *photometer* (or *light meter*) is used for measuring light in terms of its perceived brightness to the human eye. Photometers measure such quantities as luminous flux (lumen), luminous intensity (candela), and illuminance (lux). In photometry, illuminance is the total luminous flux incident on a surface per unit area. In SI derived units, illuminance is measured in *lux (lx)* or *lumens per square meter (cd·sr/m²)*. Photometers are most sensitive in the middle of the visible spectrum (555 nm) where the human eye is also the most sensitive.

A *photon meter* (or *quantum meter*) is intended for measuring solar radiation in the spectral range from 400 to 700 nm, called *photosynthetically active radiation (PAR)*.

The spectral sensitivity of the photon sensor is similar to that of the photosynthetic system in plants.

*Methods of Measuring Environmental Parameters*, First Edition. Yuriy Posudin.
© 2014 John Wiley & Sons, Inc. Published 2014 by John Wiley & Sons, Inc.

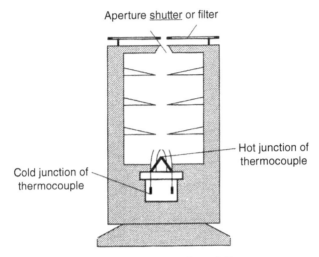

Aperture <u>shutter</u> or filter

Hot junction of thermocouple

Cold junction of thermocouple

**FIGURE 12.1**    Diagram of a pyrheliometer.

## 12.2   MEASUREMENT OF DIRECT SOLAR RADIATION—PYRHELIOMETER

Direct solar radiation is measured using a *pyrheliometer*, an instrument that directly measures solar irradiance. The field of view of this instrument is limited to 5°. The pyrheliometer consists of a collimation tube with precision apertures and a detector (Figure 12.1). Solar radiation enters the instrument through a window, apertures, and reaches a thermopile that converts heat into an electrical signal that is proportional to solar irradiance.

## 12.3   MEASUREMENT OF GLOBAL RADIATION—PYRANOMETER

Global radiation is the sum total of direct solar radiation and diffuse sky radiation received by a horizontal surface per unit area.

A *pyranometer* is a radiometer for measuring the solar irradiance (radiation flux density, W/m$^2$) on a plane surface, that results from direct and diffuse solar radiation incident from the hemisphere above in a solid angle of $2\pi$ sr (spectral range 0.3–3 μm). A diagram of a pyranometer is presented in Figure 12.2.

A pyranometer consists of a thermopile sensor with a black and white coating. The hot junctions are located beneath a black coating absorbing all radiation, whilst the cold junctions are beneath a white coating. A temperature difference between the black and white sectors occurs when the sensor is exposed to solar radiation. The voltage output signal of the thermopile sensor is proportional to the solar radiation.

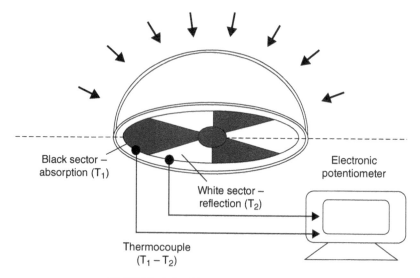

**FIGURE 12.2**    Diagram of a pyranometer.

## 12.4    MEASUREMENT OF DIFFUSE RADIATION—PYRANOMETER WITH A SUN-SHADING RING

Diffuse solar radiation is measured using a pyranometer that is equipped with a narrow sun-shading ring with its axis parallel to the Earth's. The orientation of the axis takes into account the variation of the Sun's declination and follows its position (Figure 12.3). Such a design makes it possible to exclude the measurement of direct solar radiation.

## 12.5    MEASUREMENT OF LONG-WAVE RADIATION—PYRGEOMETER

A *pyrgeometer* is an instrument used to measure long-wave (LW) radiation, either terrestrial or atmospheric.

The pyrgeometer includes a thermopile which detects the net radiation balance between the incoming and outgoing LW radiation flux and converts it into voltage. It consists of a silicon dome or window with a solar blind filter coating (Figure 12.4) and has a transmittance between 4.5 and 50 μm that eliminates solar short-wave (SW) radiation.

Net LW radiation balance can be determined as

$$E_{net} = E_{in} - E_{out}, \tag{12.1}$$

where $E_{net}$ is the net radiation at sensor surface (W/m$^2$); $E_{in}$ is the LW radiation received from the atmosphere (W/m$^2$); $E_{out}$ is the LW radiation emitted by the Earth

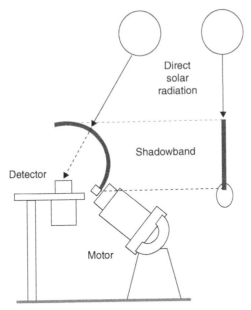

**FIGURE 12.3**  The rotating shadowband radiometer.

surface (W/m$^2$). A diagram that explains measuring the net radiation balance is presented in Figure 12.5.

## 12.6  MEASUREMENT OF ALBEDO—ALBEDOMETER

*Albedo* is the surface reflectivity of the Sun's radiation. It is determined as the ratio of solar radiation of all wavelengths reflected by a surface to the amount incident upon it.

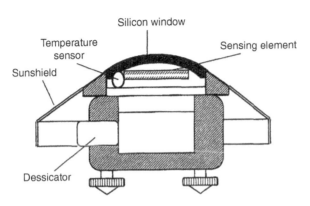

**FIGURE 12.4**  Pyrgeometer.

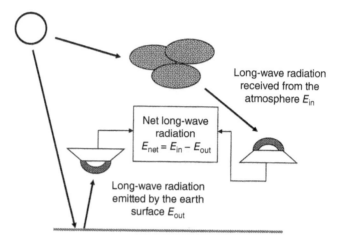

**FIGURE 12.5**   The net radiation balance measurement.

An *albedometer* is an instrument that measures both global and reflected solar irradiance. It consists of two pyranometers: the upper pyranometer measures incoming global solar radiation and the lower sensor measures solar radiation reflected from the surface below (Figure 12.6).

## 12.7   MEASUREMENT OF TOTAL RADIATION—A 4-COMPONENT NET RADIOMETER

A net *pyrradiometer* is designed to estimate the energy balance between incoming SW (0.3–3.0 μm) and LW infrared (3.0–100 μm) radiation versus surface-reflected SW and outgoing LW infrared radiation.

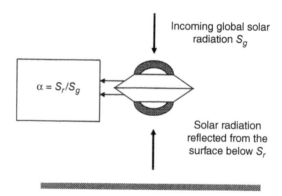

**FIGURE 12.6**   Measurement of albedo.

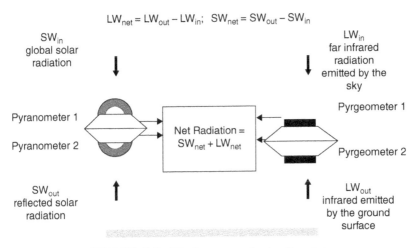

$$LW_{net} = LW_{out} - LW_{in}; \quad SW_{net} = SW_{out} - SW_{in}$$

SW$_{in}$
global solar
radiation

LW$_{in}$
far infrared
radiation
emitted by the
sky

Pyranometer 1

Pyrgeometer 1

Net Radiation =
SW$_{net}$ + LW$_{net}$

Pyranometer 2

Pyrgeometer 2

SW$_{out}$
reflected solar
radiation

LW$_{out}$
infrared emitted
by the ground
surface

**FIGURE 12.7**    The 4-component net radiometer.

Although there are many types of net radiometers, the 4-component design at present is most popular for scientific applications (Figure 12.7).

It allows measuring the following components of the surface radiation balance: SW$_{in}$ is the global solar radiation; SW$_{out}$ is the reflected solar radiation; LW$_{in}$ is the infrared emitted by the sky; and LW$_{out}$ is the infrared emitted by the ground surface.

The SW solar radiation spectrum occupies the region from 300 to 2800 nm while the long wave (LW) or far infrared (FIR) extends from 4500 to 50,000 nm. This part of the spectrum is covered by a pyrgeometer.

## 12.8   PHOTOMETER

The most popular instrument that is used in photometry is a *lux meter*, which is a small, portable device for measuring *illuminance.*

*Semiconductor photodiode* is a semiconductor device which changes its state under the influence of the flow of optical radiation for estimation of this radiation. The principle of operation of this photodetector is based on the inner photoelectric effect. If a photon of sufficient energy strikes the diode, it extracts an electron from the valence band and transfers it to the conduction band (Figure 12.8). Consequently, the number of conduction electrons and holes is increased. The total current through the photodiode is proportional to the flow of optical radiation.

*Photomultiplier tube* (or *photomultiplier*) presents the class of vacuum tubes that are sensitive detectors of radiation in the ultraviolet, visible, and near-infrared ranges of the electromagnetic spectrum. Such a photomultiplier consists of several electrodes (dynodes), whose potentials are increased in succession along the length of the tube (Figure 12.9). The external photon extracts electrons from a photocathode due to the photoelectric effect. A transmission type of photocathode is performed as a coating

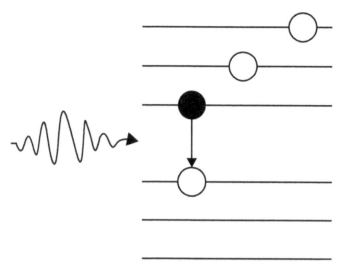

**FIGURE 12.8**    Semiconductor photodiode.

upon a glass window where the light enters one surface and electrons exit through the opposite surface. These emitted electrons strike the dynodes and produce about one million of other electrons that reach the last dynode (anode). In such a way, photomultiplier is characterized by extremely high sensitivity.

## 12.9    PHOTON METER

A *silicon-cell quantometer* is a photon meter which is based on combining a photodiode with an interference filter with a sharp cut-off at 700 nm and colored glass filters (Figure 12.10). Together they create a sensor with a spectral response similar to that

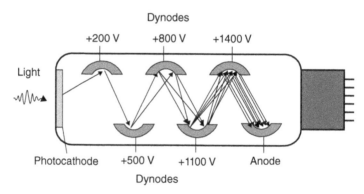

**FIGURE 12.9**    Photomultiplier tube.

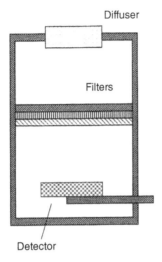

**FIGURE 12.10**   Silicon-cell quantometer.

of the principal plant pigments such as chlorophylls, carotenoids, phycoerythrin, and phycocyanin.

## 12.10   CONVERSION OF LIGHT ENVIRONMENT UNITS

Conversion of light environment units can be found at site: http://www.fb.u-tokai.ac.jp/WWW/hoshi/env/light.html.

| Item | PRFD | PPFD | Illuminance |
|------|------|------|-------------|
| Unit | $Wm^{-2}$ | $\mu mol\ m^{-2}s^{-1}$ | lx |

The ratio among units of optical radiation, measured in natural conditions, is presented in Table 12.1.

**TABLE 12.1   The relationship between $\mu mol/m^2 \cdot s$, $W/m^2$, and lx**

| $E$, $\mu mol/m^2 \cdot s$ | $E$, $W/m^2$ | $E$, lx |
|---|---|---|
| 10,000 | 2500 | 5,95,238 |
| 5000 | 1250 | 2,97,619 |
| 3000 | 750 | 1,78,571 |
| 2500 | 625 | 1,48,809 |
| 2000 | 500 | 1,19,047 |
| 1500 | 375 | 89,285 |
| 1000 | 250 | 59,523 |
| 500 | 125 | 29,761 |

# PRACTICAL EXERCISE **6**

# PARAMETERS OF OPTICAL RADIATION

## 1  PARAMETERS OF ELECTROMAGNETIC RADIATION

Electromagnetic waves are typically described by the following parameters: the frequency $v$, wavelength $\lambda$, or photon energy $E$.

*Frequency $v$* is defined as a number of cycles per unit time. In SI units, the unit of frequency is the hertz (Hz): 1 Hz means that an event repeats once per second.

The *wavelength $\lambda$* of a sinusoidal wave is the spatial period of the wave—the distance over which the wave's shape repeats. The wavelength is inversely proportional to frequency:

$$\lambda = \frac{c}{v}, \tag{P6.1}$$

where $c$ is the speed of light: $c = 2.998 \times 10^8$ m/s.

A *photon* is an elementary particle, the quantum of light and all other forms of electromagnetic radiation. There is a relationship between the energy of a photon $E$, the frequency $v$, and wavelength of the light $\lambda$ given by the equation:

$$E = hv = \frac{hc}{\lambda}, \tag{P6.2}$$

where $h$ is Planck's constant: $h = 6.626 \times 10^{-34}$ J·s

*Methods of Measuring Environmental Parameters*, First Edition. Yuriy Posudin.
© 2014 John Wiley & Sons, Inc. Published 2014 by John Wiley & Sons, Inc.

**Example**    Calculate the energy of a photon of green light whose frequency is 5.77 × 10¹⁴ Hz.

**Solution**    By Equation P6.2 the energy of this photon is $E = hv = 6.626 \times 10^{-34}$ J·s $\times 5.77 \times 10^{14}$ 1/s $= 38.23 \times 10^{-20}$ J.

**Control Exercise**    Find the energy of a photon of red light whose frequency is 4.41 × 10¹⁴ Hz.

The *wavenumber* $\sigma$ is the number of wavelengths per unit distance. It is the reciprocal of the wavelength $\lambda$:

$$\sigma = \frac{1}{\lambda}. \tag{P6.3}$$

The *angular* (or *circular*) *wavenumber* $k$ is the number of wavelengths per $2\pi$ units of distance:

$$k = \frac{2\pi}{\lambda}. \tag{P6.4}$$

SI unit of the wavenumber is m⁻¹; in addition, the wavenumber can be measured in such reciprocal distance unit as cm⁻¹ : 1 m⁻¹ = 0.01 cm⁻¹; 1 cm⁻¹ = 100 m⁻¹.

It is possible to convert inverse centimeters in wavenumber to micrometers (microns) or nanometers using the following formulas:

$$\sigma(cm^{-1}) = \frac{10,000,000}{\lambda(nm)}; \sigma(cm^{-1}) = \frac{10,000}{\lambda(\mu m)}; \lambda(nm) = \frac{10,000,000}{\sigma(cm^{-1})}; \lambda(\mu m) = \frac{10,000}{\sigma(cm^{-1})}$$
$$\tag{P6.5}$$

**Example**    Calculate the wavenumber $\sigma$ if the wavelength $\lambda$ is 500 nm.

*Solution*

$$\sigma(cm^{-1}) = 10^7/500 = 20000 \text{ cm}^{-1}. \tag{P6.6}$$

## 2   THE INVERSE-SQUARE LAW

**Example**    The illuminance of a surface that is placed directly below a lamp is 300 lx. What is the intensity of the lamp if the distance between the lamp and the surface is 3 m? Assume that the lamp can be considered as the point source.

*Solution*    Using Equation P6.5, the intensity of the lamp can be determined as

$$I = Ed^2 = 300 \text{ lx} \times 9 \text{ m}^2 = 2700 \text{ cd}.$$

**Control Exercise**    A guidance for recommended light level in different work spaces recommends the illuminance 500 lx of the table during normal office work, PC work, and laboratory.

A standard incandescent lamp that has luminous intensity of 100 cd is placed above the table surface at the distance 2 m. Is the illuminance of the table satisfied to the guidance for recommended light level?

## 3    THE COSINE LAW

Illuminance decreases with the angle of incidence, as described by the cosine law:

$$E = \frac{I \cos \theta}{d^2},$$    (P6.7)

where $\theta$ is an incidence (the angle between a ray incident on a surface and the line perpendicular to the surface at the point of incidence, called the normal).

**Example**    The central part of the highway is illuminated by two light sources of $I = 2000$ cd, located at the height of $h = 14$ m above the ground and at a distance of $L = 60$ m from each other. Determine the illuminance at the center of the road.

**Solution**    Distance $d$ from a given point to a single lamp (left or right) is found from the relationship: $d^2 = (L/2)^2 + h^2 = 30^2 + 14^2 = 900 + 196 = 1096$ m$^2$. From this it follows that $\cos\theta = h/d = h/[(L/2)^2 + h^2]^{1/2} = 14/33 = 0.42$. Illumination, created by the left or right lamp is

$$E = I \cos\theta/d^2 = 2000 \times 0.42/1096 = 0.77 \text{ 1x}.$$

Illumination, created by two lamps is $E = 1.54$ lx.

**Control Exercise**    Find the illuminance at the table which is illuminated by two lamps of 75 cd, located at the height of 1.5 m above the surface of the table and at a distance of 2 m from each other.

**Example**    Determine the luminance $L_v$ of the Sun, if the illuminance $E_v$ of the Earth's surface is $1.3 \times 10^5$ lx, resulting from light incident perpendicular (normal) to the surface plane.

**Solution**    Luminous intensity $I_v$ is related to the illuminance $E_v$ by the expression $I_v = E_v d^2$, where $d$ is the mean distance between the Sun and the Earth ($1.5 \times 10^{11}$ m). The luminance $L_v$ can be found from the expression:

$$L_v = \frac{dI_v}{dA \cos\theta},$$    (P6.8)

where $dA$ is the area of surface element; $\theta$ is the angle that determines the direction of radiation.

For normal incidence

$$L_v = I_v/A. \tag{P6.9}$$

The luminance $L_v$ of the Sun is given by

$$L_v = I_v/A = E_v d^2/\pi R_S^2, \tag{P6.10}$$

where $R_S$ is radius of the Sun ($6.96 \times 10^8$ m).
Using numerical values of $E_v$ and solar parameters, the last equation gives

$L_v = 1.3 \times 10^5$ lx $\times$ ($1.5 \times 10^{11}$ m)$^2/3.14 \times$ ($6.96 \times 10^8$m)$^2 = 1.92 \times 10^9$ cd/m$^2$.

**Control Exercise**    Determine the luminance of the Moon, if the illuminance of the Earth's surface is 0.2 lx.
    *Answer:* $3.1 \times 10^3$ cd/m$^2$.

## 4   THE WIEN'S DISPLACEMENT LAW

From Wien's displacement law, it follows that there is an inverse relationship between the wavelength of the peak of the emission of a black body and its absolute temperature:

$$\lambda_{max} = b/T, \tag{P6.11}$$

where $\lambda_{max}$ is the peak wavelength, $T$ is the absolute temperature of the black body, and $b$ is a constant of proportionality called Wien's displacement constant, equal to $2.8977685 \times 10^{-3}$ m·K

**Example**    The temperature of the green leaf is 280 K. What is the wavelength at which the peak occurs in the radiation emitted from the leaf?

*Solution*    From Wien's displacement law, we have:

$$\lambda_{max} = 2.8977685 \times 10^{-3} \text{m} \cdot K/280\, K = 10.3 \times 10^{-6} \text{m} = 10.3\ \mu\text{m}.$$

This radiation is in the infrared region of the spectrum.

**Control Exercise**    The planetary nebula BV-1 radiates with a peak wavelength of $\lambda_{max} = 500$ nm. Using Wien's displacement law, calculate the surface temperature of BV-1.

## 5   THE STEFAN–BOLTZMANN LAW

The Stefan–Boltzmann law states that the total energy radiated per unit surface area of a black body per unit time is directly proportional to the fourth power of the black body's thermodynamic temperature $T$:

$$B = \sigma T^4, \tag{P6.12}$$

where $B$ is the emitted flux density (W/m$^2$); $T$ is the absolute temperature; $\sigma$ is the Stefan–Boltzmann constant ($\sigma = 5.67 \times 10^{-8}$ W/m$^2 \cdot$K$^4$).

**Example**   The Sun can be presented as a black-body radiator emitting at 5785 K. Find the average emitted flux density of the Sun.

*Solution*   The emitted flux density of the Sun can be determined according to the Stefan–Boltzmann law as

$$B = \sigma T^4 = 5.67 \times 10^{-8} \text{ W/m}^2 \cdot \text{K}^4 \times (5785\text{K})^4 = 63.5\text{W/m}^2.$$

**Control Exercise**   Find the average emitted flux density of the Earth, if it is emitting at 288 K.

## 6   THE PHOTOSYNTHETIC PHOTON FLUX DENSITY

**Example**   The irradiance of a leaf surface in the 400–700 nm waveband is 200 W/m$^2$. What is the photosynthetic photon flux density (PPFD)? The source is solar radiation.

*Solution*   It was previously determined in Section 12.1 that for solar radiation at sea level in the waveband 400–700 nm the photon energy is $2.17 \times 10^5$ J/mol. The PPFD is therefore:

$$\text{PPFD} = 200 \text{ W/m}^2 \cdot \frac{1 \text{ mol}}{2.35 \times 10^5 \text{J}} = 85 \times 10^{-5} \text{mol/m}^2 \cdot \text{s or } 8500 \text{ } \mu\text{m/m}^2 \cdot \text{s}.$$

**Control Exercise**   The energy flux density in the 400–700 nm spectral region is 500 W/m$^2$. What is the photon flux density of PAR?

## 7   THE LABORATORY EXERCISE "THE INVERSE-SQUARE LAW"

**Theory**   *Illuminance* is the total luminous flux incident on a surface, per unit area. The SI unit of illuminance is the lux (symbol: lx).

*Lux meter* is a portable instrument for measuring illumination. It consists of a selenium photocell that converts luminous energy into the energy of an electric current, which is measured by a measuring instrument with scales calibrated in lux (lx).

*The inverse-square law:* the intensity (or illuminance or irradiance) of light or other linear waves radiating from a point source (energy per unit of area perpendicular to the source) is inversely proportional to the square of the distance from the source.

The illuminance that is produced by the point source of light is

$$E = \frac{I}{d^2},$$ (P6.5)

where $I$ is the intensity of the source and $d$ is distance from the point source.

If the light is emitted through the aperture whose diameter is not negligible in comparison with the distance $d$, the illuminance that is produced by the source is determined as

$$E = \pi R^2 \frac{I}{d^2 + R^2},$$ (P6.6)

where $R$ is radius of aperture.

## Objectives

1. To explore the principle of operation of the lux meter.
2. To understand use of the lux meter for measurement of illuminance.
3. To verify the inverse-square law of illumination for point sources.

**Lab Summary**   Students gain experience conducting the following procedures:

1. Using a lux meter to experience the concept of light intensity varying inversely as the square of the distance from a point source of light;
2. Measuring the illuminance of the screen as the distance between the lux meter and the light source is increased;
3. Graphing measured illuminance against the square of the distance between the lux meter and the light source;
4. Linearizing a graph;
5. Examining the shape of the graph and drawing conclusions from graphed data.

## Materials Supplied

1. Source of light.
2. Circular aperture.
3. Screen.
4. Lux meter.
5. Optical bench.
6. Ruler.

## Experimental Procedure

1. Dispose the light source and aperture at the optical bench.
2. Position the screen 0.4 m from the light source.
3. Adjust the lux meter until it is pointed directly at the light source.
4. Take a reading of the lux meter.
5. Move the screen to 20 cm and record the reading of the lux meter.
6. Repeat the steps above by moving the screen at 20 cm intervals until you reach 1 m.
7. Construct a graph of illuminance $E$ versus the square of the distance $d^2$, and perform a linear fit.
8. Draw conclusions from graphed data.

## QUESTIONS AND PROBLEMS

1. Explain how the solar constant is measured.
2. What is a radiometer?
3. What is principle of operation of a pyrheliometer?
4. Explain the principle of the device for measuring diffuse radiation.
5. What is the difference between a pyranometer and a pyrgeometer?
6. What instruments are used for measuring SW radiation? LW radiation?
7. What is photosynthetic active radiation?
8. What is photosynthetic photon flux density?

## FURTHER READING

Meteorological Office. 1982. *Handbook of Meteorological Instruments,* Volume 6 Measurement of Sunshine and Solar and Terrestrial Radiation, 2nd edition. Stationery Office Books.

Nobel, P.S. 2005. *Physicochemical and Environmental Plant Physiology.* Elsevier Academic Press, Burlington.

Posudin, Y.I. 1998. *Lasers in Agriculture.* Science Publishers, Inc. Enfield, New Hampshire.

Posudin, Y.I. 2007. *Practical Spectroscopy in Agriculture and Food Science.* Science Publishers, Enfield.

Posudin Y.I., Massjuk N.P., and Lilitskaya G.G. 2010. *Photomovement of Dunaliella Teod.* Vieweg + Teubner Research.

Van der Leun, J., Tang X., and Tevini M. 1994. Assessment. Ambio. Environmental Effects of Ozone depletion: *AMBIO. A Journal of the Human Environment,* XXIV(3):138–142.

Tchijevsky, A.L. 1924. *Physical Factors of the Historical Process.* A short sketch. Kaluga.

Tchijevsky, A.L. 1995. *Space Pulse of Life.* Moscow, Misl.

# ELECTRONIC REFERENCE

Posudin Y.I 2011. *Environmental Biophysics.* Fukuoka-Kiev. http://www.ekmair.ukma.kiev.ua/handle/123456789/951 (accessed May 21, 2013).

# 13

# EDDY COVARIANCE

## 13.1 TURBULENCE

*Turbulence* is defined as irregular atmospheric motion that is accompanied by upward and downward currents and rapid special and temporal variations in pressure, temperature, density, and velocity.

## 13.2 BOUNDARY LAYER

Turbulence appears regularly in a relatively thin layer of the atmosphere which is called the *boundary layer*. It is characterized by heat, moisture, or momentum transfer to or from the ground surface. The thickness of the turbulence boundary layer varies from 100 m at night up to 4000 m during the day in the summer at mid-latitudes. Turbulence induces the formation of *eddies*—currents of air that move contrary to the direction of the main current, especially in a circular motion. Air flow can be presented as a horizontal flow of numerous rotating eddies that consist of three-dimensional components including vertical movement of the air. In spite of the apparent chaotic character of the air flow, its parameters can be estimated quantitatively using covariance techniques.

*Methods of Measuring Environmental Parameters*, First Edition. Yuriy Posudin.
© 2014 John Wiley & Sons, Inc. Published 2014 by John Wiley & Sons, Inc.

## 13.3   EDDY COVARIANCE

An eddy is a current of air, water, or other material capable of flowing that moves against the main current and with a circular or whirlpool motion. Eddies appear in the boundary layer due to the wind, the roughness of terrestrial surface, and convective heat flows.

*Covariance* is a statistical measure of the correlation of fluctuations in two different quantities. Covariance quantifies the degree to which two variables vary together.

*Eddy covariance* in environmental biophysics is a method used to measure atmospheric fluxes of $H_2O$, $CO_2$, momentum, and sensible and latent heat transferred due to the turbulence within the atmospheric boundary layer (Prandtl, 1920; Desjardins, 1972; Ohtaki and Matsui, 1982; Shimizu, 2007). While these fluxes consist of horizontal and vertical components that are chaotic, it is possible to estimate the vertical components using a meteorological tower.

*Flux* is the value that characterizes how much of a substance or material moves through a unit area per unit time. Vertical flux can be presented as a covariance of the vertical component and the concentration of the material that is being studied.

For example, the wind speed with horizontal component $u$ and vertical component $v$ can be presented as

$$u = \bar{u} + \delta u; \quad v = \bar{v} + \delta v, \tag{13.1}$$

where $\bar{u}$ and $\bar{v}$ are mean values of the wind speed components; $\delta u$ and $\delta v$ are fluctuations in the wind speed components (Figure 13.1).

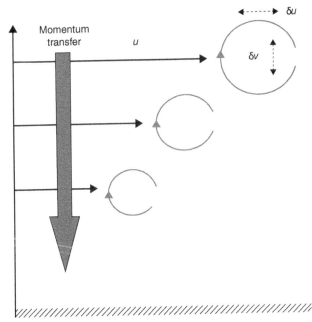

**FIGURE 13.1**   Mean values and fluctuations of the wind velocity components.

The covariance between two random variables $u$ and $v$ is defined as:

$$\text{Cov}_{uv} = \frac{\sum uv}{n},$$

(13.2)

where $n$ is the number of random variables.

By substituting (13.1) into (13.2) the covariance can be written as:

$$\text{Cov}_{uv} = \frac{\sum (u - \delta u)(v - \delta v)}{n}.$$

(13.3)

It is necessary to take into account Reynolds averaging rules. If $A$ and $B$ are variables, and $c$ is constant, then

So

$$\left.\begin{aligned}
A &\equiv \bar{A} + \delta a; \\
B &= \bar{B} + \delta b. \\
\\
c &= \bar{c}; \\
\overline{(cA)} &= c\bar{A}; \\
\overline{(\bar{A})} &= \bar{A}; \\
\overline{(\bar{A}B)} &= \bar{A}\bar{B}; \\
\overline{(A + B)} &= \bar{A} + \bar{B}; \\
\overline{\left(\frac{dA}{dt}\right)} &= \frac{d\bar{A}}{dt}; \\
\overline{\delta a} &= 0; \overline{\delta a \delta b} \neq 0.
\end{aligned}\right\}$$

(13.4)

According to the Reynolds averaging rules the average of any fluctuating components is zero.

## 13.4  TURBULENT VELOCITY FLUCTUATIONS

The time-averaged values of the fluctuations, according to Reynolds averaging rules, equals zero:

$$\overline{\delta u} = 0; \quad \overline{\delta v} = 0.$$

(13.5)

The average value of vertical wind speed also equals zero:

$$\bar{v} = 0.$$

(13.6)

The boundary layer is turbulent through its depth and all physical entities such as mass (e.g., concentration of $H_2O$ or $CO_2$, aerosols), heat, and momentum are transported in the vertical direction. One of the methods of flux measurement is eddy covariance.

## 13.5   VERTICAL MOMENTUM FLUX

The main principle of vertical flux estimation is the presentation of the covariance between measurements of vertical momentum (or vertical momentum density) and vertical wind velocity which can be calculated as

$$F = (\rho u)\delta v, \tag{13.7}$$

where $\rho$ is the density of air.

The mean momentum of a large number of air particles can be determined as

$$\bar{F} = \overline{\rho(\bar{u} + \delta u)(\bar{v} + \delta v)} = \rho\bar{u}\bar{v} + \overline{\bar{u}\delta v} + \overline{\delta u} = (\bar{v} + \overline{\delta u \delta v}). \tag{13.8}$$

As $\bar{v} = 0$ and $\overline{\delta u} = \overline{\delta v} = 0$, the last equation can be transformed as:

$$\bar{F} = \rho(\overline{\delta u \delta v}) \tag{13.9}$$

So, the mean momentum is approximately equal to air density multiplied by the mean covariance between deviations in instantaneous horizontal speed and instantaneous vertical speed.

## 13.6   SENSIBLE HEAT FLUX

The direct transfer of heat from the surface to the atmosphere through conduction and convection is called *sensible heat flux*. This heat flux results from a temperature gradient between the Earth's surface and the atmosphere.

Sensible heat flux is equal to the mean air density multiplied by the covariance between fluctuations in instantaneous vertical wind speed and temperature which can be calculated as

$$H = \rho_a C_p \overline{\delta v \delta T_a}, \tag{13.10}$$

where $\rho_a$ is the air density; $C_p$ is the specific heat capacity of air at a constant pressure; $\delta v$ is the instantaneous fluctuation of vertical wind speed; and $\delta T$ is the instantaneous fluctuation in air temperature.

## 13.7   LATENT HEAT FLUX

The energy flux to the atmosphere, transferred by water vapor through evaporation and transpiration from the surface is called *latent heat*. This heat causes a change in the physical state of the substance (e.g., water → vapor) without a change in temperature of the substance.

Latent heat flux $\lambda E$ can be defined as the covariance between fluctuations in instantaneous vertical wind speed and water vapor density and calculated as

$$\lambda E = L_v \frac{\rho_a}{\rho_w} \overline{v \delta v \delta \rho_v},$$    (13.11)

where $L_v$ is the latent heat (hidden heat) of vaporization (corresponding to the liquid-to-gas phase change); $\rho_a$ is air density; $\rho_w$ is water vapor density; and $\overline{\delta \rho}$ is the fluctuations of the water vapor density.

## 13.8  CARBON DIOXIDE FLUX

*Carbon dioxide flux* can be determined using the mean covariance between fluctuations of vertical wind speed and density of $CO_2$ in the air as

$$F_C = \overline{\delta v \delta \rho_c}.$$    (13.12)

Therefore, the transfer of heat, mass, or momentum from one level to another one is characterized by the flow of a certain magnitude that can be estimated as a product of fluctuations in temperature, and the horizontal component of wind speed or mass on the vertical component of wind speed as

$$\left. \begin{array}{l} F \sim \delta T \delta v; \\ F \sim \delta u \delta v; \\ F \sim \delta m \delta v. \end{array} \right\}$$    (13.13)

## REFERENCES

Desjardins, R.L. 1972. A study of carbon dioxide and sensible heat fluxes using the eddy correlation technique. PhD thesis, Cornell University.

Ohtaki, E. and Matsui, Y. 1982. Infrared device for simultaneous measurement of fluctuations of atmospheric carbon dioxide and water vapour. *Boundary-Layer Meteorol.* 24:109–119.

Prandtl, L. 1920. *Theory of lifting surfaces.* NACA TN 9.

Shimizu, T. 2007. Practical applicability of high frequency correction theories to $CO_2$ flux measured by a closed-path system. *Boundary-Layer Meteorol.* 122:417–438.

# 14

# MEASUREMENT OF EDDY COVARIANCE

## 14.1 METEOROLOGICAL TOWERS

The primary principle of the eddy covariance method requires measuring the vertical speed, the up-and-down movement, and changes in temperature, speed, or density induced by fast turbulent fluctuations within the atmospheric boundary layer.

The equipment required to assess these parameters is routinely installed on meteorological towers.

The eddy covariance system consists of three principal sensors that are used for measuring water vapor density, air temperature, and wind speed. This system includes an open-path infrared gas analyzer (or a hygrometer) for measuring carbon dioxide ($CO_2$) density (or water vapor density); and a three-dimensional ultrasonic anemometer-thermometer for measuring the wind speed in vertical and two mutually perpendicular horizontal directions and the temperature.

The Kahoku Meteorological Station's (Japan) eddy covariance system consists of a closed-path system (LI-7000, LI-COR, USA) for measuring $CO_2$ flux, an open-path system (LI-7500, LI-COR, USA) for measuring $H_2O$ flux, and a three-dimensional sonic anemometer-thermometer (DA600-3T, KAIJO, Japan) for measuring wind speed and air temperature.

Measurement height on the tower is 51 m, the sampling frequency is 10 Hz, and the system is equipped with data logger and data storage capability.

A list of all instruments installed on such a tower includes pyranometers, infrared radiometers, net radiometer, quantum sensor for measurement of global solar radiation, long-wave and net radiation, direct and diffuse radiation, photosynthetic photon

*Methods of Measuring Environmental Parameters*, First Edition. Yuriy Posudin.

**169**

flux density (PPFD); platinum resistance thermometer and thermistor for air and soil temperature, heat flow transducer for soil heat flux; cup and sonic anemometers for wind speed and direction; barometric pressure sensor for atmospheric pressure; rain gauge for precipitation; capacitive hygrometer for humidity; closed-path and open-path gas analyzers for water vapor and $CO_2$ concentration.

The technique is mathematically complex and requires computation of elaborate data. The advantages of eddy covariance method are many and the level of precision is excellent.

## 14.2  GAS ANALYZERS

*Closed-Path Systems.* For example, the LI-COR is a $CO_2/H_2O$ gas analyzer that comprises a source of infrared radiation, a chopper, an optical system (lens, mirror, filters), a dual cell (reference cell A and sample cell B), a beam splitter, and two detectors (for $CO_2$ and $H_2O$) (Figure 14.1). Absorption at wavelengths centered at 4.26 and 2.59 μm allow measurement of $CO_2$ and water vapor, respectively.

The instrument's software provides continuous measurement of the absolute concentration in the sample cell as well as the differential concentration.

*Open-Path Systems.* The LI-COR LI-7500 is based on the measurement of $CO_2$ and water vapor density *in situ* in turbulent air structures and interfacing the data with sonic anemometer turbulence data to determine the fluxes of $CO_2$ and $H_2O$ using eddy covariance techniques.

The principle of operation is based on measuring the absorption of infrared radiation at different wavelengths—one at a wavelength that is absorbed by each gas

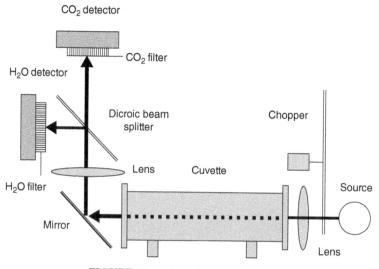

**FIGURE 14.1**   Closed-path gas analyzer.

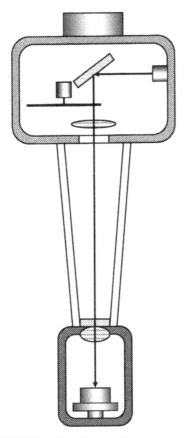

**FIGURE 14.2**   Open-path gas analyzer.

(4.26 μm for $CO_2$ and 2.59 μm for $H_2O$) and the other two at non-absorbing reference wavelengths (3.95 μm for $CO_2$ and 2.40 μm for $H_2O$).

The LI-7500 sensor head has a 12.5 cm open path; a source of infrared radiation
that emits an infrared beam of 1 cm diameter that is modulated at 150 Hz to avoid
the effect of ambient light on the operation of the head. This beam goes through a
12.5 cm open path and enters a cooled lead selenide detector (Figure 14.2).

## 14.3   QUANTUM CASCADE LASER SPECTROSCOPY FOR ATMOSPHERIC GASES: EDDY COVARIANCE FLUX MEASUREMENTS

A new approach in eddy covariance technique is a quantum cascade laser absorption
spectrometer (Sturm et al., 2012) which is characterized by high sensitivity, versatility,
ruggedness, and portability. Also, it can analyze the samples very rapidly.

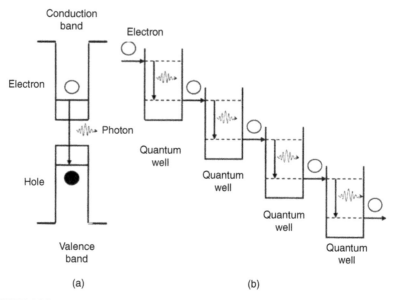

**FIGURE 14.3** Principle of operation of lasers. (a) Conventional diode laser: conduction electrons and holes are annihilated once recombining across the band gap. (b) Cascade laser: the population inversion and laser emission takes place between sub-bands of the superlattice.

A conventional diode laser emits a single photon during the transition of an electron from the conduction band to the valence band where the recombination of the electron with the hole takes place. The emission wavelength is determined by the band-gap width of the semiconductor material (Figure 14.3a).

Quantum cascade lasers (QCL) are constructed as a periodic series of thin layers of various materials that form a superlattice. The lasers utilize only a single type of carrier (electron), hence the name *unipolar* lasers. The population inversion and laser emission takes place between two sub-bands of the superlattice.

It is possible to tune the emission wavelength of a QCL over a wide range in the same material by changing the thicknesses of the quantum wells and barriers in the superlattice.

The primary peculiarity of a QCL is that a single electron causes the emission of multiple photons during the transition through the multilayer structure (Figure 14.3b); as a consequence, they are called *cascade lasers*.

## 14.4    STABLE ISOTOPES OF CARBON DIOXIDE

There are two stable isotopes of $CO_2$, $^{12}C$ and $^{13}C$ among three naturally occurring isotopes $^{12}C$, $^{13}C$, and $^{14}C$. The abundance of stable isotopes is about 98.88%.

The quantities of the different isotopes can be measured by mass spectrometry and compared to standards (see Formula 6.5). The ratio of the two most abundant isotopes ($^{13}$C and $^{12}$C) in a sample is compared to the same ratio in an international standard, using the "delta" ($\delta$) notation. The result (e.g., the delta of the $^{13}$C $= \delta^{13}$C) is expressed as parts per thousand (‰):

$$\delta^{13}C = \left[ \frac{\left(\frac{^{13}C}{^{12}C}\right)_{sample}}{\left(\frac{^{13}C}{^{12}C}\right)_{std}} - 1 \right] 1000‰. \tag{14.1}$$

## 14.5   QUANTUM CASCADE LASER ABSORPTION SPECTROMETRY

Isotopic ratio measurements can be determined using infrared mass spectrometry (IRMS); however, deficiencies in this method include complex sample preparation and chemical processing or purification. In addition, commercial IRMS requires a permanent installation and the transfer of samples to laboratory often occurs long after they are collected under field conditions. Likewise, diode lasers require liquid nitrogen cooling that results in changes in the laser parameters.

QCL can operate at near room temperature in a single-mode continuous tuning regime and with good power output. As a consequence, QCL have distinct advantages.

The wavelength range of QCL is 3.4–17 μm where the rotational–vibrational absorption bands of many atmospheric gases occur. The frequency tuning of QCL is realized through changes in the thicknesses of the quantum wells and barriers in the superlattice. A typical QCL absorption spectrometer consists of quantum cascade laser, two multi-pass optical cells and liquid-nitrogen-cooled detectors. The laser frequency is scanned over the three main $CO_2$ isotopologues $^{12}C^{16}O_2$, $^{13}C^{16}O_2$, and $^{18}O^{12}C^{16}O$. A pulsed near-room-temperature quantum cascade laser operates at a wavelength near 4.3 μm or 2310 cm$^{-1}$. The intensity of the spectral line is proportional to the concentration of certain isotope.

## 14.6   EDDY COVARIANCE MEASUREMENT OF CARBON DIOXIDE ISOTOPOLOGUES

$CO_2$ flux $F^x$ can be calculated as the mean covariance between fluctuations of vertical wind speed $v$ and the isotopologue mole fraction $c_x$ of $CO_2$ in the air (see Formula 13.12):

$$F^x = \frac{\overline{\delta v \delta c_x}}{V_m} \tag{14.2}$$

The flux ratios $F^{13}/F^{12}$ and $F^{18}/F^{16}$ can be expressed in delta notation (Sturm et al., 2012) as:

$$\delta_F^{13} = \frac{F^{13}/F^{12} - R_{st}}{R_{st}}; \quad \delta_F^{18} = \frac{F^{18}/F^{16} - R_{st}}{R_{st}}. \tag{14.3}$$

The flux ratio multiplied by $F^{12}$ yields the isotopic flux:

$$\delta_F^{13} F^{12} = \overline{\delta v \delta_F^{13} \delta c_{12}}. \tag{14.4}$$

Here the values are measured in the following units: eddy flux $F^x$ in $\mu mol/m^2 \cdot s^1$; vertical wind speed $v$ in m/s[1]; molar volume of air $V_m$ in $m^3/mol$; isotopologue mole fraction $c_x$ in $\mu mol/mol^1$; isotopic flux $\delta_F^{13} F^{12}$ in $\mu mol/m^2 \cdot s\%_0$.

Data analysis and eddy covariance flux calculations can be performed with custom software developed with MATLAB (The MathWorks Inc.).

## 14.7   MEASUREMENT OF EDDY ACCUMULATION

*Eddy accumulation* (or *conditional sampling*) is an alternative to the eddy covariance method that can be used to measure the vertical turbulent trace gas fluxes.

This method provides the sampling air into two containers. One container is filled when air goes up, and the other one is filled when air goes down. The flux is estimated from the difference in mass collected in each container divided by the collection time (generally 30 minutes):

$$F = \delta v \delta c = \frac{M_u - M_d}{2kt}, \tag{14.5}$$

where $\delta v$ is the fluctuation of vertical wind speed components; $\delta c$ is the fluctuation of concentration; $M_u$ and $M_d$ are the average concentration of the updraft and downdraft air mass correspondingly; $t$ is the time interval; $k$ is a semi-empirical coefficient.

This method does not require fast-response sensors. The disadvantage of this method is slow deposition of pollutants in sampling systems.

## 14.8   INTERACTION OF CLIMATIC FACTORS

The Earth's climate is characterized by the complex interaction of all climatic factors which were discussed in previous chapters: atmospheric pressure, wind, temperature, humidity. The spatial distribution of atmospheric pressure and temperature, intensity and direction of wind, variations in solar radiation, precipitation and humidity—all these factors have significant environmental impacts.

If the temperature and atmospheric pressure are distributed uniformly across the Earth's surface, the wind does not occur.

In a real situation, the excessive heating of separate areas of the Earth's surface provokes the non-uniform distribution of air temperature, causing the spatial distribution of atmospheric pressure.

The areas where the atmospheric pressure is lower than that of the area surrounding it are called the centers of low pressure. These areas are responsible for the occurrence of cyclones—rapid circulation of air masses (up to several hundreds or thousands of miles in diameter) with a reduced air pressure in the center. Wind in cyclone is directed from the periphery to the center. The central part of the air rises up where it is cooled by the expansion. Water vapor condenses and forms clouds; thus the precipitation occurs. Cyclones accompanied by rainy, windy, and cloudy weather.

The areas where the atmospheric pressure exceeds the surrounding pressure are called the centers of high pressure. These areas are associated with anticyclones. Wind in anticyclone is directed from the area of high pressure to outer space because air density is higher at the center. The air that leaves the center is replaced by air that is flowing in the upper atmosphere. As the air falls, it warms and the clouds are dissipated. Anticyclones are accompanied by clear cloudless weather.

The distribution of water vapor over the globe depends on the rate of evaporation and water vapor transfer by air currents from one place to another. The distribution of the partial pressure of water vapor associated with the distribution of temperature: the maximum values (25 hPa) are observed at the equator, while these values are reduced (to 4–5 hPa) in the polar regions.

The partial pressure of water vapor corresponds to the temperature in the surface layer of atmosphere, while the relative humidity is inversely proportional to the temperature.

Maximum rainfall is observed in the equatorial and subequatorial zones where the most powerful upward lifting of air saturated with moisture takes place. The adiabatic expansion of the air, its cooling, and condensation are accompanied by the formation of clouds and precipitation. The rainfall in these zones is 2000–3000 mm or more.

The high intensity of solar radiation leads to extreme dry air (relative humidity can reach 30%) and extremely high (up to +30–+50°C) temperature in tropical zones. Rainfall does not exceed 100–200 mm per year.

The main factors determining the geographical distribution of rainfall is the zonal distribution of solar radiation, temperature, and atmospheric pressure which affect the circulation of air masses and the presence or absence of clouds. Furthermore, the impact of latitude, humidity, time of year (season), nature of the Earth's surface (texture, color, content, soil moisture), the presence of particles of natural or anthropogenic origin in the atmosphere, ocean currents, relative continental location must be added.

Thus, all the environmental factors participate in complex interaction. The nature and intensity of this interaction determines the climate and its changes.

## 14.9    AUTOMATIC WEATHER STATIONS

Modern automatic meteorological stations are equipped with computers and microprocessors that are designed to perform various tasks. Observations can be sent immediately to the central institutions, or accumulated for future study of the climatic conditions in the region.

Let us consider as an example of an automatic climatological station Cimel ENERCO CE 411 (Paris, France). Power for the station is provided by solar batteries. The electronic components for the station are located in a polyester box equipped with thermal insulation and ventilation. Some sensors (temperature, wind, solar radiation, humidity) are placed on a bar, the remainder (ground temperature) is located in the ground or above the soil surface. The station is connected via telephone network or radio transmitter for direct connection with the Meteosat satellite.

Another type of automated weather station is Model 5081 (HydroLynx Systems, Inc., CA, USA) that is intended for the detection of storms, weather forecasting, the impact of wind, and the implementation of irrigation and agricultural programs. The station has wind, temperature, humidity, barometric pressure, and precipitation sensors; a data transmitter; solar panel; and interconnection cables and antenna.

A system for ground-based observations is the Automated Meteorological Data Acquisition System (AMeDAS), which was developed by the Japanese Meteorological Agency (Japan Meteorological Agency, JMA). The system consists of 1300 automated stations located at a distance of about 17 km apart throughout Japan.

The AMeDAS provides information about wind direction and speed, the types and amount of rainfall, type and height of clouds, visibility, temperature and humidity, atmospheric pressure, and weather conditions.

The data are sent to the AMeDAS Center, located at the headquarters in Tokyo, after which the information is distributed across the country. Nearly 700 stations operate without staff. About 280 stations are located in regions of possible snowfall and monitor the depth of snow cover.

The system also provides a rapid alert and warning means for natural disasters such as earthquakes and volcanic eruptions.

Overall the Japanese weather bureau consists of 100 synoptic stations, 62 weather services for aviation, 18 radiosonde stations and has 11 m wind profile, 30 radar and 4 marine meteorological centers.

## REFERENCE

Sturm, P., Eugster, W., and Knohl, A. 2012. Eddy covariance measurements of CO2 isotopologues with a quantum cascade laser absorption spectrometer. *Agricul. for Meteorol.* 152:73–82.

# PRACTICAL EXERCISE 7

# EDDY COVARIANCE MEASUREMENT

The fluctuations in the horizontal component $u$ and the vertical component $v$ of wind are depicted in Figure P7.1. Calculate the covariance between the horizontal component $u$ and the vertical component $v$ of wind speed according to the numerical data in Table P7.1.

There are similarities and distinctions between coefficients of covariance and correlation.

Coefficient of covariance is positive if $y$ increases with increasing $x$ and negative if $y$ decreases as $x$ increases. Positive coefficient of correlation numbers indicate they correlate directly and negative numbers indicate they correlate inversely. The coefficient of covariance has no upper or lower limits, while coefficient of correlation can vary from +1, through zero, to –1. Usually the coefficient of correlation values between 0.00 and 0.30 indicate a weak correlation; between 0.30 and 0.70 indicate a moderate correlation; and between 0.70 and 1.00 indicate a high correlation.

The results of our calculations ($r_{xy} = \text{cov}.xy/\sigma_x\sigma_y = 0.35$) mean that there is a moderate correlation between the horizontal $u$ and vertical $v$ component of wind speed.

**Control Exercise**    Imagine that at the Department of Environmental Sciences a question arose whether students who like to eat hamburgers also like to eat sushi. The Dean of this Department decided to resolve this important problem. So, he

*Methods of Measuring Environmental Parameters*, First Edition. Yuriy Posudin.
© 2014 John Wiley & Sons, Inc. Published 2014 by John Wiley & Sons, Inc.

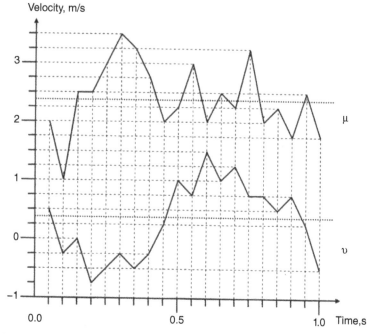

**FIGURE P7.1**    The fluctuations in horizontal component $u$ and vertical component $v$ of wind.

administered a questionnaire to his four students asking them to rate their liking of hamburgers ($H$) and sushi ($S$). Responses from the students are presented in the following matrix:

| Student | H | S |
|---------|---|----|
| 1 | 6 | 8 |
| 2 | 2 | 9 |
| 3 | 6 | 16 |
| 4 | 9 | 7 |

The question to be answered is whether the liking hamburgers and sushi are related.

## QUESTIONS AND PROBLEMS

**1.** Define turbulence.

**2.** Characterize the atmospheric boundary layer.

**3.** What are fluctuations?

**TABLE P7.1**  Results of calculation the covariance between horizontal component $u$ and vertical component $v$ of wind speed.

| N | x | Y | $x - \bar{x}$ | $y - \bar{y}$ | $(x - \bar{x})(y - \bar{y})$ |
|---|---|---|---|---|---|
| 1 | 2 | 0.5 | −0.4 | 0.19 | −0.07 |
| 2 | 1 | −0.25 | −1.4 | −0.56 | 0.78 |
| 3 | 2.5 | 0 | 0.1 | −0.31 | −0.03 |
| 4 | 2.5 | −0.75 | 0.1 | −1.06 | 0.10 |
| 5 | 3 | −0.5 | 0.6 | −0.81 | 0.81 |
| 6 | 3.5 | −0.25 | 1.1 | −0.56 | −0.61 |
| 7 | 3.25 | −0.5 | 0.85 | −0.81 | −0.68 |
| 8 | 2.75 | −0.25 | 0.35 | −0.56 | −0.19 |
| 9 | 2 | 0.25 | −0.4 | −0.06 | 0.02 |
| 10 | 2.25 | 1 | −0.15 | 0.69 | −0.10 |
| 11 | 3 | 0.75 | 0.6 | 0.44 | 0.26 |
| 12 | 2 | 1.5 | −0.4 | 1.19 | −0.47 |
| 13 | 2.5 | 1 | 0.1 | 0.69 | 0.07 |
| 14 | 2.25 | 1.25 | −0.15 | 0.94 | −0.14 |
| 15 | 3.25 | 0.75 | 0.85 | 0.44 | 0.37 |
| 16 | 2 | 0.75 | −0.4 | 0.44 | −0.17 |
| 17 | 2.25 | 0.5 | −0.15 | 0.19 | −0.02 |
| 18 | 1.75 | 0.75 | −0.65 | 0.44 | −0.28 |
| 19 | 2.5 | 0.25 | 0.1 | −0.06 | 0.00 |
| 20 | 1.75 | −0.5 | −0.65 | −0.81 | 0.52 |
| $\bar{M}$ | 2.4 | 0.31 | | | $\sum (\bar{x} - x)(\bar{y} - y) = 0.17$ |
| $\sigma^2$ | $\sigma^2{}_x = 0.35$ | $\sigma^2{}_y = 0.41$ | | | $\text{cov } xy = \frac{1}{N} \sum (\bar{x} - x)(\bar{y} - y)$ |
| | | | | | $= 0.17/20 = 0.0085$ |
| $\Sigma$ | $\sigma_x = 0.59$ | $\sigma_y = 0.64$ | $\sigma_x \sigma_y = 0.3776$ | | $r_{xy} = \text{cov } xy/\sigma_x \sigma_y$ |
| | | | | | $= 0.0085/0.3776 = 0.0032$ |

**4.** What is the essence of the method of eddy covariance?

**5.** Explain the principle of operation of an infrared gas analyzer.

## FURTHER READING

Campbell, G.S. and Norman, J.M. 1998. *Environmental Biophysics*, 2nd edition. Springer Verlag, New York.

Cieslik, S., Omasa, R., and Paoletti, E. 2009. Why and how terrestrial plants exchange gases with air. *Plant Biol.* 11:24–34.

Faist, J., Capasso, F., Sivco, D.L., Sirtori, C., Hutchinson, A.K., and Cho, A.C. 1994. Quantum cascade laser (abstract). *Science*, 264(5158):553–556.

Guyot, G. 1998. *Physics of the Environment and Climate*. John Wiley & Sons, Chichester.

Hicks, B.B. and McMillen, R. 1984. A simulation of the eddy accumulation method for measuring pollutant fluxes. *J. Appl. Meteorol.* 23(4):637–643.

Kazarinov, R.F. and Suris, R.A. 1971. Possibility of amplification of electromagnetic waves in a semiconductor with a superlattice. *Fizika i Tekhnika Poluprovodnikov.* 5(4):797–800.

Monteith, J.L. and Unsworth, M. 1990. *Environmental Physics*, 2nd edition. Edward Arnold, London.

Nelson, D.D., McManus, J.B., Herndon, S.C., Zahniser, M.S., Tuzson, B., and Emmenegger, L. 2008. New method for isotopic ratio measurements of atmospheric carbon dioxide using a 4.3 μm pulsed quantum cascade laser. *Appl. Phys. B-Lasers Opt.* 90(2):301–309.

Ruddell, B.L., Oberg, N., Kumar, P., and Garcia, M. 2010. Using information-theoretic statistics in MATLAB to understand how ecosystems affect regional climates. MATLAB Digest Academic Edition.

Saleska, S.R., Shorter, J.H., Herndon, S., Jiménez, R., Mcmanus, J.B., Munger, J.W., Nelson, D.D., and Zahniser, M.S. 2006. What are the instrumentation requirements for measuring the isotopic composition of net ecosystem exchange of CO2 using eddy covariance methods? *Isoto. Environ. Health. Stud.* 42(2):115–133.

## ELECTRONIC REFERENCE

Automated Meteorological Data Acquisition System (AMeDAS), http://en.wikipedia.org/wiki/AMeDAS. (accessed June 14, 2013).

# PART II

# ATMOSPHERIC FACTORS

# 15

# ATMOSPHERE

## 15.1  COMPOSITION OF THE ATMOSPHERE

*Atmospheric environmental factors* include the structure and composition of the atmosphere, as well as the physical and chemical properties of the atmosphere that affect living organisms.

The atmosphere is like a shell that surrounds the Earth; it consists of many gases and suspended liquid and solid particles. At sea level, the Earth's atmosphere contains 78.08% nitrogen, 20.95% oxygen, 0.93% argon, traces of other gases, and water vapor (about 1%).

## 15.2  AIR POLLUTION

There are a number of hazardous pollutants in the atmosphere according to the United States Environmental Protection Agency. Pollutants can be separated into two general classes: primary pollutants that are emitted directly to the atmosphere and secondary pollutants that are formed during reactions with primary pollutants within the atmosphere.

Six categories of primary pollutants are considered to be the most widespread health threats: nitrogen dioxide ($NO_2$), sulfur dioxide ($SO_2$), carbon monoxide (CO), lead (Pb), particulates ($PM_{10}$ and $PM_{2.5}$), and volatile organic compounds (VOCs). Ozone is not considered to be a primary pollutant but it is included in the list of critical pollutants. Ozone is an example of secondary pollutants that are formed as

*Methods of Measuring Environmental Parameters*, First Edition. Yuriy Posudin.
© 2014 John Wiley & Sons, Inc. Published 2014 by John Wiley & Sons, Inc.

a result of photochemical reactions in the presence of CO, $NO_x$, and VOCs (Artiola et al., 2004).

## 15.3    AIR QUALITY

*Air quality* is defined as a measure of the condition of air relative to the requirements of one or more biotic species or to any human need or purpose.

The EPA established an Air Quality Index (AQI) for principal air pollutants regulated by the Clean Air Act which indicates the daily air quality. AQI values vary from 0 to 500; the higher the AQI value, the greater the level of air pollution. An AQI value of 50 represents good air quality, while an AQI value of 500 represents hazardous air quality.

## REFERENCE

Artiola, J.F., Pepper, I.L., and Brusseau, M.L. 2004. *Environmental Monitoring and Characterization*. Elsevier Academic Press, San Diego, CA.

# 16

# MEASUREMENT OF AMBIENT AIR QUALITY

## 16.1 MEASUREMENT OF NO$_2$

Nitrogen oxides (NO$_x$), particularly nitrogen dioxide (NO$_2$) occur during high-temperature combustion, which may be accompanied by the appearance of brown haze dome above the cities. NO$_2$ is a gas with a nasty odor; it participates in the formation of photochemical smog—the chemical reaction of sunlight and NO$_x$ in the atmosphere, which provides serious effects on human health. Possible sources of NO$_2$ in the air include road vehicles, non-road equipment, electricity generation, fossil fuel combustion, and industrial processes.

### 16.1.1 Chemiluminescence

A *chemiluminescent method* is used for the determination of NO$_2$. This is reference method for the measurement of NO$_2$ and oxides of nitrogen (NO$_x$) that is based on the reaction between nitrogen oxide (NO) and ozone (O$_3$), which leads to the formation of the excited form of NO$_2^*$:

$$NO + O_3 \rightarrow NO_2^* + O_2; \tag{16.1}$$

$$NO_2^* \rightarrow NO_2 + h\nu. \tag{16.2}$$

*Methods of Measuring Environmental Parameters*, First Edition. Yuriy Posudin.
© 2014 John Wiley & Sons, Inc. Published 2014 by John Wiley & Sons, Inc.

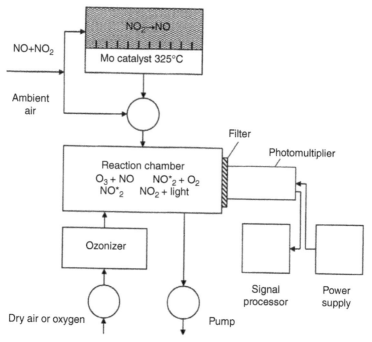

**FIGURE 16.1**    Schematic of chemiluminescent method of measuring NO and $NO_2$ concentrations.

The transition of excited $NO_2$ back to the ground state is accompanied with the emission of light. Typically the spectrum of chemiluminescence occupies the region around 1200 nm.

Air entering the chemiluminescent analyzer is divided into two streams (Figure 16.1). In one stream, the $NO_2$ is transformed into NO using a molybdenum (Mo) catalyst heated to 325°C. The NO reacts with ozone that is generated by an ozonizer in the low-pressure chamber, which leads to the formation of $NO_2$ molecules. Some of these molecules are excited ($NO_2^*$) and emit light during their transition to the ground state. The light is measured by a photodetector and converted into an electrical signal that is proportional to the light intensity and thus, the concentration of $NO_2$.

The second stream is passed through the Mo catalyst but not reacted with ozone; the control air concentration of $NO_2$ is computed as the difference between the converted and non-converted forms of $NO_2$.

For example, the sensitivity of the chemiluminescent method for measuring $NO_2$ using a Model 42 Analyzer (Thermo Scientific) is $5 \times 10^{-10}$ $NO_2$.

The another gas analyzer MODEL SIR S-5012 $NO_x$ is characterized by the following parameters:

Measurements: (NO), ($NO_x$), ($NO$-$NO_2$-$NO_x$).
Ranges 0–50, 500 ppb; 20 ppm.

Lower detectable limit: 0.4 ppb.

Precision: $\pm 0.5\%$ of reading.

*The advantages of a chemiluminescent method:*

- The specificity of analyzer to NO$_x$.
- High sensitivity at low air pressures (5–25 mbars) within the reaction chamber.
- Low detection limit (about 1 $\mu g/m^3$).
- Real-time measurement.
- Short-time resolution (<1 hour).
- It requires no excitation source, monochromator, and filter.

*The disadvantages of a chemiluminescent method:*

- A risk of error when measuring the concentrations of NO and NO$_2$ vary faster than the duration of a measurement cycle.
- Effect of other nitrogen species (peroxyacetyl nitrate PAN, nitrous acid HNO$_2$, and ozone) which interfere with the measurement of NO$_2$.
- High cost.
- Unreliability of ozone generator.
- Presence of finicky NO$_2$ converter.

## 16.1.2    The Automatic Cavity Attenuated Phase Shift NO$_2$ Analyzer

The principal shortcoming of the chemiluminescence method is the presence of finicky NO$_2$ converter and unreliable ozone generator. The chemiluminescence method suffers with the possibility to overestimate or underestimate NO$_2$, especially during fast NO$_x$ transitions. The concentrations of NO and NO$_2$ vary faster than the duration of a measurement cycle. In addition, errors of measurement can appear due to the effect of other species that are converted to NO$_2$ in the NO catalytic oven.

The automatic cavity attenuated phase shift (CAPS) NO$_2$ analyzer (Environnement S.A.) is an example of the advanced type of analytic instrument which can overcome these disadvantages (Kebabian et al., 2008). The CAPS analyzer can provide direct, continuous, and fast measurements of the NO$_2$, detect short events with high precision. It requires no catalyst, no ozonizer.

Principle of operation of the CAPS analyzer is based on an optical absorption spectrometry and provides the comparison of the phase shift between amplitude-modulated radiation at the output and the input of the object.

This analyzer consists of a light source (a modulated broadband incoherent 430-nm LED), a sample cell which is equipped with two high reflectivity mirrors at 450 nm and a vacuum phototube detector (Figure 16.2). The air sample passes through the dryer and enters the cell. The light beam participates in multiple reflections from two

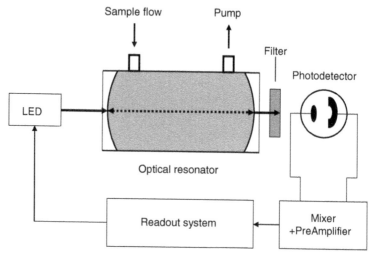

**FIGURE 16.2**   The automatic cavity attenuated phase shift (CAPS) nitrogen dioxide analyzer.

mirrors that form the optical cavity and provide the concentration measurement. The LED radiation leaves the optical cavity and reaches the detector. The electric signal of detector is elaborated by the system of readout.

The optical scheme of the CAPS is similar to optical heterodyne detection system, which uses interaction of light waves that propagate in opposite direction. The mixing product can be obtained by detecting the superimposed waves with a square-law photodetector. The presence of $NO_2$ in the cell induces a phase shift in the signal received by a photodetector that is proportional to the $NO_2$ concentration. Typical levels of detection is about 0.1 parts per billion.

This technology can be applied for ambient air monitoring, estimation of roadside air pollution and street canyon effect, traffic pollution control, and continuous indoor and outdoor air continuous monitoring.

*The advantages of a CAPS nitrogen dioxide analyzer:*

- Direct measurement of the sample—no chemical conversion required.
- High sensitivity (measurement of ambient $NO_2$ concentrations to 1000 ppbv).
- Fast and accurate measurements.
- Insensitive to presence of varying levels of peroxyacetyl nitrate PAN, nitrous acid $HNO_2$, ozone, aerosols, humidity, and other trace atmospheric species.
- Linear response (0–1000 ppbv).
- No toxic gas emissions.
- Compactness, easy to use, and to maintain.
- Completely autonomous, automatic, and capable of detecting short events with high precision.

### 16.1.3    Micro-Gas Analysis System for Measurement of NO$_2$

A team of scientists from the University of Kumamoto, Japan (Toda et al., 2007), proposed an original design of the micro-analysis system for measuring NO$_2$. The principle of operation of such a system is based on the collection of the test gas in the microchannel scrubber—a honeycomb-shaped microchannel array formed on polydimethylsiloxane (PDMS), which was covered with a porous hydrophobic membrane.

The test gas NO$_2$ passed through the membrane and was collected in an absorbing solution; after going through the scrubber this solution was mixed with a Griess–Saltzman reagent solution, which is able to change its color depending on the concentration of the test gas. Changes in the absorption of the mixed solution were fixed by the detector.

Such a system was approbated for mobile atmospheric monitoring. Its total weight was 5.3 kg; the observed maximum and minimum concentrations were 114 and 21.2 ppbv, respectively.

### 16.1.4    Measurement of NO$_2$ in a Liquid Film/Droplet System

The main idea of this method is based on the interaction of a green (555 nm) radiation of light-emitted diode with a liquid film, composed of Griess–Saltzman reagent (Cardoso and Dasgupta, 1995). This film was supported on a U-shaped wire guide. The transmitted light was measured by photodetector; the output signal of it was proportional to the NO$_2$ concentration. The sensitivity of such a system reached 10 ppb.

### 16.1.5    Electrochemical Sensor

The principle of operation of electrochemical gas sensor is based on measuring the concentration of the gas of interest by oxidizing or reducing it at an electrode, producing an electrical signal proportional to the gas concentration, and measuring the resulting current (Chou, 1999).

A typical electrochemical sensor consists of a capillary diffusion barrier, hydrophobic membrane (thin, low-porosity Teflon), sensing electrode, reference electrode, counter electrode, and electrolyte (Figure 16.3).

Gas (e.g., NO$_2$) enters into the sensor through a small capillary-type opening, passes through a hydrophobic membrane, and get into contact with the sensing electrode. This gas participates in chemical reaction that leads to the production of electric current. Electrochemical NO$_2$ sensor involves a reduction reaction of NO$_2$, which results the production of water at the cathode:

$$NO_2 + 2H^+ + 2e^- \rightarrow NO + H_2O. \tag{16.3}$$

The reference electrode produces the stable constant potential which is applied to the sensing electrode. The current flow during reaction is measured between the

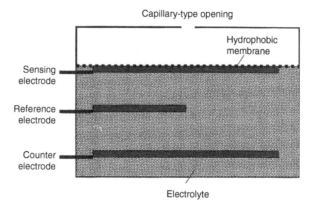

FIGURE 16.3    A typical electrochemical sensor.

sensing and the counter electrodes. This current is proportional to the concentration of $NO_2$.

The sensitivity of $NO_2$ electrochemical sensor is 20–50 ppm. It can be used also for measuring concentration of such gases as $NH_3$, $H_2S$, $SO_2$, $Cl_2$, $H_2$, CO, and HCl.

*The advantage of $NO_2$ electrochemical sensor:*

- Compactness, portable sensors can be applied in the field conditions.

*The disadvantage of $NO_2$ electrochemical sensor:*

- Lower detection limit of some samplers (about 200 $\mu g/m^3$).

### 16.1.6    Passive Diffusive Samplers

*Passive diffusive sampling* is based on the molecular diffusion of gas of interest (analyte) through a diffusive surface onto an adsorbent. Then the adsorbed analyte is extracted from the adsorbent by solvent or thermal desorption, and is transported to the laboratory equipped with analytical instrumentation (e.g., ion chromatography).

The gas molecules diffuse into the cylindrical sampler through either its end (axial sampler) or the lateral surface (radial sampler).

Let us discuss the classical Palmes diffusion tube for passive sampling of $NO_2$ (Palmes et al., 1976). It has length 71 mm and internal diameter 9.5 mm. The sampler body is an acrylic plastic that is sealed at both ends (Figure 16.4). One end of the sampler contains a Teflon filter to stop aerosols and stabilize diffusion path and the inlet cup. The other end of the sampler is equipped with stainless steel meshes coated with triethanolamine (TEA) and colored cap for accommodating the meshes. When the inlet cup is removed a concentration gradient occurs and molecular diffusion toward the other end of the tube takes place. After a period of time, $t$ the certain

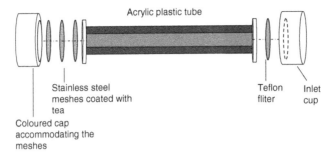

**FIGURE 16.4**   Palmes diffusion tube for passive sampling of NO$_2$.

amount $Q$ of NO$_2$ is transferred along the sampler. The average gas concentration $C$ during exposure is determined as

$$C = \frac{4Ql}{D\pi d^2 t},$$ (16.4)

where $D$ is the diffusion coefficient; $l$ is the length of the tube; $d$ is the internal diameter of the tube.

The analytical procedure provides the realization of reaction of NO$_2$ with TEA that leads to the formation of a product that liberates nitric (NO$_2-$) ions. Ion chromatography makes it possible to estimate the efficiency of this reaction and to detect low concentrations of NO$_2$.

Typical parameters of passive diffusive sampler (e.g., SKC UMEX 200 Passive Sampler) are

| | |
|---|---|
| Validated Concentration Range | 0.4–8 ppm |
| Lower Detection Limits | 15 minutes: 2 ppm |
| | 8 hours: 0.1 ppm |
| Accuracy | ± 30% |

Passive samplers for NO$_2$ can be applied for ambient air monitoring; indoor air monitoring; workplace monitoring; odor investigations; landfill perimeter monitoring; and air monitoring in remote locations where electrical power is absent.

This technology can be applied also for measurement of SO$_2$ and O$_3$.

*The advantages of a passive diffusive sampling:*

- Lower detection limit about 2–3 µg/m$^3$ for a month exposure period.
- Accurate and reliable sampling of ppm-level nitrous oxide.
- Simplicity, reliability, compactness.
- Portable design makes it possible to study spatial distribution of NO$_2$ concentrations over wide areas.

- Low cost.
- Requires minimal training of personnel.
- No sampling pump or power supply are required.
- Site calibrations are not required.

*The disadvantages of a passive diffusive sampling:*

- Only provide weekly or longer averages.
- Unproven for some pollutants.
- Laboratory analysis required.
- Only provide weekly or longer averages.
- Effect of air currents, humidity, and temperature on the results of measurements.

### 16.1.7 Thick Film Sensors

It is shown that a group of semiconductor oxides $SnO_2$, $ZnO$, $In_2O_3$, $WO_3$ is characterized by an extreme sensory effect due to their high reactivity and sensitivity of the electrical properties of a semiconductor surface to the composition of the surrounding atmosphere.

Semiconductor gas sensors are devices that convert the information about the change of composition of gas phase into an electrical signal. Thick film semiconductor gas sensors are intended for recognition of target gas from others and transduction of the gas recognition into a gas concentration (Guidi, 2002).

Chemisorption of molecules from the gas phase and chemical reactions on the surface lead to significant changes in the band structure of semiconductor in a narrow surface layer, the formation of the energy barriers at the solid–gas interface that affects the surface conductivity of semiconducting materials.

Thick film semiconductor gas sensor is a metal-oxide film which is fabricated from oxides of the transition metals or heavy metals and deposited on a silicon substrate between two electrodes. The surface of the sensor is heated to a constant temperature of about 200–250°C to increase the rate of reaction and to avoid the effects of ambient temperature changes.

According to modern ideas, sensor material consists of individual grains that are in mechanical contact. In this case, the grain boundaries present the potential barriers. However, the width of such a barrier is small, which makes the possibility of electron tunneling through the barrier.

If a sensor crystal is heated, molecules of $O_2$ are adsorbed on the crystal surface with a negative charge because these molecules tie up free electrons from the surface of the crystal. These electrons are transferred to the adsorbed oxygen, leaving positive charges in a spatial charge layer and creating a potential barrier against electron flow (Figure 16.5a).

Exposure of the film to $NO_2$ (oxidizing gas) is accompanied with the chemical reaction of $NO_2$ with surface oxygen ions of the film. The oxidation of the film

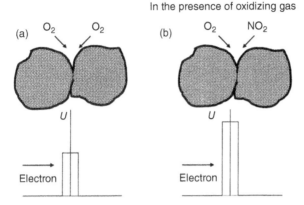

**FIGURE 16.5**   Thick film semiconductor gas sensor.

leads to a decrease in the number of free carriers and a corresponding increase of the potential barrier in the grain boundary and the resistance of the film (Figure 16.5b). This change in resistance which is proportional to the NO$_2$ concentration is measured.

*The advantages of a thick film sensor:*

- High stability.
- High sensitivity. Lower detection limit is about 4 µg/m$^3$.
- Simplicity and sufficient reliability.
- Long-term measurements.
- Portable samplers can be used in the field conditions.
- Several pollutants can be measured simultaneously.

*The disadvantages of a thick film sensor:*

- Needs temperature controlled heater.
- Expensive.
- Loss of sensitivity with age.

### 16.1.8   Open-Path Differential Optical Absorption Spectrometer

The differential optical absorption spectrometer (DOAS) technique is now widely used for the measurement of trace gases in the atmosphere. It is based on measuring specific narrow band absorption structures of the gases in the UV and visible region of the spectrum (Platt, 1994).

A differential absorption spectrometer measures the absorption at two wavelengths—one $\lambda$ corresponds to the absorption band of the object (gas or pollutant), while the another $\lambda_0$ coincides with the base absorption band.

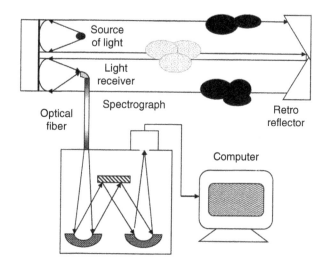

**FIGURE 16.6**    Open-path differential absorption spectrometer.

The absorption of the object is compared with the absorption of the pollutant at a known concentration, which is located in the data bank of the spectrometer.

A typical DOAS instrument consists of a continuous light source (i.e., a Xe-arc lamp, direct, or scattered sun light), an optical telescope that is used as the collimator, a retro-reflector, that reflects optical radiation in the opposite direction, a quartz fiber to transmit radiation from the telescope to the spectrograph, a monochromator with diffraction grating, and a system for recording the information (Figure 16.6). The typical open-path length in the atmosphere varies from several hundred meters to many kilometers.

Differential absorption spectrometers can measure the components of smog in the air of cities (e.g., $NO_x$, $SO_x$, $O_3$, formaldehyde), pollutants in the air of industrial areas, airports and highways, and monitor air quality in the troposphere and stratosphere.

*The advantages of a differential optical absorption spectrometer:*

- Several different trace-gas species can be measured simultaneously.
- High sensitivity. Lower detection limit is about 1 $\mu g/m^3$.
- Gives an average concentration that is integrated over the open-path length.
- Any inlet manifold.

*The disadvantages of a differential optical absorption spectrometer:*

- Expensive equipment.
- Effect of weather conditions (rain, fog, snow) on the results of measurements.

## 16.2   EFFECT OF NITROGEN DIOXIDE ON HUMAN HEALTH

Emissions of nitrogen in the atmosphere, converting them into nitrates and nitric acid causes acid rain. $NO_2$ dissociates under short-wave solar radiation; the series of photochemical reactions lead to the formation of such a powerful oxidant as ozone.

$NO_2$ exposure is especially dangerous near roadways, railroad, and airports. The EPA states that near-roadway (within about 50 m) concentrations of $NO_2$ have been measured to be approximately 30–100% higher than concentrations away from roadways. It is necessary to mention that about 16% of US housing units are located within 300 ft (about 90 m) of the roadways.

$NO_2$ is poisonous gas. It provokes severe irritation of the respiratory organs. It is shown that short-term $NO_2$ exposure ranging from 30 minutes to 24 hours causes inflammation of a healthy persons and enhancement of respiratory symptoms in people with asthma.

Long-term $NO_2$ exposure decreases lung function and increases the risk of respiratory symptoms. Such diseases as acute bronchitis and chronic respiratory symptoms (cough and phlegm) that are frequent among children can be explained by $NO_2$ exposure.

The US primary air quality standard for $NO_2$ is 0.053 ppm (100 μg/m$^3$) annually.

## 16.3   MEASUREMENT OF SO$_2$

$SO_2$ is a colorless gas with a pungent irritating odor. It comes in the air due to fossil fuel combustion at power plants, industrial processes, and volcanic eruptions. When released into the atmosphere, sulfur dioxide ($SO_2$) moves into the air, where it can be converted to sulfuric acid, and thus to acid rain.

### 16.3.1   Fluorescence Spectroscopy

*Fluorescence spectroscopy* (or spectrofluorometry) is a technique that is based on the excitation of the electrons in molecules of a sample, typically using ultraviolet (UV) light and the subsequent induction of fluorescence (emission of optical radiation typically in visible range). The emitted radiation is proportional to the concentration of the analyte in the sample being measured.

UV fluorescence spectroscopy can be used for quantitative measurement of $SO_2$ in the air. $SO_2$ exhibits a strong UV absorption spectrum between 200 and 240 nm, while the spectrum of emission of fluorescence is in the 300–400 nm range:

$$SO_2 + h\nu_1 \rightarrow SO_2^*; \qquad (16.5)$$

$$SO_2^* \rightarrow SO_2 + h\nu_2 \text{ (fluorescence).} \qquad (16.6)$$

A spectrofluorometer is an instrument used to measure fluorescence of $SO_2$ (Figure 16.7). It consists of a source of fluorescence excitation (discharge lamp and 214 nm UV bandpass filter), a fluorescence cell, and a system of registration.

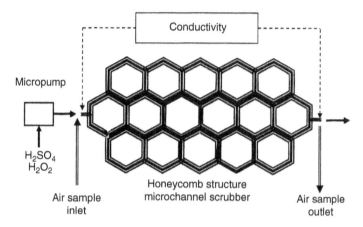

**FIGURE 16.7** The microchannel honeycomb scrubber (Ohira and Toda, 2005).

Fluorescence emission at 350 nm passes through the filter and enters the photomultiplier, producing a signal that is proportional to the concentration of $SO_2$. The fluorescence emission is detected by a photomultiplier at the angle 90° to the excitation radiation. Direct excitation radiation passes through the cell and enters a reference detector which is used for correcting fluctuations in lamp intensity.

Florescence is an active process. That is why the spectrofluorometers are characterized by high sensitivity. For example, the range of $SO_2$ measurement by Serinus 50 Sulfur Dioxide analyzer (Ecotech Pty Ltd) is 25–500 ppb in air.

*The advantages of a fluorescence spectroscopy:*

- High sensitivity.
- It is possible to analyze low quantities of the compound in question.
- Easy and fast procedure of measurement.

*The disadvantages of a fluorescence spectroscopy:*

- Time-consuming sample preparation.
- Not useful for sample identification.
- Requires expensive and sophisticated equipment.
- Not all compounds fluoresce.

### 16.3.2 Micro-Gas Analyzers for Environmental Monitoring

Mobile monitoring systems that contain portable gas analyzers are characterized by high sensitivity and resolution. These systems have been available during recent years for studying the environment, air pollution, and detecting various gases.

Liquid drops and liquid films have a large surface-area-to-volume ratio and therefore are suitable for collecting dissolved gases.

When the gas stream passes through a liquid drop, soluble components that are contained within the gas diffuse into and are dissolved within the drop.

It is often necessary to separate gases and particles. Because the diffusion coefficients of gases are about four times larger than that of small particles of atmospheric aerosols, the possibility of separating gases and particles is evident.

Water drops are natural collectors of soluble gases. A striking example is the fresh feeling of air after a rain. It is interesting to also note that during the evaporation of gases from the surface of a drop, the flux of the molecules leaving the surface prevents the approach of additional particles to the drop.

*Droplet-based sensor* utilizes drops of a reagent solution that is formed at the tip of a tube in a cylindrical chamber (Liu and Dasgupta, 1995). The droplet is used not as collector but as a reactor for chromogenic reactions. The color change of the solution depends on the concentration of the gases that partition into it.

For determination of SO$_2$, the following reagent was used: a solution of 5.0 mM NaHSO$_3$ in 0.2 M phosphate buffer (pH $\sim$ 4) was prepared by dissolving 0.30 g of NaHSO$_3$ in 500 mL of 0.2 M potassium hydrogen phthalate aqueous solution (Liu and Dasgupta, 1995).

Such systems can be used for the analysis of SO$_2$, NH$_3$, Cl$_2$, H$_2$O$_2$, H$_2$S, and HCHO with various reagents.

*Miniature-membrane-based diffusion scrubber.* Such systems are characterized by high sensitivity and time resolution. These advantages were achieved due to a microchannel scrubber with a honeycomb structure that is used to collect the gases (Ohira and Toda, 2005).

The sensitivity of such a system is inversely proportional to the channel depth $d$ which is related to the parameters of the gas and membrane:

$$C_s = kTC_g/d, \tag{16.7}$$

where $C_s$ is the analyte concentration collected by the solution; $k$ is the gas permeation rate constant of the membrane; $T$ is the length of time the gas is collected; and $C_g$ is the analyte gas concentration.

The system consists of a miniature pump, honeycomb scrubber, and a fluorescent detector. The microchannel scrubber contains 500 sets of hexagonal elements with sides 600 μm long (Figure 16.7). These elements were deposited by photolithography on a plastic plate (26 mm × 76 mm). The length of a single cell was 600 μm, and the thickness of each channel was 100 μm.

The honeycomb-like microchannel scrubber was fabricated as a sandwich which contained the glass plate, PDMS microchannel scrubber, and porous polytetrafluoroethylene (pPTFE) membrane sheet. The solution sickness was 70 μm—an optimal value from the point of view of the high sensitivity of thin solution layer and the prevention of water evaporation from such a thin layer.

The absorbing solution for SO$_2$ contained 1 μM fluorescein mercuric acetate (FMA) with 5 μM H$_2$SO$_4$ + 0.006% H$_2$O$_2$. The analyte dissolved in H$_2$SO$_4$ + H$_2$O$_2$

solution and was oxidized to be $H_2SO_4$. The concentration of $SO_2$ was estimated by measuring the electrical conductivity of the solution due to the application of a pair of platinum microelectrodes.

The sensitivity of honeycomb structure microchannel scrubber was 1 ppbv for $SO_2$.

*The advantages of a honeycomb structure microchannel scrubber:*

- Compactness, portability.
- High sensitivity (about 1 ppbv).
- High absorbing area (five times larger than a single-channel scrubber).
- High permeation rate (hundred times higher compared to the solid PDMS film).
- Effect of small defects in microchannels is negligible.

*The disadvantages of a honeycomb structure microchannel scrubber:*

- Reagent required.
- Expensive.

The system can be used for the analysis of $SO_2$, $H_2S$, $CH_3SH$, and $NH_3$ in a natural environment.

## 16.4   EFFECT OF SULFUR DIOXIDE ON HUMAN HEALTH

High concentrations of $SO_2$ causes wheeze, shortness, and limitation of breath even in healthy people; provokes lung and heart problems.

Long-term effects of $SO_2$ cause respiratory and cardiovascular diseases. People with asthma, chronic lung or heart disease are especially sensitive to $SO_2$.

In 1952, about 4000 deaths occurred in London due to 4-day $SO_2$ air pollution. About 4000 excess deaths were attributed to air pollution during the 4-day period. The maximum $SO_2$ concentration measured during this period of time was 1.34 ppm ($3510\ \mu g/m^3$).

For comparison, the US primary air quality standard for $SO_2$ is 0.03 ppm or $80\ \mu g/m^3$ during the year; 0.14 ppm or $365\ \mu g/m^3$ over a 24-hour period.

## 16.5   MEASUREMENT OF CO

Carbon monoxide (CO) is a colorless, odorless, and tasteless gas that is produced by the incomplete combustion of hydrocarbon fuel. Motor vehicle exhaust, industrial processes, and natural sources such as fire are responsible for high concentration of CO in the atmosphere.

### 16.5.1   Infrared Photometry

This method is based on the ability of air pollutants to absorb optical radiation at certain wavelengths. Particularly, it is used for measuring the concentration of CO, which absorbs infrared radiation at a wavelength of 4.7 μm.

The amount of CO in the sample is determined according to the Beer–Lambert law.

The level of transmission $T$ (absorption) is directly proportional to the concentration $C$ of CO based on the Beer–Lambert law:

$$T = (I/I_0) = \exp(-\alpha Cl),\qquad(16.8)$$

where $\alpha$ is the coefficient of absorption of the pollutant; $C$ is the concentration of pollutant; and $l$ is the optical path length.

There are two possible configurations of photometers for the estimation of CO. In the first version of photometer (Figure 16.8a) an infrared radiation passes alternately through two identical cells, one filled with carbon dioxide ($CO_2$) to be analyzed, and the other filled with reference gas. Both beams are followed by the detector. The intensity of the signal that is received by the detector is proportional to the amount of CO in the sample.

The second version of photometer (Figure 16.8b) contains a rotating gas-filled filter wheel. The detector responds to the modulation of intensity that is inversely proportional to the amount of CO in the sample cell Such a technique is called *gas filter correlation (GFC)*.

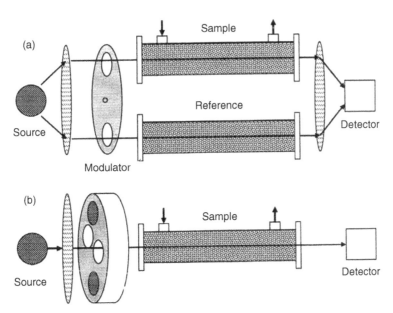

**FIGURE 16.8**   Two possible configurations of photometers for estimation of carbon monoxide: (a) two-cell version; (b) one-cell version.

The method of infrared photometry is sensitive. For example, the Ecotech Serinus 30 Carbon Monoxide analyzer is used to measure CO in ambient air in the range of 0–200 ppm to a sensitivity of 0.05 ppm.

*The advantages of an infrared photometer:*

- This system has the simplistic design.
- A very quick process of measurement.
- This technique is non-destructive to the sample.
- High sensitivity.
- Can be used for cross-stack monitoring of CO.

*The disadvantages of an infrared photometer:*

- It requires frequent calibrations to retain the accuracy and precision of the instrument.
- It requires very sensitive and properly tuned instruments.

### 16.5.2   Open-Path Fourier Spectrometry

This analytical technique is based on the measurements of the temporal coherence of a radiative source using time-domain or space-domain measurements of the electromagnetic radiation.

It is known that any complex fluctuation $y(t)$ can be represented by Fourier transformation as a combination of a large number of sine and cosine waves, which form a *Fourier series*:

$$y(t) = \sum_{n} (A_n \sin 2\pi v_n t + B_n \cos 2\pi v_n t), \qquad (16.9)$$

where $A_n$ and $B_n$ are amplitudes of harmonic oscillations; and $v_n$ is the frequency of the $n$th oscillation.

One of the devices commonly used for measuring the temporal coherence of the light is the *Fourier transform spectrometer*, which consists of a *Michelson interferometer* that comprises a source of light, beam splitter, movable and fixed mirrors, and a detector. A typical optical setup of Michelson interferometer is described by Hariharan (2007).

If the source is monochromatic and the movable mirror is moved at a constant rate, the detector signal oscillates with a single frequency. The radiant power can be recorded as a function of time as the cosine oscillation (*time domain*) or as a function of frequency as a spectral line (*frequency domain*). A plot of the output power from the detector versus the mirror displacement is called an *interferogram*.

If the source is polychromatic, each input frequency can be considered to produce a separate cosine oscillation. The resulting interferogram is a summation of all cosine oscillations caused by all the frequencies of the source.

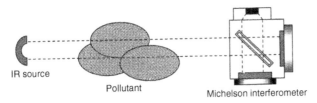

**FIGURE 16.9**    Open-path Fourier spectrometer.

The recorded signal is mathematically manipulated using the Fourier transform technique to produce a spectrum that can be used to identify the specific contaminants and their concentrations.

Direct Fourier transform is

$$S(t) = \int_{-\infty}^{\infty} I(v)e^{-iv2t}dv, \tag{16.10}$$

while the inverse Fourier transform is described as

$$I(v) = 2Re \int_{-\infty}^{\infty} S(t)e^{2i\pi vt}dt. \tag{16.11}$$

Here, a signal $S(t)$ presents a sum total of all harmonics in the frequency domain and $I(t)$ in the time domain.

An *open-path Fourier transform spectrometer* can be used for air quality monitoring in industrial areas. It consists of a source of infrared radiation that passes through the gas that is analyzed and the Michelson interferometer (Figure 16.9).

This Fourier spectrometer can be used to identify and quantify molecules found in industrial emissions with a high level of sensitivity and resolution.

Open-path Fourier spectrometry can be applied also for the determination of carbon dioxide ($CO_2$), methane ($CH_4$), acetic acid ($CH_3COOH$), methanol ($CH_3OH$), formaldehyde ($HCHO$), ethene ($C_2H_4$), ammonia ($NH_3$), acetylene ($C_2H_2$), nitric oxide ($NO$), ethane ($C_2H_6$), phenol ($C_6H_5OH$), propene ($C_3H_6$), formic acid ($HCOOH$), nitrogen dioxide ($NO_2$), sulphur dioxide ($SO_2$), hydroxyacetaldehyde ($HOCH_2CHO$), and furan ($C_4H_4O$).

*The advantages of a Fourier transform spectrometer:*

- Requires few optical elements; that is, why the power of radiation reaching the detector at once is much greater than that in dispersive spectrometers.
- Extremely high wavelength accuracy, resolution ($<0.1$ cm$^{-1}$) and precision.
- Improved signal-to-noise ratio.
- All elements of the source reach the detector simultaneously; it is possible to obtain an entire spectrum in a brief (1 second or less) period.
- Can be used for the identification of the samples.

*The disadvantages of a Fourier transform spectrometer:*

- Requires only a single beam, so the environment can easily affect it.
- Cannot detect atoms or individual ions, which contain no chemical bonds, do not possess vibrational motion and do not absorb infrared radiation.
- The analysis and interpretation of spectra of complex sample mixtures are difficult.
- It is difficult to analyze the aqueous solutions using infrared spectroscopy, because water is a strong infrared absorber.

### 16.5.3    Effect of Carbon Monoxide on Human Health

CO is a toxic gas which enters through the lungs to the blood vessels, where it binds to hemoglobin reducing the amount of oxygen that reaches the organs and tissues. Sensitive to CO people are suffering from heart and respiratory diseases (such as angina). Low concentrations of CO provoke such symptoms as headaches, nausea, and dizziness. High concentrations of CO lead to loss of mental control, impaired vision, and death.

The US primary air quality standard for CO is 9 ppm or 10 mg/m$^3$ during 8 hours; 35 ppm or 40 mg/m$^3$ during 1-hour period.

## 16.6    PARTICULATE MATTER SAMPLING

*Particulates* (or particulate matter (PM)), are tiny portions of solid or liquid matter suspended in the Earth's atmosphere. The size of atmospheric particles ranges from a few nanometers to several micrometers.

Particles come in a wide range of sizes and can be separated into categories based on size. Particles less than 2.5 μm in diameter are called "fine" particles (PM$_{2.5}$). These particles are so small they can only be detected using an electron microscope. Particles between 2.5 and 10 μm in diameter are referred to as "coarse" particles (PM$_{10}$).

Particles PM$_{2.5}$ and PM$_{10}$ are emitted from primary combustion processes such as transportation, industrial, and domestic activity, solid waste disposal, while PM$_{10}$ can arise additionally from the formation of aggregates in the atmosphere, road dust, volcanic eruptions, wind erosion, forest and grassland fires, and other natural phenomena.

A *PM sampler* is a device for measuring the mass concentration or chemical composition of particulates in the atmosphere.

There are two principal approaches to particulate sampling. The first one (called gravimetric) is filter-based manual sampling which provides the laboratory weighting of the filters before and after sampling followed by analysis of the sample.

The second method is based on continuous monitoring for PM$_{2.5}$ and PM$_{10}$. It provides an *in situ* automated sampling and weighting in the field conditions.

## 16.7   GRAVIMETRIC METHODS

### 16.7.1   High-Volume Samplers

A high-volume $PM_{10}$ sampler provides a passing of certain volume of air at a constant flow rate through the hole of small diameter (inlet) and the filter. As a result, the particles are collected in the filter during the 24-hour period, after which the weight of the filter before and after deposition of the particles is compared. The concentration of the particles is determined as the ratio of the total weight of particles and the total volume of air, which is defined on the basis of known volumetric flow rate and the sampling time. The unit of measurement of particle concentration is $\mu g/m^3$.

Each high-volume $PM_{10}$ sampler consists of inlet and flow-control systems. There are two principal types of high-volume $PM_{10}$ inlet systems.

### 16.7.2   Impaction Inlet

Ambient air is passing through the buffer chamber, system of acceleration nozzles, and impaction chambers, where the separation and fractionation of particles takes place. Then the air flow enters a sample filter. Each particle size that is greater than 10 μm is captured on the collector shims, while the particles of size less than 10 μm are collected on the quartz filter (Figure 16.10).

### 16.7.3   Cyclonic Inlet

The movement of particles in this inertial system takes place due to centrifugal and gravitational forces (Figure 16.11). Cyclone collector uses cyclonic action to separate particles from the gas stream, which enters at an angle and is spun rapidly.

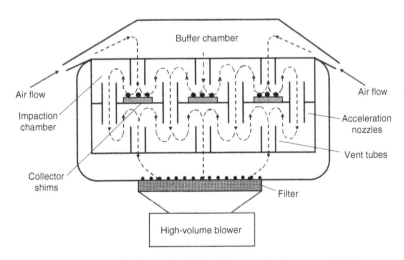

**FIGURE 16.10**   Impaction high-volume particulate matter inlet.

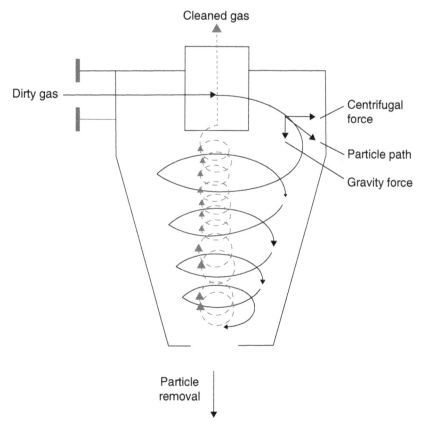

**FIGURE 16.11**    Inertial cyclone collector.

The centrifugal force created by the circular flow throws the particles toward the wall of the cyclone. After interacting with the wall, these particles are removed.

### 16.7.4    Low-Volume Samplers

Low-volume samplers are characterized with a flow rate (2.3 m³/h) considerably smaller in comparison with high-volume samplers (68 m³/h). In spite of the same principle of operation, there are differences in parameters as the inlet size, flow rate, and filter size. Let us discuss one of the types of low-volume samplers.

### 16.7.5    Dichotomous Sampler

The principle of operation of a dichotomous system is based on the separation of particles using two channels with different cross-sectional areas and therefore different air flow speeds. The particles smaller than 2.5 μm follow the higher air flow

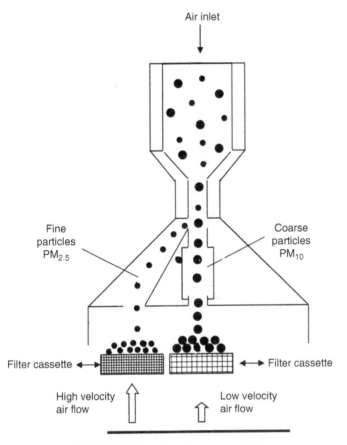

**FIGURE 16.12**   Dichotomous chamber.

path and are collected on a fine filter. The coarse particles (2.5–10 μm) demonstrate sufficient inertia and are collected on the coarse filter (Figure 16.12).

*The advantages of a gravimetric method:*

- High accuracy.
- Does not require expensive equipment.
- Can be used for calibration of other instruments.

*The disadvantages of a gravimetric method:*

- Can lead to loss of PM, especially volatile compounds, due to heating.
- Suffers with formation of particles from other gases due to chemical reactions that are accompanied by overestimation of PM mass.
- Sampling is complicated and expensive.

## 16.8  CONTINUOUS METHODS

Continuous method of atmospheric particles monitoring provide useful information in real time concerning the PM concentration, extreme pollution situations, diurnal variations, etc. We shall discuss here two continuous methods that are currently available.

### 16.8.1  Beta Attenuation Monitor

This method is based on the interaction of ionizing $\beta$-radiation with a PM that is deposited on the filter. A beta attenuation monitor consists of a low-volume size selective inlet, a beta radiation source (isotopes $^{14}$C or $^{85}$Kr), a filter type, flow controller, and a timer. A compensation chamber is used as a reference to compare the sample measurement in the sampling chamber with $\beta$-radiation transmitted through the clean filter type. Heating the ambient air stream makes it possible to avoid emission of semi-volatile compounds. The filter material is designed as a roll that moves automatically (Figure 16.13). When PM is located between the source and a detector (photomultiplier connected with a scintillation counter), the $\beta$-radiation is attenuated. The level of the reduction in detected $\beta$-radiation depends on the concentration $C$ of PM, which can be determined as follows (Gobeli et al., 2008):

$$C(\text{kg/m}^3) = \frac{A(\text{m}^3)}{Q(\text{m}^3/\text{s})t(\text{s})}x(\text{kg/m}^3), \qquad (16.12)$$

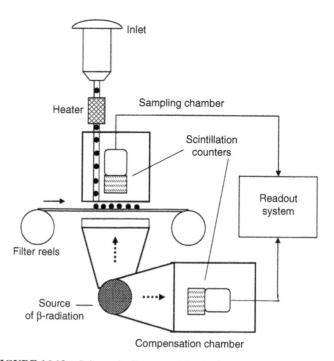

**FIGURE 16.13**  Schematic diagram of the Andersen beta gauge monitor.

where $A$ is the cross-sectional area of the tape spot over which particles are deposited; $Q$ is the rate of air sampling; $t$ is the sampling time; $x$ is the mass density of collected particles.

The procedure of measurements provides the estimation of the intensity $I_0$ of $\beta$-radiation that passes across the clean filter type, the intensity $I$ of attenuated $\beta$-radiation (filter with PM) and determination of the ambient concentration of PM:

$$C(\mu g/m^3) = \frac{D}{\mu} \ln \frac{I_0}{I}, \qquad (16.13)$$

where $D$ is a constant, which makes it possible to express concentration in $\mu g/m^3$, $\mu$ is the absorption cross section of the material absorbing $\beta$-radiation (kg/m$^3$).

The commercial Andersen FH621-N Beta Gauge Sampler (Newterra) is characterized by the following parameters:

Measuring Range 5–10,000 $\mu g/m^3$;
Minimum Detectable Limit less than 2 $\mu g/m^3$ (24 hour);
Accuracy 2 $\mu g/m^3$ (24 hour).

### 16.8.2    Tapered Element Oscillating Microbalance

The principle of operation of the tapered element oscillating microbalance (TEOM) particulate sampler is based on the measurement of the frequency of oscillation of a cylindrical glass tube with a filter attached to one of its ends on the mass of the particulate material deposited on this filter due to the air flow.

The lower end of the tube is fixed to a base; the upper end with a filter that oscillates due to an electrical circuit. The monitor consists of two principal units: sensor unit (tapered element, mass transducer, supporting electronics) and control unit (onboard microprocessor for data processing and storage, frequency counter electronics, mass flow controller, interface) (Figure 16.14). The temperature of the sample chamber is about 50°C to eliminate the effects of ambient temperature changes and avoid the thermal expansion of the tapered element.

The resonance frequency $f$ of the tube is proportional to the square root of the mass $m$ of PM:

$$f = \sqrt{\frac{k}{m}}, \qquad (16.14)$$

where $k$ is a constant determined during calibration of the TEOM.

The change of mass $\Delta m$ from the initial condition can be determined as (Patashnick et al., 2002)

$$\Delta m = k \left( \frac{1}{f_f^2} - \frac{1}{f_i^2} \right), \qquad (16.15)$$

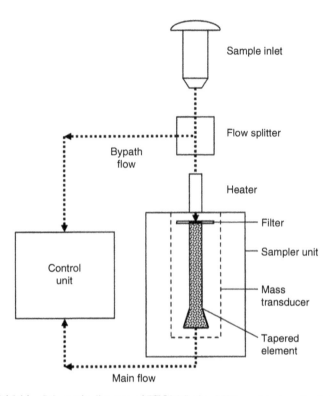

**FIGURE 16.14**    Schematic diagram of TEOM Series 1400a ambient particulate monitor.

where $f_i$ is the initial oscillation frequency of the system; $f_f$ is the oscillation frequency of the system after the loading filter by PM. A minimum mass detection limit can reach 0.01 μg.

*The advantages of a continuous method:*

- Real-time measurements with short time resolution (<1 hour).
- TEOM samplers operate continuously and do not need filter changes as frequently as high-volume air samplers do.

*The disadvantage of a continuous method:*

- Can lead to loss of PM, especially volatile compounds, due to heating in TEOM technology.

## 16.9    EFFECT OF PARTICULATE MATTER ON HUMAN HEALTH

$PM_{2.5}$ enters the bronchial tubes, lungs, and blood stream and can cause serious health problems.

$PM_{10}$ is $\leq 10$ μm in diameter and can enter the nasal cavity.

The US primary air quality standard for PM, $PM_{2.5}$ is 15 μg/m$^3$ annually; 65 μg/m$^3$ during 24 hours; for PM, $PM_{10}$ is 50 μg/m$^3$ annually; 150 μg/m$^3$ during 24 hours.

## 16.10 NANOPARTICLES

*Nanoparticles* are sized between 1 and 100 nm. Nanoparticles can be highly reactive due to their large surface-area-to-volume ratio and the presence of a greater number of reactive sites. For example, surface-area-to-volume ratio $S/V$ of different particles is: $S/V = 3 \times 10^3$ (radius $R = 1$ mm); $S/V = 3 \times 10^6$ ($R = 1$ μm); $S/V = 3 \times 10^9$ ($R = 1$ nm).

## 16.11 EFFECT OF NANOPARTICLES ON HUMAN HEALTH

Nanoparticles provide a specific nanopollution which presents residues and waste generated by nanodevices or during the nanomaterial's manufacturing process. These nanoparticles can be very dangerous because of small size, extreme mobility, and reactivity, just as the ability to penetrate living cells. The mechanisms of the interaction of nanoparticles with living organisms are not known yet. Besides, the living organisms have not elaborated effective means of protection from nanoparticles which were fabricated artificially.

## 16.12 BIOAEROSOLS

Organic particles of biological origin such as bacteria, fungi, fungal spores and cells, very small insects and their fragments, pollen, and viruses suspended in the air are called *bioaerosols*. A branch of biology that studies the nature, movement, and distribution of these organic particles is defined as *aerobiology* (Mar Trigo Pérez et al., 2007).

The principal bioaerosols are: bacteria (*Legionella pneumophila*, *Micropolyspora faeni*, *Mycobacterium tuberculosis*, *Pseudomonas* spp., *Staphylococcus* spp., *Streptococcus* spp., *Bacillus* spp.) and bacterial products or components; fungi (principally molds and yeasts) *Histoplasma capsulatum*, *Alternaria* spp., *Penicillium* spp., *Aspergillus fumigatus*, *Stachybotrys atra*, *Fusarium* spp., *Cladosporium* spp.; protozoa *Naegleria fowleri*, *Acanthamoeba* spp.; anthropods such as mites *Dermatophagoides farinae*, *Dermatophagoides pteronyssinus* and insects *Blattella germanica*, *Periplaneta Americana*, *Blatta orientalis*; animals (cats, dogs, ferret, guinea pigs, hamster, rabbit, rat, mouse), particularly dermal and urine antigens from pets or vermin; plants, especially pollen (*Betuta*, *Alnus*, *Pinus*, *Artemisia*, *Platanus*, *Ambrosia*).

The size of bioaerosols varies from 0.015 to 0.45 μm (viruses), 0.3 to 10 μm (bacteria), 1 to 100 μm (fungal spores), 0.5 μm to several cm (algae), 10 to 100 μm (pollen), more than 100 μm (plant and animal fragments, seeds, insects).

Airborne particles can be easily transported, transferred, and deposited in the environment. The concentration of bioaerosols depends strongly on the ambient temperature and humidity.

## 16.13   BIOAEROSOL SAMPLING AND IDENTIFICATION

*Bioaerosol samplers* are intended for collection of airborne particles, spores, and pollen. There are several types of bioaerosol sampling devices such as gravitational samplers and inertial samplers (see Section 16.5), spore traps (particles impacted on to coated glass slide or adhesive tape), impactors (collection on slide or agar plates), impingers (air drawn through liquid and particles are removed by impingement), non-inertial samplers (filtration, electrostatic precipitation, thermal precipitators, and condensation traps) (Meschke, Bioaerosol Sampling).

Identification of bioaerosols is realized due to such analytical methods as microscopy and cultivation, flow cytometry, polymerase chain reaction technique, ATP-bioluminescence, spectroscopic techniques (Matrix assisted laser desorption/ionization time of flight-mass spectrometry (MALDI-ToF-MS)), Raman spectroscopy, laser-induced fluorescence).

### 16.13.1   Bioaerosol Sampler Spore-Trap

Let us discuss the most traditional and effective bioaerosol sampler spore trap. The inventor of this method, J.M. Hirst created in 1952 the first aspiration sampler which consisted an air admission chamber with a slit and petroleum jelly-smeared slide, that was mounted on a weather vane (Hirst, 1952). Later this device was modified by the Burkard company as "Burkard Seven-Day Volumetric Spore-Trap®."

The airflow enters a 14 × 2-mm slit and is directed onto a 345-mm adhesive band (Figure 16.15). The slit is oriented toward the direction of wind due to the

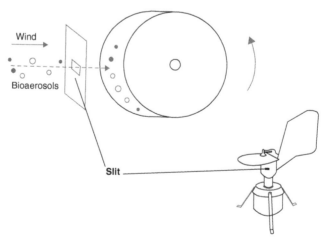

**FIGURE 16.15**   Bioaerosol sampler spore-trap.

weather vane. The sampler is accompanied with a drum and clock mechanism, which provides an uninterrupted sampling period during 7 days. The pieces of tape with a sample are analyzed qualitatively through the microscope and quantitatively due to the densitometer.

### 16.13.2    Matrix Assisted Laser Desorption/Ionization Time of Flight Mass Spectrometry

*MALDI-ToF-MS* is a soft ionization technique that is based on the formation of gas-phase ions without extensive fragmentation.

*Desorption* is the process opposite of sorption (adsorption, absorption), which is accompanied by a release of substance from a surface.

*Ionization* is the process of converting an atom or molecule into an ion by adding or removing charged particles such as electrons or ions.

The term matrix-assisted laser desorption ionization (MALDI) belongs to Franz Hillenkamp, Michael Karas and their colleagues (Karas et al., 1986, 1987). Koichi Tanaka of Shimadzu Corp. and his coworkers shared the Nobel Prize in Chemistry in 2002 for developing a novel method for mass spectrometric analyses of biological macromolecules and demonstrating that a process of ionization of these molecules can be realized with the proper combination of laser wavelength and matrix (Tanaka et al., 1988).

MALDI-ToF-MS technique can be used for analysis of biomolecules, large organic molecules, and bioaerosols.

Main principles of MALDI-ToF-MS are based on irradiation of a sample of a nanosecond-pulsed UV laser. Usually such UV lasers as nitrogen lasers (337 nm) and frequency-tripled and quadrupled Nd:YAG lasers (355 nm and 266 nm, respectively) are used in MALDI technique. Duration of laser pulses is a few nanoseconds, intensity is about $10^6$–$10^7$ W/cm$^2$. The sample is introduced in a matrix, forming a solid solution or a mixture of the matrix material and the analyte. Intense laser radiation induces the *laser ablation*—the process of removing material from a surface in the form of microparticles that can reach the size of several hundred micrometers.

As a result of the interaction of laser radiation with the matter, the so-called plume is created. This plume consists mainly of neutral particles, but a small ($10^{-4}$–$10^{-5}$) portion of the charged particles is also presented. The spatial expansion of the plume in the first nanoseconds provokes the decay of conglomerates into separate fragments and molecules as well as charged particles.

Ionization of molecules that occurs directly with the release of material from the condensed state can be considered as primary ionization. In addition, there are numerous collisions of particles in the expanding plume that lead to the ion–molecule reactions between the matrix charged particles and the analyte molecules and to the secondary ionization of the latter.

At first the matrix is ionized with a single positive charge; later, this charge is transferred from matrix to sample molecules through the collisions in the gas phase.

In general, the absorption of laser radiation by matrix is accompanied by such processes as ionization (addition of a proton), desorption, transfer of proton from matrix

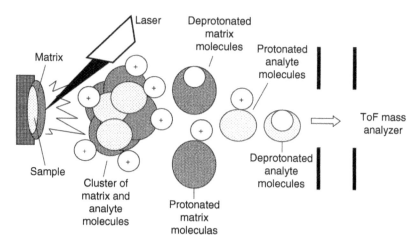

**FIGURE 16.16**    Matrix assisted laser desorption/ionization time of flight mass spectrometry.

to analyte, charging the analyte molecules, and formation of protonated molecules (Figure 16.16).

The ionized sample molecules are directed then through a positively charged electrostatic field which accelerates these molecules into the time of flight (TOF) mass analyzer which measures the mass and charge of each molecule and depicts the information as a mass spectrum.

*The advantages of a MALDI-ToF-MS technique:*

- Ability to ionize and desorb high molecular weight biomolecules into the gas phase while preserving their intact state.
- Broad applicability to bioaerosols.
- Cost effective.

*The disadvantages of a MALDI-ToF-MS technique:*

- Identification is limited by reference spectra in database.
- Inability to differentiate among closely related organisms.
- Repeat analysis is required.

## 16.14    MEASUREMENT OF ATMOSPHERIC OZONE

### 16.14.1    Radiosondes

The ozone layer is located in the stratosphere at altitudes of 10–50 km with a maximum at about 20–25 km.

It should be noted that despite the small amount of ozone, it is capable of absorbing a significant amount of UV solar radiation, that is, almost all radiation in the UV-C region (200–300 nm) and part of the UV-B region (280–320 nm).

*Balloon-borne radiosondes* are equipped with instrumentation to measure the vertical distributions of ozone.

Ozonesondes have a lower quantifiable limit of less than 15 ppb with a precision of ± 5 ppb or ± 10%.

The method of detection of an ozonesonde is based on the reaction of ozone with potassium iodide in an aqueous solution.

When ozone in the air enters the sensor, iodine is formed in the cathode half of the cell according to the relation:

$$2KI + O_3 + H_2O \rightarrow I_2 + O_2 + 2KOH. \tag{16.16}$$

The cell converts the iodine to iodide according to

$$I_2 + 2e^- \rightarrow 2I^-, \tag{16.17}$$

during which the two electrons flow into the cell's external circuit.

### 16.14.2  Dobson and Brewer Spectrophotometry

*Dobson spectrophotometer* can be used to measure both total column ozone and profiles of ozone in the atmosphere. Total ozone measurements are made by determining the relative intensities of selected pairs of UV wavelengths (Figure 16.17). One wavelength (305 nm) is strongly absorbed by ozone and the other (325 nm) is not.

The Dobson method has its drawbacks. It is strongly affected by aerosols and pollutants in the atmosphere, which also absorb some of the light at the same wavelength.

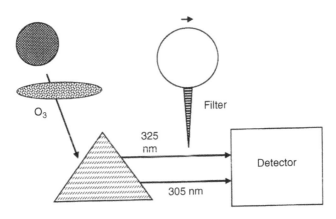

**FIGURE 16.17**  Dobson spectrophotometer.

In addition, measurements are made over only a small area. Today this method is often used primarily to calibrate data obtained by other methods, including satellites.

The *Brewer spectrophotometer* is devoid of those shortcomings. Total ozone is determined by comparing direct sunlight signals measured at the wavelengths of 306.3, 310.1, 313.5, 316.8, and 320.1 nm.

## 16.15   MEASUREMENT OF GROUND-LEVEL OZONE

Photometry method can be used also for measuring the concentration of ozone, which absorbs UV radiation at a wavelength of 254 nm.

Chemiluminescence is also used for measuring ground-level ozone and is based on the emission of light energy as the result of chemical reactions (16.1–16.2).

EPA reference method is based on the reaction of ozone with ethylene which is accompanied with the emission of photons. The intensity of photon flow is transferred into electric signals of photomultiplier, which is proportional to the concentration of ozone in air. The disadvantage of this method is inflammability, toxicity of ozone, and its ability to be polymerized in the gas stream.

## 16.16   EFFECT OF OZONE ON HUMAN HEALTH

Children and elderly people with respiratory diseases such as asthma can react to ground-level ozone. Healthy people suffered from irritation, wheezing, coughing exhaustion, and headache.

The US primary air quality standard for ground-level ozone is 0.08 ppm ($155 \, \mu g/m^3$) annually; 0.12 ppm ($232 \, \mu g/m^3$) during 1 hour.

## 16.17   MEASUREMENT OF LEAD

*Lead* is a bluish-white soft, malleable, and ductile metal. Native lead can be found in ore. Lead can enter the atmosphere through soil erosion, volcanic eruptions, sea spray, and forest fires. But the most widespread sources of lead, lead salts ($PbSO_4$, $PbCO_3$) and lead-containing dust in the atmosphere are the exhausts of cars and fuel combustion, solid waste combustion, industrial processes, mining operations, battery recycling, and the production of lead fishing sinkers. The small particles of air-borne lead can travel long distances through air and remain in the atmosphere.

### 16.17.1   Atomic Spectrometry of Lead

The determination of lead in PM in ambient air is based on active sampling and analysis that is realized by atomic absorption spectrometry. There are two principal methods of atomic spectrometry: flame atomic absorption spectroscopy (FAAS) and graphite furnace atomic spectroscopy (FDAA). We shall consider here the GFAA method which has the detection limit about two orders better than FAAS (Compendium method IO-3.2).

## 16.17.2   Graphite Furnace Atomic Absorption Spectroscopy

GFAAS method is based on the application of high (about 3000°C) temperature to the sample which is deposited in a small graphite tube, its vaporization and atomization—conversion of analyte to a free gaseous atom. An optical radiation containing the wavelength that is required to excite atoms of analyte is directed through the flame of furnace, monochromator, and enters the detector (Figure 16.18). Electric signals from the output of the detector are proportional to the attenuation of the intensity of the radiation that is absorbed by the element in the furnace and in such a way to the concentration of the metal in the sample.

This instrument can measure the same metals at concentrations 100–1000 lower, in the parts per billion (µg/L) range.

*The advantages of a GFAA spectrometry:*

- Greater sensitivity and detection limits than other methods.
- Direct analysis of some types of liquid samples.
- High conversion efficiency of sample into free atoms.
- Very small sample volume (about 20 µL).

*The disadvantages of a GFAA spectrometry:*

- Longer analysis time than flame AA or ICP analysis.
- Limited dynamic range.
- No true field-portability, with a mobile laboratory setup usually required.

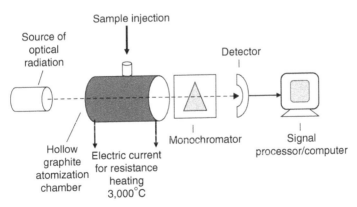

**FIGURE 16.18**   Graphite furnace atomic absorption spectrometer.

## 16.18    EFFECT OF LEAD ON HUMAN HEALTH

Lead can provoke some diseases and health impairments such as rise of blood pressure, disruption of nervous systems, mental retardation and behavioral disorders, anemia, nausea, gastric problems, and brain damage.

In children, the symptoms of lead exposure can be behavioral disruptions of children, poor development of motor abilities, and memory, colic, and gastric problems.

The US primary air quality standard for leas is 1.5 $\mu g/m^3$ during 3 months.

### Constructive Tests

1. What is the principal difference between the "absorption coefficient" and the "attenuation coefficient?"
2. What does "Fourier synthesis" mean?
3. Define the terms "mass concentration" and "molar concentration."
4. Explain the properties of Griess–Saltzman reagent solution.
5. What are the advantages of liquid drops and liquid films for analysis of air?
6. Cuvette of photometer has two windows that are fabricated from high quality optical material. What optical material can be used for measurement of the concentration of ozone: calcium fluoride $CaF_2$ or borosilicate crown glass BR-7? Explain your choice.
7. Why the concentration of CO in urban air is higher during early morning and early evening hours?

## REFERENCES

Cardoso, A.A. and Dasgupta, P.K. 1995. Analytical chemistry in a liquid film/droplet. *Anal. Chem.*, 67(15):2562–2566.

Chou, J. 1999. Electrochemical sensors. In: *Hazardous Gas Monitors: A Practical Guide to Selection, Operation, and Applications*, 1st edition. McGraw-Hill Professional, pp. 27–35.

Gobeli, D., Schloesser, H., and Pottberg, T. 2008. Met One Instruments BAM-1020 Beta Attenuation Mass Monitor US-EPA PM$_{2.5}$ Federal Equivalent Method Fild Test Results. *Paper #2008-A-485-AWMA*.

Guidi, V., Butturi, M.A., Carotta, M.C., Cavicchi, B., Ferroni, M., Malagu, C., Marinelli, G., Vincenzi, D., Sacerdoti, M., and Zen, M. 2002. Gas sensing through thick film technology. *Sens. Actuators B*, 84:72–77.

Hariharan, P. 2007. *Basics of Interferometry*, 2nd edition. Elsevier.

Hirst, J. 1952. An automatic volumetric spore trap. *Ann. Appl. Biol.* 39:257–265.

Karas, M., Bachmann, D., and Hillenkamp, F. 1985. Influence of the wavelength in high-irradiance ultraviolet laser desorption mass spectrometry of organic molecules. *Anal. Chem.* 57 (14):2935–2939.

Karas, M., Bachmann, D., Bahr, D., and Hillenkamp, F. 1987. Matrix-assisted ultraviolet-laser desorption of nonvolatile compounds. *Int. J. Mass Spectrom. Ion Proc.* 78:53–68.

Kebabian, P.L., Wood, E.C., Herndon, S.C., and Freedman, A. 2008. A practical alternative to chemiluminescence detection of nitrogen dioxide: cavity attenuated phase shift spectroscopy. *Environ. Sci. Technol.* 42:6040–6045.

Liu, S. and Dasgupta, P.K. 1995. Liquid droplet. A renewable gas sampling interface. *Anal. Chem.* 67(13):2042–2049.

Mar Trigo Pérez, M., Caballero, M.M., Sanchez, J.G., Criado, M.R., Fernández, S.D., and Artero, B.C. 2007. *El polen en la atmósfera de Vélez-Málaga.* Concejalia de Medio Ambiente. Ayuntamiento de Vélez-Málaga.

Ohira, S.-I. and Toda, K. 2005. Micro gas analysis system for measurement of atmospheric hydrogen sulfide and sulfur dioxide. *Lab Chip.* 5:1374–1379.

Palmes, E.D., Gunnison, A.F., DiMattio, J., and Tomczyk, C. 1976. Personal sampler for nitrogen dioxide. *Am. Ind. Hyg. Assoc. J.* 37:570–577.

Patashnick, H., Meyer, M., and Rogers, B. 2002. Tapered element oscillating microbalance technology. Mine Ventilation, edited by Euler De Souza. Proceedings of the North American/Ninth U.S. Mine Ventilation Symposium. pp. 625–631. Kingston, Ontario, June 8–12, 2002.

Platt, U. 1994. Differential optical absorption spectroscopy (DOAS). *Chem. Anal. Series,* 127:27–83.

Tanaka, K., Waki, H., Ido, Y., Akita, S., Yoshida, Y., and Yoshida, T. 1988. Protein and polymer analyses up to m/z 100 000 by laser ionization time-of flight mass spectrometry. *Rapid Commun. Mass Spectrom.* 2 (20):151–153.

Toda, K., Hato, Y., Ohira, S.-I., and Namihira, T. 2007. Micro-gas analysis system for measurement of nitric oxide and nitrogen dioxide: Respiratory treatment and environmental mobile monitoring. *Anal. Chem. Acta.* 603:60–66.

# PRACTICAL EXERCISE 8

# FUNDAMENTALS OF SPECTROSCOPY

## 1 BEER–LAMBERT–BOUGER LAW

Let us discuss the main terminology which is related to the passing of light through the object.

*Absorption* is the process in which incident radiated energy is retained without reflection or transmission on passing through a medium.

*Absorbance* is the ability of a solution or a layer of a substance to absorb radiation that is expressed mathematically as the negative common logarithm of the transmittance of the substance or solution:

$$A = -\lg\left(\frac{I_t}{I_0}\right) = \varepsilon Cl,$$ 

(P8.1)

where $\tau = \frac{I_t}{I_0}$ is transmittance; $I_0$ and $I_t$ are the intensity of the incident and the transmitted light, respectively; $\varepsilon$ is the *molar absorptivity* (molar absorption coefficient, molar extinction coefficient) of the absorber; $C$ is the molar concentration of the absorbing species; $l$ is the thickness of the sample (the path length).

Analogue expressed in terms of natural logarithm can be defined as

$$A' = -\ln\left(\frac{I_t}{I_0}\right) = kCl.$$ 

(P8.2)

*Methods of Measuring Environmental Parameters*, First Edition. Yuriy Posudin.
© 2014 John Wiley & Sons, Inc. Published 2014 by John Wiley & Sons, Inc.

The *molar absorptivity* $\varepsilon$ (or $k$) is a measurement of how strongly a chemical species absorbs light at a given wavelength.

The SI unit for $\varepsilon$ (or $k$) is $m^2/mol$; off-system unit is $L/mol \cdot cm$ or $M^{-1} \cdot cm^{-1}$ (here $1M = 1$ mol/L).

Absorbance is also called *optical density D*, but absorbance is related to absorption inside the substance only, while optical density is determined either by absorption or by scattering.

It is possible to obtain from Equations P8.1 and P8.2 the important relationship that is known as *Beer–Lambert–Bouger law* in terms of natural logarithm:

$$I_t/I_0 = e^{-kCl}, \tag{P8.3}$$

and in terms of common logarithm:

$$I_t/I_0 = 10^{-\varepsilon Cl}. \tag{P8.4}$$

## 2  PHOTOMETRY OF OZONE IN GAS PHASE

The concentration of ozone can be estimated in terms of *ppm* or *percentage (%) by volume*. Such units as ppm in terms of mass means parts per million, or $10^{-6}$. Volume percent takes into account the dependence of gas volume on the pressure, temperature, and number of molecules of the gas.

Let us use the photometer with the quartz cell of 1 cm path length. The absorbance which was measured during experiment is $A = 0.36$. The extinction coefficient for gas phase ozone is given as 308.8 atm$^{-1}$·cm$^{-1}$ (Ozone Results. EPA Method), atmospheric pressure is 750 mmHg; absolute temperature is 273 K.

The ozone concentration in terms of *ppm* is determined as

$$C(\text{ppm}) = -\frac{10^6}{\varepsilon l} \ln \left( \frac{I}{I_0} \right). \tag{P8.5}$$

The ozone concentration in terms of percentage (%) by volume can be determined as follows:

$$C(\% \text{ by volume}) = \frac{100\%}{\varepsilon} \frac{T_a}{293} \frac{760}{p} \frac{A}{l}. \tag{P8.6}$$

Using this equation we can find

$$C(\% \text{ by volume}) = \frac{100\%}{308.8 \text{ atm}^{-1} \text{ cm}^{-2}} \times \frac{273}{293} \times \frac{760}{750} \times \frac{0.36}{1 \text{ cm}} = 0.11\%.$$

The ozone concentration in terms of mass per volume can be calculated as

$$C(\text{mg/L}) = 21.427 \frac{293}{T_a} \cdot \frac{p}{760} C(\% \text{ by volume}). \tag{P8.7}$$

## 3  FOURIER TRANSFORM SPECTROMETRY

Wavenumber $\bar{\nu}$ is the number of wavelengths per unit distance $\bar{\nu} = \frac{1}{\lambda}$, where $\lambda$ is the wavelength.

Wavenumber $\bar{\nu}$ can be measured in the following units:

$$1\ \text{cm}^{-1} = 100\ \text{m}^{-1}; 1\ \text{m}^{-1} = 0.01\ \text{cm}^{-1}; 1\ \text{mm}^{-1} = 10\ \text{cm}^{-1}.$$

It is possible to convert inverse centimeters in wavenumber to micrometers (microns) or nanometers using the following formulas (See Practical Exercise 6, Section 1):

$$\sigma(\text{cm}^{-1}) = \frac{10,000,000}{\lambda(\text{nm})}; \quad \sigma(\text{cm}^{-1}) = \frac{10,000}{\lambda(\mu\text{m})};$$

$$\lambda(\text{nm}) = \frac{10,000,000}{\sigma(\text{cm}^{-1})}; \quad \lambda(\mu\text{m}) = \frac{10,000}{\sigma(\text{cm}^{-1})}.$$

The wavelength $\lambda$ of a sinusoidal waveform travelling at constant speed $c$ is given by

$$\lambda = \frac{c}{f}, \tag{P8.8}$$

**Example**    The mirror of Michelson interferometer is moved at a constant velocity $\upsilon_M = 1$ mm/s. Find the frequency $f$ of oscillation of the detector signal, if the wavenumber of infrared source is $\bar{\nu} = 1000$ cm$^{-1}$.

**Solution**    The frequency $f$ of oscillation of the detector signal is

$$f = 2\upsilon_M \bar{\nu}, \tag{P8.9}$$

where $\bar{\nu}$ is the wavenumber of the infrared source.

Using the data of this example and Equation P8.9, we find

$$f = 2\upsilon_M\bar{\nu} = 2 \times 10^{-3}\ \text{m/s} \times 1000 \times 100\ \text{m}^{-1} = 200\ \text{s}^{-1} = 200\ \text{Hz}.$$

**Control Exercise**    Calculate the frequency range of a modulated signal from a Michelson interferometer with a mirror velocity of 0.25 cm/s for infrared radiation of 15 μm.

**Control Exercise**    The Michelson interferometer is used for the analysis of yellow light of sodium ($\lambda = 569$ nm). Calculate the velocity of movement of the mirror if the frequency of a modulated signal is 5000 Hz.

## QUESTIONS AND PROBLEMS

**1.** Define the term "air quality."

**2.** Explain the method of infrared absorption.

**3.** What is the principle of the chemiluminescent method?

**4.** Explain the principle of operation of dichotomous system.

**5.** Describe the principle of differential optical absorption spectroscopy.

**6.** What is the principle of Fourier transform spectroscopy?

**7.** Describe principle of operation of miniature gas analysis systems.

## FURTHER READING

Choi, U.S., Sakai, G., Shimanoe, K., and Yamazoe, N. 2004. Sensing properties of $SnO_2$-$Co_3O_4$ composites to CO and $H_2$. *Sens. Actuators B.*, 98:166–173.

Gerboles, M., Buzica, D., and Plaisance, H. 2007. Examples of equivalence for the $NO_2$ membrane-closed Palmes tube and $O_3$ radial diffusive sampler. Workshop on demonstration of equivalence of ambient air monitoring methods, 2–4 May 2007, Ispra, Italy.

Johnson, D.L., Ambrose, S.H., Bassett, T.J., Bowen, M.L., Crummey, D.E., Isaacson, J.S., Johnson, D.N., Lamb, P., Saul, M., and Winter-Nelson, A.E. 1997. Meanings of environmental terms. *J. Environ. Qual.* 26:581–589.

Mandal, J. and Brandl, H. 2011. Bioaerosols in indoor environment – a review with special reference to residential and occupational locations. *The Open Environmental & Biological Monitoring Journal*, 4:83–96.

McCartney, E.J. 1983. *Absorption and Emission by Atmospheric Gases*. John Wiley & Sons, New York.

Toda, K. and Dasrupta, P.K. 2008. New applications of chemiluminescence for selective gas analysis. *Chem. Eng. Comm.* 195:82–97.

Yoon, J.-W., Son, K.C., Yang, D.S., and Kays, S.J. 2009. Removal of tobacco smoke under light and dark conditions as affected by foliage plants. *Kor. J. Hort. Sci. Technol.*, 27:312–318.

## ELECTRONIC REFERENCES

Air monitoring methods - Inorganic (IO) compendium methods. Determination of PM10 in ambient air using the Andersen continuous beta attenuation monitor. 1999. Center for Environmental Research Information. Office of research and development U.S. EPA Cincinnati, OH 45268, 1999, http://www.epa.gov/ttn/amtic/inorg.html (accessed July 1, 2013).

Air quality index (AQI) - A guide to air quality and your health, http://airnow.gov/index.cfm?action=aqibasics.aqi (accessed July 2, 2013).

Borse, R.Y. Synthesis and characterization of nanostructured ZnO thick film gas sensors prepared by screen printing method sensors & transducers. December 1, 2010, http://www.readperiodicals.com/201012/2240337141.html (accessed July 5, 2013).

Chapter XVII: Light-Scattering and Molecular Spectrophotometry. Ozone: Gas phase, http://www.ecs.umass.edu/cee/reckhow/courses/572/572bk17/572BK17.html (accessed July 8, 2013).

Compendium method IO-3.2. Determination of metals in ambient particulate matter using atomic absorption (AA) spectroscopy. Center for Environmental Research Information Office of Research and Development U.S. Environmental Protection Agency, Cincinnati, OH 45268. June 1999, http://www.epa.gov/ttn/amtic/files/ambient/inorganic/mthd-3-2.pdf (accessed July 12, 2013).

Locations for global ozone passive monitoring project, http://www.thesalmons.org/ozone/countries.html (accessed July 15, 2013).

McElroy, F., Mikel, D., and Nees, M. 1997. Determination of ozone by ultraviolet analysis. A new method for volume ii, ambient air specific methods, quality assurance handbook for air pollution measurement systems. Final draft, http://mattson.creighton.edu/Ozone/OzoneEPAMethod.pdf (accessed July 17, 2013).

Nitrogen dioxide in the United Kingdom. 2001. Main report: nitrogen dioxide in the United Kingdom. Chapter 4 Measurement methods and UK monitoring networks for $NO_2$, http://archive.defra.gov.uk/environment/quality/air/airquality/publications/nitrogen-dioxide/chapter4.pdf (accessed July 20, 2013).

Ozone Results. Creighton University. Part 6. EPA method, Determination of Ozone by Ultraviolet Analysis. Results of 3/4 December 2002, http://mattson.creighton.edu/Ozone/Ozone6.html (accessed July 22, 2013).

Poehler, D., Rippel, B., Stelzer, A., Mettendorf, K., Hartl, A., Platt, U., and Pundt, I. 2005. Tomographic DOAS measurements of the 2D trace gas distribution above the city centre of Heidelberg, Germany, http://www.cosis.net/abstracts/EGU05/08790/EGU05-J-08790.pdf (accessed July 24, 2013).

Quality Assurance Handbook, Vol. II, Part II 2.11–PM 10 High Volume. 2.11 Introduction September 1997, P.1–4, http://www.epa.gov/ttn/amtic/files/ambient/qaqc/2-11meth.pdf (accessed July 25, 2013).

Sulfur dioxide measurement, http://www.serinus-gas-analyzer.com/index.php/technical-information/sulfur-dioxide-so2-measurement (accessed July 27, 2013).

# 17

# INDOOR AIR QUALITY

## 17.1 INDOOR AIR

Humans spend almost all of their lifetime in closed spaces—at home and in nonresidential buildings (offices, commercial establishments, universities, schools, hospitals, enterprises) or inside motor vehicles, trains, ships, and airplanes. According to investigations in the United States and Europe, the population of industrialized countries spend more than 90% of their time indoors where air quality is often inferior to that outside (Snyder, 1990; Indoor Air Pollution, ... , 1994). This indoor environment is often contaminated with various air pollutants and the concentration of these pollutants may reach high levels due to the small spatial volume. Indoor air in cities has been reported to be as much as 100 times more polluted than that outdoors (Brown et al., 1994; Godish, 2001; Brown, 1997).

*Indoor air quality* takes into consideration the content and nature of interior air that affects the health and well-being of building occupants.

According to the World Health Organization (WHO's Programme on Indoor Air Pollution, 2002), every year, indoor air pollution is responsible for more than 1.6 million annual deaths and 2.7% of the global burden of disease.

The main pollutants found in indoor air include inorganic pollutants (carbon dioxide, carbon monoxide, nitrogen dioxide, sulfur dioxide, ozone), organic pollutants (volatile organic compounds, formaldehyde, pesticides, polynuclear aromatic hydrocarbons, polychlorinated biphenyls), physical pollutants (particulate matter, asbestos, man-made mineral fibers), environmental tobacco smoke, combustion-generated, microbial and biological contaminants, and radioactive pollutants (radon).

---

*Methods of Measuring Environmental Parameters*, First Edition. Yuriy Posudin.
© 2014 John Wiley & Sons, Inc. Published 2014 by John Wiley & Sons, Inc.

## 17.2   VOLATILE ORGANIC COMPOUNDS

One class of hazardous pollutants in indoor air is *volatile organic compounds (VOCs)*. VOCs are gases that are emitted from certain solids or liquids and include a variety of chemicals hazardous to human health; they have high enough vapor pressures under normal conditions to significantly vaporize and enter the atmosphere. The term "volatile" relates to the tendency of these compounds to vaporize at normal ambient temperatures and pressures due to their low boiling points.

It is possible to distinguish

*very volatile (gaseous) organic compounds* (VVPCs) which have boiling points ranging from <0°C to 50–100°C;

*volatile organic compounds* (VOCs) ranging from 50–100°C to 240–260°C;

*semivolatile organic compounds* (SVOCs) ranging from 240–260°C to 380–400°C;

organic compounds, associated with particulate matter or particulate organic matter (POM) with boiling point range >380°C.

The number of identified VOCs has increased progressively from more than 300 in 1986 to over 900 in 1989 (Maroni et al., 1995); nowadays it is possible to distinguish thousands of different VOCs.

## 17.3   SOURCES OF VOLATILE ORGANIC COMPOUNDS

Volatile organic pollutants emanate from wood panels, paint, occupants, pets, furnishings, floor covering, carpets, curtains, draperies and clothing, books, newspapers, and journals, electric shavers, portable CD players, liquid waxes, and certain adhesives (Posudin, 2010). Potential sources of VOCs include electronic devices such as copy machines, toners and printers, and heating, ventilating, and air-conditioning systems; household chemicals and materials, new construction and renovation, automobile exhausts and products, biological sources, and smoke.

Homehold materials include furnishings, floor covering, carpets, curtains, draperies and clothing, books, newspapers, and journals. Production and manufacture of wood furniture is accompanied with the technology of elaboration of wood surfaces with various chemicals that perform the protective or decorative functions. These chemicals (paints, lacquers, varnishes, and other coatings) can emit hazardous VOCs and present certain health risk.

Carpets are the potential sources of VOCs due to the materials that these carpets include synthetic fibers, latex components, and adhesives. Carpet is recognized not only as a potential source of VOCs in indoor air but can also serve as adsorptive sinks that provide reemission of VOCs over prolonged periods of time.

Homehold machines and devices as potential sources of VOCs include electronic devices such as copy machines, toners and printers, and heating, ventilating, and air-conditioning systems.

Nomenclature of homehold chemicals and materials includes waxes, room fresheners, deodorants, furniture polishes, lacquers, insecticides, cosmetics. The potential indoor air pollutants include automotive products, household cleaners, paint-related products, fabric and leather treatments, cleaners for electronic equipment, oils, greases and lubricants, and adhesive-related products, wallpaper glue, multicolor paint, floor wax, floor covering glue, polyurethane foam insulation.

New construction and renovation, newly erected and remodeled dwellings are responsible for the emission and high concentration of VOCs in the indoor air.

Automobile exhausts and products such as gasoline, oils, automobile fluids, various interior components (leather and fabric trims) of new cars are the potential sources of VOCs in indoor air, which are related to the garages that are either attached or located inside the building. Gasoline that is emitted from automobiles as uncombusted fuel and via evaporation is a source of VOCs which penetrate into automobile cabine from the roadway, thereby exposing commuters to higher levels than they would experience in other microenvironments. The levels of VOCs were related to intense traffic density and were inversely related to driving speed and wind speed. It was shown that total VOC is highly correlated with traffic air pollution.

Biological sources include building inhabitants that emit metabolic products. Human exhaled breath is also the source of VOCs. The most dangerous source of indoor pollution is smoking. The sources of VOCs in indoor air of biological origin include viruses, bacteria, molds, pollen, fungi, insects, bird droppings, cockroaches, flea, moth, rats and mouse, fungi, and animal feces. Ornamental plants can also be considered sometimes as the sources of VOCs (Yang et al., 2009a, 2009b).

Tobacco smoke contains about 4700 chemicals, including nicotine, tar, polycyclic aromatic hydrocarbons, vinyl chloride, phenols, and cadmium. The smoke can be separated into gas and particulate phases. Gaseous phase of tobacco smoke consists of carbon monoxide, carbon dioxide, nitrogen oxides, ammonia, volatile nitrosamines, hydrogen cyanide, volatile sulfur-containing compounds, volatile hydrocarbons, alcohols, and aldehydes and ketones. The environmental tobacco smoke provides a substantial contribution to concentrations of VOCs (Posudin, 2010).

It was shown that humans exhale VOCs. Major VOCs in the breath of healthy individuals include isoprene, acetone, ethanol, and methanol. Though human breath emissions are a negligible source of VOC on regional and global scales, they may become an important (and sometimes major) indoor source of VOC under crowded conditions.

Airborne biocontaminants include fungi, bacteria, dust mite allergens, cat allergens, and cockroach allergens.

## 17.4 EFFECT OF EXTERNAL FACTORS ON VOCs EMISSION IN INDOOR AIR

The concentration of volatile and semivolatile organic compounds depends in a complex manner on the interaction of sources of VOCs and internal conditions: structural situation inside the house, room climate, ventilation regime, air velocity, temperature, relative humidity, and the season (Volland et al., 2005).

Most air pollutants occur in typical concentration range of $1-1000$ $\mu g/m^3$ (Salthammer, 1999). Typically VOC concentrations in the buildings depend on a number of factors such as age of building, renovation, decoration, ventilation, season, location near traffic routes. Comparison of the concentrations of VOCs in indoor and outdoor air demonstrated that indoor air concentrations of many VOCs exceed those in outdoor air. It was shown that the average indoor VOC concentrations were higher in winter than those in summer. In winter, VOC concentration showed a trend of increasing because of closed windows. VOC concentrations in newly built houses had higher concentrations of chemicals related to solvents and building materials. Ventilation is the effective way of reducing indoor VOC concentration. There is a correlation between temperature, relative humidity, and concentration of VOCs.

## 17.5   HEALTH EFFECTS AND TOXICITY OF VOLATILE ORGANIC COMPOUNDS

### 17.5.1   Sick Building Syndrome

The most potential effects of VOCs are irritation of eyes and respiratory tract, narcotic action, and depression of the central nervous system. Many of the VOCs are known as human or animal carcinogens. The other health effects are related with the affect of heart, kidney, and liver (Hess-Kosa, 2002).

In spite of relatively low concentration, emission of VOCs can provoke *sick building syndrome* (*multiple chemical sensitivity* or *new house syndrome*) through the irritation of sensory systems, neurotoxic effects, skin irritation, nonspecific hypersensitivity reactions, odor and taste sensations (Ando, 2002).

### 17.5.2   Estimation of Health Effects of VOCs through the Questionnaires

Quite an interesting approach to assess the impact of VOCs on the human health was proposed by the investigators of Hokkaido University, Japan (Saijo et al., 2004, 2009; Takeda et al., 2009). The symptoms of sick building syndrome of 343 residents in 104 detached houses at Hokkaido region were surveyed (Takeda et al., 2009); a total of 429 dwellings in Sapporo and 135 in the environs of Sapporo were also analyzed (Saijo et al., 2004); the questionnaires were distributed to the occupants of 1240 dwellings which were all detached houses that have been newly built within 7 years (Takeda et al., 2009).

The questionnaire contained the information concerning personal characteristics and lifestyle of the residents of dwellings in Sapporo such as age, gender, current smoking, time spent in dwelling, working hours, stress levels, relation to asthma, allergies, and sick building syndrome symptoms. Simultaneously, the instrumental assessment of indoor environmental factors (VOC analysis in air samples, indoor air fungi, dust, dampness index, temperature, and humidity) was performed. The Japanese version of the program MM040EA was used to obtain statistical relations between the sick building syndrome symptoms and the environmental factors in newly built dwellings. It was concluded that a significant relation between VOCs and sick building syndrome of residents of newly built dwellings is a serious problem.

### 17.5.3  Principles of Phytoremediation

According to Prof. Stanley J. Kays, Department of Horticulture, University of Georgia, USA, and his colleagues, some of ornamental indoor plants demonstrate their ability to remove harmful VOCs from indoor air and accordingly to improve physical and psychological health of the inhabitants of closed spaces (Yoo et al., 2006; Kim et al., 2008, 2010, 2011, 2014; Yang et al., 2009a, 2009b; Yoon et al., 2009; Kays, 2011). This ability of plants to remove VOCs is called *phytoremediation*.

Researchers have proved that ornamental plants such as *Hemigraphis alternata* (purple waffle plant), *Hedera helix* (English ivy), *Hoya carnosa* (variegated wax plant), and *Asparagus densiflorus* (Asparagus fern) had the highest removal rates for all of the VOCs introduced; *Tradescantia pallida* (purple heart plant) was rated superior for its ability to remove four VOCs (November 4, 2009. Science News).

The advantages of plants that remove VOCs from indoor air are related with low cost of phytoremediation technology, which does not require the sources of electrical and thermal energy, provides continuous removing VOCs during day and night (Yang et al., 2009a).

## REFERENCES

Ando, M. 2002. Indoor air and human health. Sick-house syndrome and multiple chemical sensitivity. *Kokuritsu Iyakuhin Shokuhin Eisei Kenkyusho Hokoku*, 120: 6–38.

Brown, S. 1997. Volatile organic compounds in indoor air: sources and control. *Chem. Australia*, 64: 10–13.

Brown, S.K., Sim, M.R., Abramson, M.J., and Gray, C.N. 1994. Concentrations of volatile organic compounds in indoor air – a review. *Indoor Air*, 4(2): 123–134.

Godish, T. 2001. *Indoor Environmental Quality*. Lewis Publishers, Boca Raton-London-New York-Washington.

Hess-Kosa, K. 2002. *Indoor Air Quality. Sampling Methodologies*. Lewis Publishers, Boca Raton-London-New York-Washington.

Kays, S.J. 2011. Phytoremediation of indoor air – Current state of the art. In: *The Value Creation of Plants for Future Urban Agriculture* (ed. J.K. Kim), Nat. Inst. Hort. Herbal Science, RDA, Suwon, Korea, pp. 3–21.

Kim, K.J., Kil, M.J., Song, J.S., Yoo, E.H., Son, K.C., and Kays, S.J. 2008. Efficiency of volatile formaldehyde removal by indoor plants: Contribution of aerial plant parts versus the root-zone. *J. Amer. Soc. Hort. Sci.*, 133: 1–6.

Kim, K.J., Jeong, M.I., Lee, D.W., Song, J.S., Kim, H.D., Yoo, E.H., Jeong, S.J., Lee, S.Y., Kays, S.J., Lim, Y.W., and Kim, H.H. 2010. Variation in formaldehyde removal efficiency among indoor plant species. *HortScience*, 45: 1489–1495.

Kim, K.J., Yoo, E.H., Jeong, M.I., Song, J.S., Lee, S.Y., and Kays, S.J. 2011. Changes in the phytoremediation potential of indoor plants with exposure to toluene. *HortScience*, 46: 1646–1649.

Kim, K.J., Jung, H.H., Seo, H.W., Lee, J.A., and Kays, S.J. 2014. Volatile toluene and xylene removal efficiency of foliage plants as affected by top to root zone size. *HortScience*, 49(2): 230–234.

Maroni, M., Seifert, B., and Lindvall, T. (eds). 1995. *Indoor Air Quality. A Comprehensive Reference Book*. Elsevier, Amsterdam.

Saijo, Y., Kishi, R., Sata, F., Katakura, Y., Urashima, Y., Hatakeyama, A., Kobayashi, S., Jin, K., Kurahashi, N., Kondo, T., Gong, Y.Y., and Umemura, T. 2004. Symptoms in relation to chemicals and dampness in newly built dwellings. *Int Arch Occup Environ Health*, 77: 461–470.

Saijo, Y, Nakagi, Y., Ito, T., Sugioka, Y., Endo, H., and Yoshida, T. 2009. Relation of dampness to sick building syndrome in Japanese public apartment houses. *Environ. Health Prev. Med.*, 14: 26–35.

Salthammer, T. 1999. *Organic Indoor Air Pollutant: Occurrence-Measurement-Evaluation*. Wiley-VCH, New York.

Snyder, S.D. 1990. *Building Interior, Plants and Automation*. Prentice Hall, Englewood Cliffs, NJ.

Takeda, M., Saijo, S., Yuasa, M., Kanazawa, A., Araki, A., and Kishi, R. 2009. Relationship between sick building syndrome and indoor environmental factors in newly built Japanese dwellings. *Int. Arch. Occup. Environ. Health*, 82(5): 583–593.

Volland, G., Krause, G., Hansen, D., and Zoeltzer, D. 2005. Organic pollutants in indoor air – basics and problems. *Otto Graf Journal*, 16: 95–109.

Yang, D.S., Son, K.C., and Kays, S.J. 2009a. Volatile organic compounds emanating from indoor ornamental plants. *HortScience*, 44: 396–400.

Yang, D.S., Pennisi, S.V., Son, K.C., and Kays, S.J. 2009b. Screening indoor plants for volatile organic pollutant removal efficiency. *HortScience*, 44: 1377–1381.

Yoon, J.-W., Son, K.C., Yang, D.S. and Kays, S.J. 2009. Removal of tobacco smoke under light and dark conditions as affected by foliage plants. *Kor. J. Hort. Sci. Technol.* 27: 312–318.

# 18

# METHODS OF ANALYSIS OF VOLATILE ORGANIC COMPOUNDS

## 18.1 PRINCIPAL STAGES OF VOLATILE ORGANIC COMPOUNDS ANALYSIS

All the methods of volatile organic compound (VOC) analysis consist of several principal stages:

(1) sampling (e.g., trapping from the air and collection); (2) sample concentrating and enriching; (3) transfer of the VOCs from the air sample to the analytical device; (4) detection and identification of each compound.

There are two methods of sampling—active sampling that involves actively drawing air through a sorbent device, and passive sampling that utilizes the diffusion of the sample due to a concentration gradient (Hess-Kosa, 2002).

A sorbent is a material used to adsorb VOCs. Preliminary sample enriching permits to achieve the required sensitivity and selectivity of the analytical device. The most preferable method for enriching VOCs is application of *solid sorbents*. There are three main types of solid sorbents such as *inorganic sorbents*, *porous materials* based on carbon, and *organic polymers*. VOCs can be tested by carbon-based sorbents (activated charcoal, carbon molecular sieves, graphitized carbon black) and porous polymers (styrene polymers and phenyl-phenylene oxide polymers).

The principal ways to remove the sample from the sorbent are thermal desorption and solvent extraction. *Thermal desorption* means utilization of heat to increase the volatility of a compound that is analyzed and to transfer it from the sorbent to the analytical device. *Solvent extraction* is a method that makes it possible to separate compounds on the basis of their different solubilities in water and organic solvent.

*Methods of Measuring Environmental Parameters*, First Edition. Yuriy Posudin.
© 2014 John Wiley & Sons, Inc. Published 2014 by John Wiley & Sons, Inc.

## 18.2   GAS CHROMATOGRAPHY

*Gas chromatography (GC)* is an analytical technique that is based on vaporization of the sample and separation of mixtures by passing the mixture of gases dissolved in a mobile phase through a stationary phase.

The principle of operation of GC is based on the interaction of the gaseous component with the walls of the column that is covered with a stationary phase. The mechanisms of interaction between the components of the mixture and the stationary phase are discussed in Section 10.2.3. Each component can be extracted from the mixture at its proper *retention time* $t_R$ to come out of the gas chromatograph and to enter the detector system.

Correspondingly, the detector will give a signal of registration. With adequate separation, each component of the mixture is presented as a spectral peak. A time-based graphic record of the signals produced by all components is called a *chromatogram*.

The detector estimates the concentration of each component through the comparison of parameters (retention time and area of signal) of a sample being analyzed and standard sample with known concentration.

The typical gas chromatograph consists of a balloon with a carrier gas, an injector, an oven with a capillary column, and a detector (Figure 18.1).

GC involves the use of an inert gas as the mobile phase (nitrogen, helium, argon, and carbon dioxide are used as the so-called *carrier gas*).

The sample should be introduced into the heated injector, where the process of vaporization takes place, and then the sample is carried to the column via a syringe or valve.

There are two principal types of GC column. The *packed columns* consist of 300–500 plates of solid support material coated with liquid stationary phase. The length of the packed column is 1.5–6 m and internal diameter is 2–4 mm.

The *capillary (open tubular) columns* can be represented by the following types which differ in internal coating of the internal wall of the column: *wall-coated open tubular* (WCOT), *support-coated open tubular* (SCOT), or *fused silica open tubular*

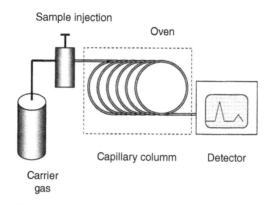

**FIGURE 18.1**   Principles of gas chromatography.

(FSOT) systems. The length of capillary columns varies from 5 to 100 m. Internal diameter is 0.15–0.53 mm. The capillary columns are more efficient than packed columns.

The temperature in the oven must be a little bit higher than the boiling temperature of the sample. The higher the temperature the shorter the retention time; the smaller the temperature the better the resolution. Column temperature must be controlled with high (±0.1–0.2°C) precision. The temperature programming can be used to separate mixtures with a broad boiling point range.

*The advantages of a gas chromatograph:*

- High sensitivity.
- The possibility to estimate simultaneously a wide number of VOCs.

*The disadvantage of a gas chromatograph:*

- Long response time (from minutes to hours).

## 18.3   DETECTION SYSTEMS

Detectors, that are used in GC, can be classified as non-selective detectors (respond to all gases except gas carrier), selective detectors (respond to gases with common physical and chemical properties), and specific detectors (respond to a single gas compound). Besides, the detectors can respond to concentration or mass flow. All detectors are characterized by different support gases, selectivity, detectability, and dynamic range.

### 18.3.1   Flame Ionization Detectors

The principle of operation of the flame ionization detector (FID) is described by Scott (1957). The gas from the chromatograph column enters the area of high temperature where the mixture of it with hydrogen and air in the gaseous state is created. This mixture is fed to the nozzle, where the process of combustion is supported by oxygen. Flame ionizes the gas in the space between the electrodes. Ionized particles reduce resistance and significantly increase the current, which is measured by a sensitive ammeter. The products of combustion are removed from the combustion zone.

*The advantages of the flame ionization detector:*

- High sensitivity.
- Low cost.
- Low maintenance requirements.
- Rugged construction.
- High linearity and wide detection ranges.

*The disadvantages of the flame ionization detector:*

- Require flammable gas.
- Destroy the sample.
- Cannot differentiate between different organic substances.
- FID flame oxidizes all compounds that pass through it.

## 18.3.2    Thermal Conductivity Detectors

Thermal conductivity detector is an electrical bridge (Wheatstone bridge) that contains four resistances. Two of the resistances are placed in the cells for measurement of thermal conductivity. The carrier gas (helium) is flowing through the first cell, while the mixture of the carrier gas and the sample is passing through the second cell. Four resistances of the bridge are fabricated with platinum or tungsten.

When both cells are filled with carrier gas, the bridge is balanced. When one of the cells is filled with the sample mixture, the bridge is unbalanced and the electric current goes through the diagonal of the bridge with a galvanometer. The carrier gas has very high thermal conductivity; the addition of the sample can change the conductivity of the sample cell (thermal conductivity of most gases is less than helium) and its resistance that depends on the temperature. The temperature change of resistance is proportional to the concentration of the compound. The sensitivity of the detector is about $10^{-6}$–$10^{-7}$ g/mL.

*The advantages of a thermal conductivity detector:*

- Easy to use and to maintain.
- Simplicity.
- Ability to estimate inorganic and organic compounds.
- Nondestructive action.
- The analyte can be collected after separation and detection.

*The disadvantages of a thermal conductivity detector:*

- Low sensitivity.
- Chemically active compounds and oxidizing substances can destroy the wires.

The flame ionization detector and the thermal conductivity detector are the most commonly used detectors. Other detection methods such as electron-capture detectors, atomic emission detectors, chemiluminescence detectors, photoionization detectors should be mentioned also.

## 18.4   MASS SPECTROMETRY

*Mass spectrometry (MS)* is an analytical technique for the separation of ionized atoms and molecules in electric and magnetic fields in vacuum according to their mass-to-charge ratio and identification of the chemical composition and structure of a sample.

A typical mass spectrometer contains an ion source (transformation of neutral molecules of a sample into ions); a mass analyzer (separation of ions by their masses and charges in applied electric and magnetic fields); and a detector (qualitative and quantitative estimation of sample compounds). There are several types of mass spectrometers.

### 18.4.1   Sector Field Mass Analyzer

This analytical device measures the mass-to-charge ratio of charged particles that are accelerated by an electric field and are separated based on their mass and charge in a magnetic field.

A beam of ions first passes through an electric field that acts as a velocity selector and then enters a uniform magnetic field ($\vec{B}$) directed into the paper (Figure 18.2). When a positively charged particle moves in a uniform external magnetic field with its initial velocity vector perpendicular to the field, the particle moves in a circle whose plane is perpendicular to the magnetic field. As the force ($\vec{F}_M$) deflects the particle, the directions of $\vec{v}$ and $\vec{F}_M$ change continuously. Therefore, the force ($\vec{F}_M$) is a centripetal force. From Newton's second law, we find that

$$F = qvB = \frac{mv^2}{r}, \tag{18.1}$$

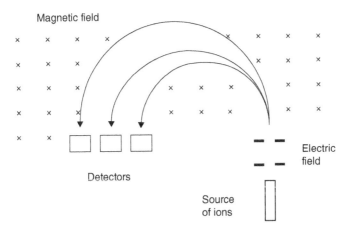

**FIGURE 18.2**   Sector field mass analyzer.

or

$$r = \frac{m\upsilon}{qB}. \tag{18.2}$$

A mass spectrometer separates atomic and molecular ions according to their mass-to-charge ratio. Upon entering the magnetic field, the ions move in a semicircle of radius $r$ before striking the target at point $P$. From Equation 18.2, we can express the ratio $m/q$ as

$$\frac{m}{q} = \frac{rB}{\upsilon} \tag{18.3}$$

Thus, the ions that pass through the system of electric and magnetic deflection, are separated in space corresponding to the mass-to-charge ratio $m/q$.

### 18.4.2   Quadrupole Mass Analyzer

This analyzer is used to separate the ions according to their mass-to-charge ratio, which is determined by the trajectories of the ions under the influence of an electric field. The quadrupole consists of four parallel electrodes of circular cross-section (Figure 18.3a). The combination of constant and a radio frequency fields $U_0 = U + V\cos\omega t$ (where $U$ is the direct voltage and $V$ is the amplitude of alternative voltage of the frequency $\omega$) applied to the electrodes. The ions are flying along the axis of the quadrupole and are beginning to oscillate under the influence of the field. If the amplitude of ion fluctuations does not exceed the transverse distance between the electrodes, those ions continue to travel in the quadrupole and can be detected (Figure 18.3b). The other ions with higher amplitude of oscillations are colliding with electrodes and are neutralized (Figure 18.3c). There are modifications of this type

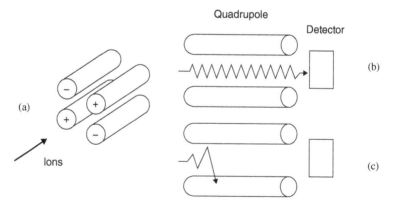

**FIGURE 18.3**   Quadrupole mass analyzer.

of mass spectrometer such as *quadrupole ion trap* and *linear quadrupole ion trap*, which are based on the same physical principles but differ by the ion trap systems.

The *time-of-flight analyzer* measures the time the ions take to reach the detector under the influence of an accelerating electric field. If the ions have the same charge, but different masses, the lighter ions will be detected first.

*The advantage of a mass spectrometer for the monitoring VOCs in the atmosphere:*

- High sensitivity and accuracy.

*The disadvantages of a mass spectrometer:*

- It requires a trained operator.
- High cost for the instrumentation.


## 18.5   COMBINATION OF GAS CHROMATOGRAPHY AND MASS SPECTROMETRY

*Gas chromatography–mass spectrometry (GC/MS)* is a combination of GC and MS methods that is used to identify different substances within a test sample. This method is effective in separating compounds into their various individual components and identification of the specific substances.

The GC/MS method combines the capabilities and advantages of both (GC and MS) analytical approaches. The gas mixture is separated into components by gas chromatograph according to the retention time of each component forming the chromatogram. After entering the mass spectrometer these components are captured, ionized, and detected. Thus, each peak of chromatogram is resolved into the consequences of mass spectrum components according to their mass-to-charge ratio (Figure 18.4). GC/MS technique combines high resolution separation of components with very selective and sensitive detection and in such a way makes it possible to

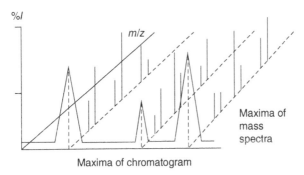

**FIGURE 18.4**   GC/MS spectra. Three-dimensional plot of scan number (time) versus mass/charge (*m/z*) versus relative intensity (%).

achieve such a high level of identification of unknown gas that cannot be realized by GC or MS alone (Grob and Barry, 2004).

Possible applications of GC/MS method in VOCs monitoring are reflected in a number of papers (Yoshida et al., 2004; Hodgson, 1995; Dai et al., 2005; Stachowiak-Wencek and Pradzynski, 2005; Zhu and Cao, 2000).

*The advantages of a GC/MS system:*

- High sensitivity.
- Fast process of measurements.

*The disadvantages of a GC/MS system:*

- It requires a trained operator.
- Not all peaks are calibrated.
- High cost for the instrumentation.

## 18.6  PHOTOACOUSTIC SPECTROSCOPY

This method is based on irradiation of the sample with a laser beam that induces localized heating and thus a pressure wave emitted from the sample.

The *photoacoustic effect* (it was discovered by Alexander Graham Bell in 1880) involves the sequence of the following steps: absorption of modulated or pulsed laser radiation by either opaque or transparent sample; temporal changes of the sample temperature at the place of irradiation; expansion and contraction of the sample which induce the corresponding pressure changes in those sample regions where absorption occurred; propagation of the pressure changes which are perceived as sound within the sample body; and measurement of these changes by a detector (microphone or piezoelectric sensor).

By measuring the dependence of the sample pressure on the wavelength, the photoacoustic spectrum of a sample can be recorded and used to identify the absorbing components within the sample. A photoacoustic spectrometer (Figure 18.5) consists of a laser, a modulator, a semitransparent plate, a camera, a detector, a microphone, an amplifier, and a synchronous detector.

The photoacoustic cell is a small vessel equipped with the transparent window. Sample inlet and outlet, and a detector (Figure 18.6). The cell is mounted on anti-vibration table to limit noise.

The photoacoustic spectroscopy is useful to estimate extremely low concentrations of substances (Solid et al., 1996; Wolff et al., 2005; Krueger et al., 1995).

*The advantages of a photoacoustic spectrometer:*

- High sensitivity (at the range ppbv–pptv).
- Nondestructive action.
- Good selectivity.

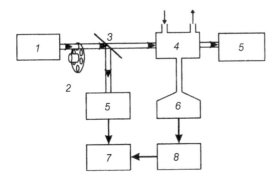

**FIGURE 18.5**    Principles of photoacoustic spectroscopy.

- Simplicity.
- Can analyze opaque samples.
- Insensitive to scattered radiation.
- No need for a photoelectric detector.
- Variety of samples.

*The disadvantages of a photoacoustic spectrometer:*

- Source of energy must be sufficient.
- The window of the sampling cell must be transparent.
- Background noise can hamper the acoustic measurements.

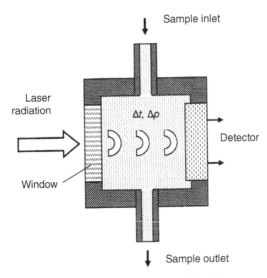

**FIGURE 18.6**    Photoacoustic cell.

## 18.7  PROTON TRANSFER REACTION MASS SPECTROMETRY

Proton transfer reaction mass spectrometry (PTR-MS) utilizes the chemical ionization that is based on proton-transfer reactions; $H_3O^+$ is used as the reagent ion. The air sample is continuously drawn into a reaction chamber of the PTR-MS where it encounters the reagent ion (Figure 18.7).

VOCs containing a polar functional group or unsaturated bonds have proton affinities larger than that of $H_2O$ and therefore will react with $H_3O^+$ in a proton transfer reaction where a proton is transferred between $H_3O^+$ and the VOC.

This reaction can be described by the following formula:

$$H_3O^+ + V \rightarrow V \cdot H^+ + H_2O, \tag{18.4}$$

where $V$ is molecule of VOC in the analyte.

PTR-MS provides real-time, online quantification of VOCs in indoor air. The instrument is characterized with a fast response time (about 1 second).

A PTR-MS system has been developed which allows online measurements of trace components with concentrations as low as a few parts per trillion by volume (pptv).

PTR-MS technology makes it possible to measure VOCs which have high proton affinity such as methanol, acetaldehyde, acetone, Me tert-Bu ether (MTBE), benzene, toluene, ethanol, xylene present in ambient air (Tani, 2003; 2007; Kato and Kajii, 2004; Lindinger et al., 1998, 2001, 2005; Hansel, 1998, 2004; Tanimoto and Inomata, 2006; Rogers et al., 2006).

*The advantages of a PTR-MS system:*

- Real-time measurements.
- Real-time quantification.

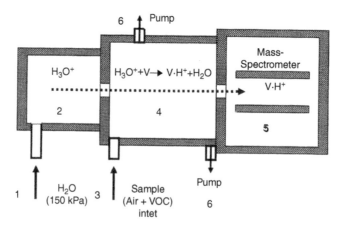

**FIGURE 18.7**    Proton transfer reaction mass spectrometer. 1, inlet of $H_2O$ vapor; 2, reaction chamber; 3, inlet of air with VOCs; 4, drift chamber; 5, entrance of mass spectrometer; 6, pump.

- Soft ionization provides low fragmentation and improves the identification capability.
- No sample preparation is necessary.
- Compact and robust setup.
- Easy to operate.
- Fast process of measurements.

*The disadvantage of a PTR-MS system:*

- Not all molecules are detectable.

The commercial IONICON PTR-MS (Ionicon Analytik Ges.m.b.H., Innsbruck, Austria) has the following parameters:

- Mass range 1–300 amu.
- Resolution <1 amu.
- Response time 100 ms.
- Limits of detection 300 pptv.
- Linearity range 10 ppmv–300 pptv.

## 18.8   FOURIER TRANSFORM INFRARED SPECTROSCOPY OF VOLATILE ORGANIC COMPOUNDS

Principles of Fourier Transform Spectroscopy are considered in Section 16.5.2. Here we shall discuss the application of Fourier transform infrared spectroscopy (FTIR) for estimation of the indoor air quality, particularly for quantitative and qualitative analysis of VOCs.

FTIR is an analytical system which is used to convert the raw data into the actual infrared spectrum of a gas through such mathematical process as Fourier transform that converts an amplitude–time spectrum to an amplitude–frequency spectrum, or vice versa.

Fourier transform infrared spectrometer consists of a source of infrared radiation, an interferometer, an enclosed sample cell of known absorption wavelength, on infrared detector, and computer system.

Detailed procedure of the application of FTIR spectroscopy for the analysis of VOCs can be found in the recommendations of the EPA (EPA Test Method 320).

The fact is that VOCs are characterized by the chemical bonds that absorb infrared radiation at specific wavelengths. The dependence of the intensity of spectral lines of VOC, which is analyzed, on the wavelength can be presented as peculiar "fingerprint" that enables to identify this compound with high accuracy.

The sensitivity of FTIR is ranging from very low parts per million (ppm) to high percent (%) levels.

The examples of the application of FTIR spectroscopy to indoor air monitoring of VOC's and semi-volatiles at very low levels are presented in a number of papers (Lee et al., 1993; Davis et al., 1998; Xu et al., 2006; Wu et al., 2007).

*The advantages of an FTIR spectrometer:*

- High sensitivity (below 1 ppm and, in many cases, down to 0.1 ppm or lower).
- Non-destructive action.
- Can measure up to 30 or more compounds simultaneously.
- Can measure most VOCs that absorb infrared radiation.
- The procedure of measurements is very fast.

*The disadvantages of an FTIR spectrometer:*

- Insufficient sensitivity for the monitoring of background concentration.
- Difficult calibration.

## QUESTIONS AND PROBLEMS

1. What compounds are the major pollutants of air?

2. Define VOCs.

3. Name the main sources of VOCs.

4. Explain the principles of the GC.

5. Explain the principles of the MS.

6. What is the advantage of combining GC and MS?

7. What is photoacoustic spectroscopy?

8. Explain the principle of PTR-MS.

## REFERENCES

Dai, H., Piao, F., Zhong, L., Asakawa, F., and Jitsunari, F. 2005. Investigation of trends and risk of indoor air pollution by VOCs and HCHO. *Dalian Yike Daxue Xuebao.*, 27(5): 337–340.

Davis, L.C., Vanderhoof, S., Dana, J., Selk, K., Smith, K., Goplen, B., and Erickson, L.E. 1998. Movement of chlorinated solvents and other volatile organics through plants monitored by Fourier transform infrared (FT-IR) spectrometry. *Journal of Hazardous Substance Research*, 4:1–26.

Grob, R.L. and Barry, E.F., eds. *Modern Practice of Gas Chromatography*. 2004. 4th edition. John Wiley & Sons, Inc., New Jersey.

Hansel, A. 2004. Proton transfer mass spectrometer. *Europhysics News*, 35(6): 197–199.

Hansel, A., Jordan, A., Warneke, C., Holzinger, R., and Lindinger, W. 1998. Improved detection limit of the proton-transfer reaction mass spectrometer: online monitoring of volatile organic compounds at mixing ratios of a few PPTV. *Rapid Communications in Mass Spectrometry*, 12(13): 871–875.

Hodgson, A.T. 1995. A review and a limited comparison of methods for measuring total volatile organic compounds in indoor air. *Indoor Air*, 5(4): 247–257.

Hess-Kosa, K. 2002. *Indoor Air Quality. Sampling Methodologies*. Lewis Publishers, Boca Raton-London-New York-Washington.

Kato, S. and Kajii, Y. 2004. Measurement of volatile organic compounds by proton transfer reaction mass spectrometry. *Shinku*, 47(8): 600–605.

Krueger, U., Kraenzmer, M., and Strindehag, O. 1995. Field studies of the indoor air quality by photoacoustic spectroscopy. *Environ Int*, 21(6): 791–801.

Lee, K.A.B., Hood, A.L., Clobes, A.L., Schroeder, J.A., Ananth, G.P., and Hawkins, L.H. 1993. Infrared spectroscopy for indoor air monitoring. *Spectroscopy*, 8(5): 24–29.

Lindinger, W. and Jordan, A. 1998. Proton-transfer-reaction mass spectrometry (PTR-MS): online monitoring of volatile organic compounds at pptv levels. *Chem Soc Rev*, 27(5): 347–354.

Lindinger, W., Fall, R., and Karl, T.G. 2001. Environmental, food and medical applications of proton-transfer-reaction mass spectrometry (PTR-MS). *Advances in Gas Phase Ion Chemistry*, 4: 1–48.

Lindinger, C., Pollien, P., Ali, S., Yeretzian, C., Blank, L., and Maerk, T. 2005. Unambiguous identification of volatile organic compounds by proton-transfer reaction mass spectrometry coupled with GC/MS. *Anal Chem*, 77(13): 4117–4124.

Rogers, T.M., Grimsrud, E.P., Herndon, S.C., Jayne, J.T., Kolb, C.E., Allwine, E., Westberg, H., Lamb, B.K., Zavala, M., Molina, L.T., Molina, M.J., and Knighton, W.B. 2006. On-road measurements of volatile organic compounds in the Mexico City metropolitan area using proton transfer reaction mass spectrometry. *Int J Mass Spectrom*, 252(1): 26–37.

Scott, R.P.W. 1957. *Vapour Phase Chromatography* (ed. D.H. Desty). Butterworths, London.

Solid, J.E., Trujillo, V.L., Limback, S.P., and Woloshun, K.A. 1996. Comparison of photoacoustic radiometry to gas chromatography/mass spectrometry methods for monitoring chlorinated hydrocarbons. Proceedings of the 89th Annual Meeting (held June 23-28 in Nashville, TN), Air and Waste Management Association, Pittsburgh, Pennsylvania, June 1996. 89th wa6508/1-wa6508/15.

Stachowiak-Wencek, A. and Pradzynski, W. 2005. Investigations of volatile organic compounds during finishing furniture surfaces as well as from furniture coating. *Pol. Ann WULS-SGGW, For and Wood Technol*, 57: 220–224.

Tani, A. 2003. Novel rapid analytical method for VOCs. Proton transfer reaction mass spectrometry. *Taiki Kankyo Gakkaishi*, 38(4): A35–A46.

Tani, A., Kato, S., Kajii, Y., Wilkinson, M., Owen, S., and Hewitt, N. 2007. A proton transfer reaction mass spectrometry based system for determining plant uptake of volatile organic compounds. *Atmospheric Environment*, 41(8): 1736–1746.

Tanimoto, H. and Inomata, S. 2006. Development and future direction of an instrument for real-time measurements of volatile organic compounds in air. Proton transfer reaction – time of-flight mass spectrometry (PTR-TOFMS). *Shigen Kankyo Taisaku*, 42(8): 61–66.

Wolff, M., Groninga, H.G., Dressler, M., and Harde, H. 2005. Photoacoustic sensor for VOCs: First step towards a lung cancer breath test. Proceedings of SPIE-The International

Society for Optical Engineering. 5862 (Diagnostic Optical Spectroscopy in Biomedicine III). 58620G/1-58620G/7. 12–16 June, 2005, Munich, Germany. doi: 10.1117/12.633019

Wu, C.-Y., Chin-Chun, C.-C., and Chung, T.W. 2007. Dynamic determination of the effluent concentrations of the acetates in an adsorption column packed with zeolite 13X using a gas phase FT-IR. *J Chem Eng Japan*, 40(10): 817–823.

Xu, L.-H., Feng, Y.-Q., and Chen, J.-Q. 2006. Study on simultaneous analysis of indoor air multi-component VOCs with FTIR. *Guangpuxue Yu Guangpu Fenxi*, 26(12): 2197–2199.

Yoo, M.H., Kwon, Y.J., Son, K.C., and Kays, S.J. 2006. Efficacy of indoor plants for the removal of single and mixed volatile organic pollutants and physiological effects of the volatiles on the plants. *J. Amer. Soc. Hort. Sci.*, 131: 452–458.

Yoshida, T., Matsunaga, I., and Oda, H. 2004. Simultaneous determination of semivolatile organic compounds in indoor air by gas chromatography-mass spectrometry after solid-phase extraction. *J Chromatogr.* 1023(2): 255–269.

Zhu, J.P. and Cao, X.L. 2000. A simple method to determine VOC emissions from constant emission sources and its application in indoor air quality studies. Air Conditioning in High Rise Buildings 2000, International Symposium, Shanghai, China, October 24–27, 2000. pp. 204–209.

## FURTHER READING

Hunter, R. and Oyama, S.T. 2000. *Control of Volatile Organic Compound Emissions. Conventional and Emerging Technologies.* John Wiley & Sons, Inc.

Kay, J.G. 1991. *Indoor Air Pollution: Radon, Bioaerosols, and VOCs.* Lewis Publishers.

Leslie, G.B. 1994. *Indoor Air Pollution: Problems and Priorities.* Cambridge University Press.

Michaelian, K.H. 2010. *Photoacoustic IR Spectroscopy: Instrumentation, Applications and Data Analysis.* 2nd edition. Wiley-VCH.

Oyama, S.T. and Hunter, P. 2000. *Control of Volatile Organic Compound Emissions: Conventional and Emerging Technologies USA.* Wiley-Interscience.

Parmar, G.R. and Rao, N.N. 2006. Recent advances in monitoring techniques for volatile organic compounds. *Journal of Indian Association for Environmental Management*, 33(3): 127–136.

Pluschke, P. 2004. *Indoor Air Pollution (Handbook of Environmental Chemistry).* Springer.

Poole, C.F., ed. 2012. *Gas Chromatography*, 1st edition. Elsevier.

Scott, R.P.W. 2012. *Gas Chromatography (Chrom-Ed Book Series).* The Reese-Scott Partnership.

Sparkman, D.O., Penton, Z., and Kitson, F.G. 2011. *Gas Chromatography and Mass Spectrometry: A Practical Guide*, 2nd edition. Academic Press.

Zhang, Y. 2005. *Indoor Air Quality Engineering.* CRC Press, Boca Raton.

## ELECTRONIC REFERENCES

*Indoor Air Pollution: An Introduction for Health Professionals.* 1994, http://www.epa.gov/iaq/pubs/hpguide.html (accessed August 1, 2013).

Posudin Yuriy, Smoking: historical, medical and social aspects. eKMAIR.University Library, National University of «Kyiv-Mohyla Academy», 2010, http://www.ekmair .ukma.kiev.ua/handle/123456789/1722 (accessed August 8, 2013).

Posudin, Yuriy. Volatile Organic Compounds in Indoor Air: Scientific, Medical and Instrumental Aspects/eKMAIR.University Library, National University of «Kyiv-Mohyla Academy», 2010, http://www.ekmair.ukma.kiev.ua/handle/123456789/885 (accessed August 15, 2013).

WHO's Programme on Indoor Air Pollution. 2002, www.who.int/indoorair/contact/en/ index.html (accessed August 10, 2013).

EPA Test Method 320: Measurement of Vapor Phase Organic and Inorganic Emissions by Extractive Fourier Transform Infrared (FTIR) Spectroscopy, http://www.midac.com/ files/Ap-139.PDF (accessed August 15, 2013).

Common plants can eliminate indoor air pollutants. November 4, 2009. Science News, http://www.sciencedaily.com/releases/2009/11/091104140816.htm (accessed August 22, 2013).

# PART III

## HYDROGRAPHIC FACTORS

# 19

# WATER QUALITY

*Hydrographic factors* are the physical and chemical properties of water as they relate to water as a habitat for living organisms.

## 19.1   WATER RESOURCES

It is known that the hydrosphere contains about 1.36 billion $km^3$ of water. About 97.2% of the world's water is saline water in the seas and oceans. The remaining part consists of fresh water which exists as surface water, groundwater, and vapor in atmosphere (0.65%) and frozen water in glaciers and ice sheets (2.15%) (Van der Leeden et al., 1990).

It is estimated by water consumption statistics (Worldometers) that 70% of world-wide water is used for agriculture, 20% for industry, and 10% for domestic purposes.

The water in the natural environment is involved in various processes of the global water cycle such as evaporation, transpiration, condensation, precipitation, and ice formation.

## 19.2   PROPERTIES OF WATER

*Water* is the simplest stable chemical compound of hydrogen and oxygen; under normal conditions water is a colorless and odorless substance which can exist in liquid, solid, and gaseous forms. Water is the richest component of the biosphere that

*Methods of Measuring Environmental Parameters*, First Edition. Yuriy Posudin.
© 2014 John Wiley & Sons, Inc. Published 2014 by John Wiley & Sons, Inc.

covers about three-quarters of the Earth's surface; it is an important component of living organisms which use water for metabolic processes.

*Mechanical properties* of water include a density which is 999.8 kg/m$^3$ at 0°C (32°F). Maximal value of water density is 1000.0 kg/m$^3$ at 4°C (39.2°F). The fact is that water is the only compound that expands when freezes and therefore the density of ice is less than the density of water. That is why the ice is on the surface of water, which makes it possible to preserve the viability of aquatic organisms in cooling.

*Hydrodynamic properties* of water are related to its ability to solve large number of compounds. This property of water contributes to a better supply of nutrients to living organisms. From the environmental point of view water is an efficient solvent that can be used for extraction and dissolution of most natural and man-made waste. Water is characterized by a high surface tension (72.75 × 10$^{-3}$ N/m at 20°C) which is responsible for transport of nutrients from soil to plant due to the capillary phenomena in soil pores and stems of the plants.

*Thermophysical properties* of water are characterized by freezing point of water (273.16 K or 0°C), boiling point (373.16 K or 100°C), high heat of vaporization (2.26 × 10$^6$ J/kg). This means that water molecules absorb large amounts of heat during the evaporation due to the solar radiation and release a large amount of heat due to the condensation and precipitation on the Earth's surface. Every day the Sun forces to leave the Earth's surface about 1230 km$^3$ of water due to evaporation from the seas, lakes, rivers, soil, and transpiration from the vegetation. Heating of water reservoirs and atmospheric vapor leads to climate and weather changes.

Water has an extremely high heat capacity (4186 J/kg·K), which is three to four times higher than the heat capacity of the soil. As the water should get more heat than the soil for uniform heating, it is heated and cooled very slowly. Thus, significant massifs of water are not affected by rapid changes in temperature; this situation leads to a moderate climate on Earth.

There is a layer of dry air (thermal conductivity is 0.0257W/m K at 20°C) between the atmospheric water vapor (0.599 W/m·K at 20°C) and soil (0.125–0.209 W/m·K at 20°C), which performs the function of a heat-insulating layer and plays an essential role in the greenhouse effect.

Water particles are in the state of continuous movement and turbulent mixing. This specific feature of water provides the transfer of heat into the deep layers.

*Electric properties* of the water include its high dielectric constant. This feature indicates that the electric forces between dissolved substances in water are weak. The dielectric constant of water is almost a constant ($\varepsilon = 80$), while the soil dielectric constant is very sensitive to the volumetric soil moisture. This property is used during the determination of soil moisture.

In addition, non-polar water molecules are characterized by dipole moment, which determines the ability of water to absorb long-wave radiation of the Earth's surface and participate in the greenhouse effect.

*Optical properties* of water are based on its ability to interact with solar radiation. Water is a transparent body for light emission in the visible range. The absorption coefficient of distilled water is characterized by a minimum in the region of 400–500 nm, the scattering coefficient shows a decline from short-wave to long-wave parts of the spectrum. Therefore, optical radiation penetrates to the considerable depth, which leads to heating water within the layer thickness of several meters. The dissolved salts and organic compounds ($SiO_2$, $Fe_2O_3$, $Al_2O_3$), bacteria, phyto- and zooplankton, and muddy substances affect optical properties such as absorption and scattering.

## 19.3 CLASSIFICATION OF WATER

*Drinking water* is water that is not harmful to human health and corresponds to the requirements of the quality standards.

*Surface water* is water that collects on the ground or in various water basins and streams. The main disadvantage of surface water is its ability to be contaminated by industrial and municipal emissions and runoff, soil erosion. Excess supply of nutrients provokes algal outbreaks, leading to eutrophication of water bodies. Excessive turbidity and pollution cause numerous unwanted taste, odor, and color of water.

*Groundwater* is water located beneath the Earth's surface. Groundwater penetrates through soil pores of the surface layer. Soil or mineral formations are saturated with water, and create an aquifer—underground layer of water-bearing permeable rock or unconsolidated materials. Typically, water is pumped from these horizons through the wells. Groundwater can also be contaminated, but the process of groundwater remediation is long-term and complex. Most pathogens are excluded from the groundwater by soil particles due to their filtering action. That is why it is not necessary to apply a lot of effort to bring the quality of groundwater to drinking water standards. Well water, despite its limited amount, is characterized by uniform quality and transparency, but requires a water-softening procedure.

## 19.4 QUALITY OF WATER

*Water quality* is based on its physical, chemical, and biological characteristics that are considered in comparison with existing standards.

Requirements for the quality of water depend on its destination. Naturally, the most serious quality criteria relate to drinking water, as the quality of this water is related to human health. The main sources of freshwater in natural conditions include groundwater, surface water, and precipitation. That is why the quality of groundwater and surface water affects the quality of drinking water. For environmental water attention focuses on the protection of natural water bodies from pollution and eutrophication. Water intended for agricultural purposes must ensure the safety of crops and domestic animals.

## 19.5 WATER QUALITY PARAMETERS

*Water quality parameters (indicators)* are numerical values that provide information about the trends or changes of water quality over time that can be measured. Water quality parameters include physical, chemical, and biological measures of water quality. The normal presence, appearance, or absence of such parameters can be used for the identification of a change in water quality.

### 19.5.1 Drinking Water Quality Parameters

Water monitoring in the United States began with the Clean Water Act (1948) and the Safe Drinking Water Act (1974). These acts have established the so-called Primary and Secondary Drinking Water Standards (USEPA, 2000).

*Primary Standards* include standards for inorganic and organic compounds, by-products of disinfectants, radionuclides, microorganisms, and volatile and synthetic organic compounds.

*Secondary Standards*, according to USEPA, are non-enforceable, maximum contaminant levels that take into account public welfare criteria (e.g., taste, color, corrosivity, odor) rather than health effects.

The Standards of the European Union (1998) include chemical, quality, and microbiological parameters.

The Standards of the World Health Organization (1993) for drinking water emphasize cationic and anionic composition of drinking water, microbiological, and other parameters.

Drinking water quality parameters can be classified as follows:

- Physical parameters:

Total suspended solids (TSS); turbidity; color.

- Chemical parameters:

Inorganic chemicals (e.g., nitrates, heavy metals); organic chemicals (e.g., insecticides, herbicides); solvents, disinfectants, and disinfection by-producers; volatile and synthetic organic compounds.

- Radionuclides.
- Microorganisms:

Viruses; pathogenic bacteria; protozoa; toxic algae.

### 19.5.2 Groundwater Quality Parameters

Groundwater has its own specificity. On the one hand, groundwater is filtered by the soil components, on the other hand, it is considerably contaminated from industrial, agricultural, and domestic chemicals that enter from the surface. As a result, the level of dissolved substances in the groundwater is higher than in the surface waters.

Groundwater is characterized by the following quality parameters:

- Physical parameters:

Total dissolved solids (TDS); turbidity; pH.

- Chemical parameters:

Hardness ($Ca^{+2}$ and/or $Mg^{+2}$); iron ($Fe^{+2}$, $Fe^{+3}$); ammonia ($NH_3$); nitrate ($NO_3-$) and nitrite ($NO_2^{-2}$); silica ($SiO_2$); sulfide ($S^{-2}$) and sulfate ($SO_4^{-2}$); pesticides.

- Radionuclides:

Radon and radium.

- Microorganisms:

Bacteria.

### 19.5.3    Surface Water Quality Parameters

Surface water quality should be assessed directly *in situ* and in real time. Such a technology makes it possible to measure very precisely water quality parameters and changes of its physical and chemical properties by inserting sensors of the measuring device directly into the water. The following water quality parameters that can be measured directly in water are used (Artiola et al., 2004):

- pH.
- Dissolved oxygen.
- Redox potential.
- Turbidity.
- Electric conductivity.
- Temperature.
- Stream flow.

### 19.6    EFFECT OF WATER QUALITY ON HUMAN HEALTH

According to the UN "Water quality facts and statistics" (2010), worldwide, 2.5 billion people live without adequate sanitation (UNICEF WHO, 2008); unsafe or inadequate water, sanitation, and hygiene cause approximately 3.1% of all deaths worldwide (WHO 2002); unsafe water causes 4 billion cases of diarrhea each year, and results in 2.2 million deaths, mostly of children under five (WHO and UNICEF, 2000).

More than a billion people—almost one-fifth of the world's population—lack access to safe drinking water, and 40% lack access to basic sanitation, according to the second UN World Water Development Report (WWDR2), 2006 "Water, a Shared Responsibility."

Clean water provides the necessary human existence. Contamination of groundwater, surface water, and precipitation provokes a corresponding deterioration of the quality of drinking water. Hazardous microorganisms are transferred into drinking water and cause dangerous for human health diseases. Among them we should mention cholera that is caused by the bacterium *Vibrio cholerae*; typhoid fever that is induced by the bacteria *Salmonella enterica* and serovar *Salmonella Typhi*; *Escherichia coli* infections that can happen after drinking water that has been contaminated by feces; dysentery (bacteria of the genus *Shigella*, or amoeba *Entamoeba histolytica*); cryptosporidiosis (protozoan parasite in the phylum *Apicomplexa*); hepatitis A (hepatitis A virus (HAV)); and giardiasis (protozoa *Giardia lamblia*).

Some toxins (chemicals or minerals) that are present in water are responsible for such diseases as cancer, kidney damage, nervous disorders, tooth decay, skin irritation, birth defects, and fertility problems.

According to the World Health Organization, waterborne diseases that are caused by pathogenic microorganisms provoke about 1.8 million human deaths annually. The World Health Organization estimates that 88% of such a situation is attributable to unsafe water supply, sanitation, and hygiene.

## REFERENCES

Artiola, J.F., Pepper, I.L. and Brusseau, M.L. 2004. *Environmental Monitoring and Characterization.* Elsevier Academic Press, San Diego, CA.

United States Environmental Protection Agency (USEPA). 2000. Drinking Water Standards and Health Advisories. U.S. Environmental Protection Agency. Office of Water 4304 EPA 822-B-00-001. Summer 2000.

Van der Leeden, F., Troise, F.L., and Todd, D.K. 1990. *Water Encyclopedia,* 2nd edition. Lewis Publishers, Chelsea, MI.

# 20

# MEASUREMENT OF WATER QUALITY PARAMETERS

## 20.1 *IN SITU* MEASUREMENT OF WATER QUALITY PARAMETERS

The main water quality parameters that are measured directly in natural water are pH, dissolved oxygen (DO) concentration, redox potential, turbidity, salinity, temperature, and stream flow.

### 20.1.1   pH value

pH is a quantitative measure of acidity or alkalinity of a water solution. The level of acidity or alkalinity is determined by the relative number of hydrogen ($H^+$) or hydroxide (OH) ions. This special measure pH is defined as the negative logarithm of hydrogen ion concentration $pH = -\lg [H^+]$ or hydroxide ion concentration $pOH = -\lg [OH^-]$.

Water solution that is neutral has equal concentrations of hydrogen and hydroxide ions, and at 25°C its pH = pOH = 7. Solutions with a pH <7 are acidic, while with a pH >7 are alkaline.

### 20.1.2   Measurement of pH of Water

The operative principle of a pH-metry is described in Section 10.2.1. A typical pH meter consists of one electrode which is inserted into a liquid that has a fixed pH, and the other electrode which responds to the acidity of the water sample. The difference between the voltages of the two electrodes is measured by an electronic meter. In

*Methods of Measuring Environmental Parameters*, First Edition. Yuriy Posudin.
© 2014 John Wiley & Sons, Inc. Published 2014 by John Wiley & Sons, Inc.

such a way, the pH is determined by the transfer of hydrogen ions between chemical species. The shortcoming of measuring pH of the water is that a change of pH does not identify the specific type of contamination.

### 20.1.3   Concentration of Dissolved Oxygen

The concentration of water that is in equilibrium with the atmosphere (DO) decreases with increasing temperature from 14.5 mg to 5 mg of $O_2$ per liter. In addition, DO depends on salinity and latitude. DO enters the water from the atmosphere through the processes of diffusion, aeration, and as a product of the photosynthetic activity of vegetation. Analysis of DO is very important in that it provides information on the biological and biochemical processes occurring in water.

Level of DO (mg/L) influence on living organisms consists of those regions (Behar, 1997):

0–2 mg/L: not enough oxygen to support life;

2–4 mg/L: only a few fish and aquatic insects can survive;

4–7 mg/L: good for many aquatic animals, low for cold water fish;

7–11 mg/L: very good for most stream fish.

If the amount of DO drops below normal levels in water bodies, the phenomenon as *eutrophication* takes place. It is accompanied with an increase in the rate of supply of organic matter and deterioration of water quality, which affects the survival of aquatic organisms.

### 20.1.4   Measurement of Dissolved Oxygen

DO is commonly measured using an electrometric technique such as a *Clark electrode* (Figure 20.1).

The Clark electrode consists of a platinum cathode in the form of a disk that supports a potential of −0.6 V with respect to the annular silver anode and a thin (about 20 μm thick) Teflon membrane. The electrodes are immersed in a buffered electrolyte solution (KCl). The gap between the electrodes and the membrane, which is filled with the solution, has a thickness of about 10 μm.

When the device is immersed in water in which the quality has been analyzed, molecular oxygen diffuses through the membrane and the electrolyte film. The following electrochemical reactions occur under the influence of the applied voltage:

(a) at the cathode

$$4Ag \rightarrow 4Ag^+ + 4e^-; \qquad (20.1)$$

$$4Ag^+ + 4Cl^- \rightarrow 4AgCl; \qquad (20.2)$$

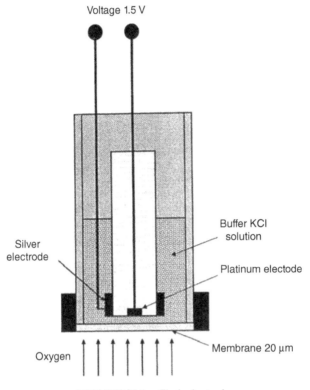

Voltage 1.5 V

Silver electrode

Buffer KCl solution

Platinum electode

Membrane 20 μm

Oxygen

**FIGURE 20.1**    Clark electrode.

(b) at the anode

$$O_2 + 2H_2O + 2e^- \rightarrow H_2O_2 + 2OH^-;$$    (20.3)

$$H_2O_2 + 2e^- \rightarrow 2OH^-.$$    (20.4)

A closed electrical circuit is formed due to the KCl electrolyte; the resulting current is directly proportional to the concentration of molecular oxygen in solution.

Modern instruments for measuring DO contain the probe which is located at the end of the cable.

### 20.1.5    Oxidation–Reduction Potential

*Oxidation–reduction potential* (ORP) or redox potential characterizes the tendency of aqueous solutions to gain or lose electrons and thereby be reduced or oxidized.

Water is a complex mixture consisting of atoms of hydrogen, oxygen, and other chemical elements that are present as impurities in the water. The reduction potential of aqueous solution is characterized by the transfer of electrons between chemical species.

### 20.1.6  Measurement of Oxidation–Reduction Potential

Reduction potentials of aqueous solutions are determined by measuring the potential difference between two electrodes: the *working electrode* is made of a noble metal (Au, Pt, Hg) or inert materials (e.g., graphite) while the *reference electrode* has AgCl or saturated calomel $Hg_2Cl_2$.

Both electrodes are combined in one body in modern ORP-meters; this combination is called "a probe."

Equilibrium potential is associated with the concentration by the Nernst equation:

$$E = E^0 + \frac{RT}{nF} \ln \frac{C_{ox}}{C_{red}}.$$

Here $E$ is the measured potential between the working and reference electrodes; $E^0$ is the standard reduction potential (at a concentration $C_{ox} = C_{red}$); it can be found in tables in handbooks and it is referred to the standard hydrogen electrode; $R$ is the universal gas constant (8.31441 J/K·mol); $T$ is the absolute temperature; $F$ is the Faraday constant (96485 C/mol); $n$ is the number of moles of electrons involved in the reaction; $C_{ox}$ is the oxidant concentration (moles/L); $C_{red}$ is the reductant concentration (moles/L).

The occurrence of various redox reactions explains why the value for the redox potential of water varies in range from −400 to +700 mV. In this way, ORP values characterize the chemical composition of water.

### 20.1.7  Turbidity

The relative transparency of water, which depends on the scattering and absorption of optical radiation by particles of clay, dirt, silicon rust, as well as algae and bacteria, is called *turbidity*. High turbidity is induced by soil erosion, emission of waste water, outbreaks of algae growth, fish activity, rainfall, human activity, and eutrofication.

Muddy water contains viruses or bacteria that provoke gastroenterological diseases in humans. Suspended particles that are absorbed by microorganisms inhibit the development of aquatic fauna and flora. Solar radiation does not pass into the deeper layers of water causing the photosynthetic activity of aquatic plants to be restricted. The presence of blue-green algae that consume oxygen can lead to adverse conditions and the inhibited growth of fish. Increased turbidity raises water temperature and lowers DO

### 20.1.8  Measurement of Turbidity

*Turbidimetry* is a method for determining the concentration of a substance in a solution by measuring the amount of transmitted light by suspended particulate matter.

*Nephelometry* is a method for estimation of the number and size of particles in a suspension by measurement of light scattered by suspended particulate matter.

The most widely used measurement unit for turbidity is the Nephelometric Turbidity Units (NTU) in the United States and the Formazin Nephelometric Units (FNU) which are used as international standards. Typical values of turbidity: drinking water −0.02 to 0.5 NTU; spring water −0.05 to 10 NTU; waste water −70 to 2000 NTU.

Turbidity of water with suspended clay particles can reach 10 units, turbidity of surface water can vary from 10 to 1000 units, and turbid rivers can reach 10,000 units.

In such a way, turbidity values can be used as parameters of water quality in water bodies.

*Turbidimeter* is an instrument for measuring the loss in intensity of a transmitted light that passes through a solution with suspended particulate matter. Attenuation meters and spectrophotometers can be used as turbidimeters. Turbidimeter can be applied for the measurement of the concentration and size distribution of the particles.

*The advantages of a turbidimeter:*

- High accuracy.
- Ability to measure very low (<5 NTU) turbidities.

*The disadvantages of a turbidimeter:*

- High cost.
- Easily damaged.
- Needs power supply.
- Requires well-collimated beam.

*Nephelometer* is an instrument that measures how much light is scattered by suspended particles in the water. A nephelometer measures scattered light at an angle of 90° using a submerged photocell (Figure 20.2). Since the heavy particles settle quickly and the suspended particles remain, nephelometry provides a unique opportunity to assess the total amount of suspended solid particles. The amount of scattered light is proportional to the level of turbidity in water.

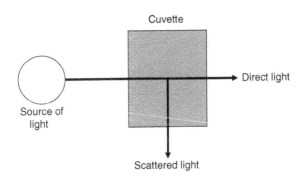

**FIGURE 20.2**   Diagram of a nephelometer.

Nephelometer can be used for the estimation of the concentration, number, and size of particles in suspension.

*The advantages of a nephelometer:*

- As the amount of scattered light is considerably higher than the transmitted light, nephelometers demonstrate higher sensitivity than turbidimeters.
- Can be used for low-level (<40 NTU) measurements (e.g., Model EPA 180.1, HF Scientific MicroTPW).

*The disadvantage of a nephelometer:*

- High cost.

*GLI-2 method.* The conventional nephelometers are suffering with the fluctuations of infrared power, photodiode sensitivity, and window fouling. The GLI-2 method (Great Lakes Instruments, Inc. 1992. Turbidity, GLI method 2 Milwaukee, Wisconsin) makes it possible to eliminate these fluctuations due to the incorporation of two source-detector pairs arranged orthogonally. Such a system was approved for drinking water measurement (0–40 NTU) (Downing, 2005).

An *optical backscatter sensor* measures turbidity by measuring the quantity of light scattered by suspended particles and reflected back to the sensor. The backscatter sensor is characterized by the response to turbidity up to 4000 NTU. The angle range of detection of scattered radiation is 140–165° (Downing, 1989).

The intensity of scattered light depends on the particle size, shape, and reflectivity. Thus, this sensor can be used for estimation of the suspended-sediment concentration (SSC).

*The advantages of an optical backscatter sensor:*

- High sensitivity to coarse-grained particles.
- Wide linear range for turbidity.
- Insensitivity to bubbles and phytoplankton.
- Ambient light rejection.
- Compactness.
- Low cost.

*The disadvantages of an optical backscatter sensor:*

- The susceptibility to color of suspended particles.
- Effect of background concentration of relatively fine sediments.

An *acoustic backscatter sensor* is based on the application of short (10 μs, 1–5 MHz) acoustic pulses that are emitted by a sonar transducer. The sound is scattered from the suspended material and is detected by a sound receptor. The intensity of the backscattered signal depends strongly on the sediment concentration, particle size, and the time delay between transmission and reception.

The acoustic backscatter sensor can be applied for the estimation of particle size distributions on the order of 10–500 µm.

*The advantages of an acoustic backscatter sensor:*

- Can be used in areas with no optical or direct measurements of SSC.
- High temporal and spatial resolution.

*The disadvantages of an acoustic backscatter sensor:*

- Calibration in uniform suspensions is required.
- Sensitivity to water bubbles.

*Laser diffractometry* is based on the *in situ* analysis of the diffraction pattern, which is obtained from a laser beam passing through the particles suspended in water. Schematic of a laser diffractometer is shown in Figure 20.3. The photodetector system consists of a specially constructed annular ring detector that responds to the diffracted radiation. The intensity and character of the diffraction pattern strongly depends on the size of particles. Such a system can be used for *in situ* measuring of the particle-size distribution of the sediments suspended in freshwater rivers and streams.

*The advantages of a laser diffractometer:*

- Reliability, flexibility, and low power consumption.

A *Secchi disk* is a device that has a circular disk (diameter of 23 cm) with black and white sectors (Figure 20.4). The disk is lowered in depth in turbid water until the difference between the white and black sectors disappear. The Secchi disk is used to measure water transparency in water reservoirs.

*The advantages of a Secchi disk:*

- Integrates turbidity over depth (where variable turbidity layers are present).
- Is quick and easy to use.
- Portable.
- Inexpensive.

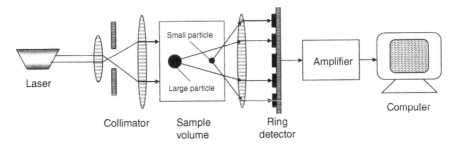

**FIGURE 20.3**  Schematic of laser diffractometer.

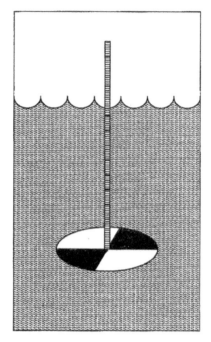

**FIGURE 20.4**    Principle of measuring turbidity of water with Secci disk.

*The disadvantages of a Secchi disk:*

- It cannot be used in shallow waters or swift currents where the disk can still be seen on the bottom.
- Not applicable to small sample size.

*Turbidity tube* contains a 1–3/4″ (4.4 cm) polycarbonate tubing which is marked in centimeters from 0 to 60 with a standard Secchi disk pattern at the bottom of the tube. The turbidity tube is filled with water and excess water is drained off using a crimp until the Secchi pattern appears. The height of the column that is scaled in units of turbidity is then recorded.

*The advantages of a turbidity tube:*

- It can measure turbidity in shallow waters.
- Low cost.
- Portable.
- Easy to learn.

*The disadvantages of a turbidity tube:*

- Less precise.
- Cannot measure turbidity <5 NTU.

## 20.1.9    Electrical Conductivity of Water

*Electrical conductivity* indicates a material's ability to conduct an electric current. The current is formed by the movement of ions through the solution. The conductivity is directly proportional to the ion concentration; it is affected by the presence of such dissolved inorganic solids in water that carry positive or negative ions. With regard to natural bodies of water, electrical conductivity is related to the concentration of salts dissolved in water, and therefore to the total dissolved solids (TDS) and total salinity.

The unit of electric conductivity is the *Siemens* (A/V); however, usually units of electric conductivity such as dS/m (deciSiemens/m) or μS/cm (microSiemens/cm) are used. Note that 1000 μS/cm = 1 dS/m. Molar conductivity has the SI unit $S \cdot m^2/mol$.

The conductivity is affected by temperature; the values of conductivity must be corrected to a standard value of 25°C. The concentration of ions in a solution affects conductivity also: below 10 μS/cm the dependence of conductivity response on increasing concentration has linear character, while above 10 μS/cm this dependence is nonlinear (Gray, 2005).

The conductivity scale varies from 0.05 μS/cm for pure water; 5 μS/cm for demineralized; 10–500 μS/cm for flowing water; $5 \times 10^4$ μS/cm for sea water; to $10^6$ μS/cm for concentrated acidic and alkaline environments.

## 20.1.10    Measurement of Electrical Conductivity

There are two methods which can be used to measure conductivity.

*Conductive conductivity* measurement is based on the immersion of metal or graphite electrodes into the water. The distance between electrodes is 1.0 cm. A constant voltage is applied across the electrodes. An electrical current that flows through the water due to this voltage is proportional to the concentration of dissolved ions in the water. The conductive conductivity analyzer uses two-electrode or four-electrode types of sensor. The four-electrode configuration is less affected by polarization of electrode surfaces and allows to avoid degradation of electrodes. Besides, such a configuration provides more precise measurements especially in water with high conductivity.

*Inductive conductivity* measurement uses an alternative current that passes through a toroidal coil and induces a current in the water of interest and in a second toroidal coil. The inductive conductivity analyzer contains the coils that are sealed in a doughnut-shaped housing and have no direct contact to water.

*The advantages of an inductive conductivity analyzer:*

- As both coils do not need to touch the sample, it is possible to use the sensor in solutions with concentrated acids and alkalis.
- The inductive conductivity analyzer can be applied to measure dirty, abrasive, or corrosive water or acid/alkaline solution with high conductivity.

*The disadvantage of an inductive conductivity analyzer:*

- Lower sensitivity in comparison with contacting measurements.

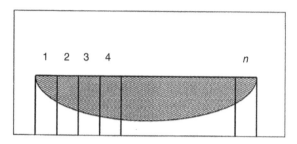

**FIGURE 20.5**   Cross-section of river.

We examined the main factors that characterize the quality of natural waters. Simultaneous measurement of several parameters of water quality such as pH, ORP, DO, conductance, salinity, TDS, seawater, specific gravity, temperature, turbidity, and water depth is possible through the application of a portable multiparameter water quality checker (Horida Ltd., Japan).

### 20.1.11   Measuring Stream Flow

A cross-section of a river is made up of several segments (Figure 20.5) that are numbered from 1 to $n$.

The amount of water passing through the first segment is less than the amount of water that passes, for example, through the fourth segment. But we are interested in the total amount of water that flows through all segments $(1 + 2 + 3 + 4 + \ldots n)$. So it is necessary to summarize the amounts of water passing through all of the segments using the following formula:

$$Q = \sum (w_1 D_1 \bar{v}_1 + w_2 D_2 \bar{v}_2 + \cdot + w_n D_n \bar{v}_n) \qquad (20.5)$$

We measure the current flow as the amount of water carried by the flow through a cross-section of the river per unit time. Water quantity $Q$, which passes through the segment equals the product of the cross-section of the segment $wD$, where $w$ is the width of the segment, and $D$ is the depth. Thus $Q = wDv$, where $v$ is the velocity of flow, or $Q = Sv = SL/t$, where $S$ is the area of a cross-section of the segment, $SL$ is the water volume, and $t$ is the time. Thus, we measure the area of the segment (approximated as a rectangle), velocity of the flow (with a device), and the amount for all of the segments and estimate $Q$.

## 20.2   LABORATORY MEASUREMENT OF WATER QUALITY PARAMETERS

There are several classical methods of water analysis which are still used today.

*Atomic absorption analysis* provides the atomization of a sample in flame or electrothermal plasma. Liquid sample is turned into an atomic gas through desolvation, evaporation, and volatilization. The optical radiation passes through the atomized sample and then the level of attenuation of optical radiation allows to determine the concentration of specific elements in the sample.

*Electrochemical methods* are based on an analysis of the processes occurring at the electrodes and in the inter-electrode space followed by the measurement of the potential and/or current in an electrochemical cell containing the analyte.

*Gravimetric analysis* involves determining the amount of analyte through the measurement of mass.

These classical methods of analysis of water are illuminated sufficiently in the scientific literature and we will not discuss them. Such modern methods of analysis as *gas chromatography (GC)*, *mass spectrometry (MS)*, and the *combination of these methods (GC/MS)* are reflected in this textbook (Sections 18.2–18.5).

We shall concentrate our attention on those methods that can be used for the analysis of water pollution by new and hazardous chemical compounds that have appeared during past years.

Development and implementation of new methods of water quality analysis from the point of view of its pollution by these chemical compounds is very actual task.

As an example, the practical application of advanced analytical methods for the extraction of volatile organic compounds (VOCs) from water samples will be presented.

### 20.2.1    Purge-and-Trap Gas Chromatography/Mass Spectrometry

The purge-and-trap process provides the following operations (Figure 20.6):

1. an inert gas is bubbled though the sample at ambient temperature;
2. the VOC is transferred from the aqueous phase to the vapor phase;
3. this VOCs vapor is swept through a sorbent column where the VOCs are trapped on an absorbent material in the sorbent column;
4. transfer of the trap to a small heating oven;
5. the sorbent column is heated until VOCs are vaporized;
6. the vaporized VOCs are transported with inert gas to the gas chromatograph column;
7. the gas chromatograph column is heated to elute the VOCs components;
8. the VOCs components are detected and identified with a mass spectrometer.

*The advantages of a purge-and-trap system:*

- High sensitivity (in the ppb range).
- By purging samples at higher temperatures, higher molecular weight compounds can be detected.
- Compactness and small dimensions.

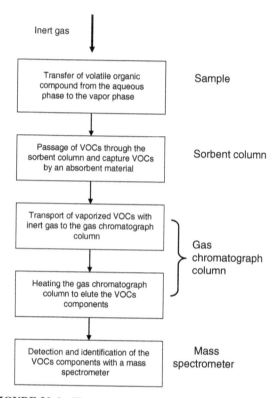

Inert gas

| | |
|---|---|
| Transfer of volatile organic compound from the aqueous phase to the vapor phase | Sample |
| Passage of VOCs through the sorbent column and capture VOCs by an absorbent material | Sorbent column |
| Transport of vaporized VOCs with inert gas to the gas chromatograph column | Gas chromatograph column |
| Heating the gas chromatograph column to elute the VOCs components | |
| Detection and identification of the VOCs components with a mass spectrometer | Mass spectrometer |

**FIGURE 20.6**   The main stages of the purge-and-trap process.

*The disadvantages of a purge-and-trap system:*

- Requires more time for sample preparation and cannot normally be automated.
- Very light volatiles and gases will not be trapped on the adsorbent resins and therefore will be missed in the analysis.

## 20.2.2   Membrane Introduction Mass Spectrometry

It was found (Westover et al., 1974) that silicone rubber was the most effective membrane material for the separation of VOCs from aqueous solutions. This method involves the introduction of the substance being analyzed into a vacuum chamber through a semipermeable membrane. Usually a thin hydrophobic silicon (polydimethylsiloxane) membrane is placed between the sample and ion source of the mass spectrometer. The method is based on separating the organic components of the sample from the water using the membrane. Organic components in the sample diffuse through the membrane and evaporate via the ion source. Since the flow of water is smaller than the flow of organic analytes, an accumulation of organic components takes place. The MIMS method is used for the identification and quantitative

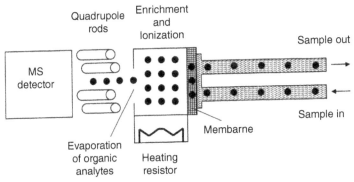

**FIGURE 20.7**    Schematic diagram of a membrane introduction mass spectrometer: separation of organic components of the sample (ooo) from water (•••).

determination of VOCs, water disinfection by-products, polyaromatic hydrocarbons, polychlorinated biphenyls, dioxines, etc. (Ketola et al., 2002).

The principle of the MIMS method is explained in Figure 20.7. The water sample is circulated next to the membrane and the analytes pass through the membrane into the ionization chamber of the mass spectrometer. As the flow of the analyte matrix (water) through the membrane is smaller than the flow of the organic analytes, the sufficient analyte enrichment can be achieved.

This method can be used also for the analysis of VOCs in soil and gases.

One of the modifications of the flow-through technique is based on the passage of the analyte solution across the inner surface of the hollow fiber membrane while the outer surface of the membrane is exposed to the vacuum of the mass spectrometer (Ketola, 2005).

*The advantages of a membrane introduction mass spectrometer:*

- The system has a VOC detection limit in the ppb level.
- This method is fast, low-cost, and does not require pretreatment of the samples nor the need for solvents; it can also be used for long-term monitoring of environmental processes.

*The disadvantages of a Membrane Introduction Mass Spectrometer:*

- High molecular weight and more polar compounds are difficult to analyze.
- Analysis of complicated mixtures is difficult.

**Constructive Tests**    What is the difference between "pure water" and "safe drinking water?"

Why do the streams that run through areas with granite bedrock tend to have lower conductivity and the streams that run through areas with clay soils tend to have higher conductivity?

What the difference between TDS and total suspended solids (TSS)?

# REFERENCES

Behar, S. 1997. *Testing the Waters: Chemical and Physical Vital Signs of a River.* River Watch Network, Montpelier, VT.

Downing, J. 1989. *Optical Backscatter Turbidimeter Sensor.* U.S. Patent Number 4, 841, 157.

Downing, J. 2005. Turbidity monitoring. In: *Environmental Instrumentation and Analysis Handbook.* (eds. R.G. Down and J.H. Lehr). John Wiley & Sons.

Gray, J.R. 2005. Conductivity analyzers and their application. In: *Environmental Instrumentation and Analysis Handbook.* (eds. R.D. Down and J.H. Lehr). John Wiley & Sons, pp. 491–510.

Ketola, R.A. 2005. Online analysis of environmetal samples by mass spectrometry. In: *Environmental Instrumentation and Analysis Handbook.* (eds. R.G. Down and J.H. Lehr). John Wiley & Sons, pp. 187–220.

Ketola, R.A., Kotiaho, T., Cisper, M.E., and Allen, T.M. 2002. Environmental applications of membrane introduction mass spectrometry. *J. Mass Spectrometry,* 37:457–476.

Westover, L.B., Tou, J.C., and Mark, J.H. 1974. Novel mass spectrometric sampling device – hollow fiber probe. *Anal. Chem.* 46:568.

# PRACTICAL EXERCISE 9

# WATER QUALITY PARAMETERS

## 1  pH-VALUE

pH means the negative log of the concentration of hydrogen ions [$H^+$] in a substance. The concentration of hydrogen is used in the form of molarity M, or moles per liter. Then, it is necessary to take the negative log of the concentration $-\log([H^+])$.

**Example**  Convert a pH of 6.4 to [$H^+$].

*Solution*  If pH $= -\log [H^+]$, then: $[H^+] = 10^{-pH} = 10^{-6.4} = 10^{-7}$ M.

**Control Exercise**  Convert the following pH values to [$H^+$]: 5.05; 9.05.

## 2  OXIDATION–REDUCTION POTENTIAL. NERNST EQUATION

ORP is a measure of the tendency of a chemical substance to oxidize or reduce another chemical substance.

To calculate the ORP in a particular solution, it is necessary to use the Nernst equation (see Formula (20.5)), which makes it possible to calculate the voltage of an electrochemical cell or to find the concentration of one of the components of the cell:

$$E = E^0 + \frac{RT}{nF} \ln Q. \tag{P9.1}$$

*Methods of Measuring Environmental Parameters*, First Edition. Yuriy Posudin.
© 2014 John Wiley & Sons, Inc. Published 2014 by John Wiley & Sons, Inc.

Here $E$ is the measured potential between the working and reference electrodes; $E^0$ is the standard reduction potential (at a concentration $C_{ox} = C_{red}$); it can be found in tables in handbooks and it is referred to the standard hydrogen electrode; $R$ is the universal gas constant (8.31441 J/K·mol); $T$ is the absolute temperature; $F$ is the Faraday constant (96485 C/mol); $n$ is the number of moles of electrons involved in the reaction; $C_{ox}$ is the oxidant concentration (moles/L); $C_{red}$ is the reductant concentration (moles/L); $Q = \frac{[C_{ox}]}{[C_{red}]}$ is the reaction quotient.

Substituting the values into the equation, we get

$$E = E^0 + \frac{RT}{nF} \ln Q = E^0 + \frac{8.31441 \text{ J/K} \cdot \text{mol} \times 298 \text{ K}}{n \times 96485 \text{ C/mol}} \ln Q = E^0 + \frac{0.026}{n} \ln Q =$$

$$= E^0 + \frac{0.026}{n} 2.303 \log Q = E^0 + \frac{0.059}{n} \log Q(V) = E^0 + \frac{59}{n} \log Q(\text{mV}).$$

**Example**   Consider the following redox reaction:

$$\text{Fe(s)} + \text{Cu}^{2+}(\text{aq}) \rightarrow \text{Fe}^{2+}(\text{aq}) + \text{Cu(s)},$$

if $[\text{Cu}^{2+}] = 1.2$ M and $[\text{Fe}^{2+}] = 0.2$ M.

*Solution*   Determine the cell reaction and the total cell potential:

| | |
|---|---|
| $\text{Cu}^{2+}(\text{aq}) + 2e^- \rightarrow \text{Cu(s)}$, | $E^0 = +0.34$ V; |
| $\text{Fe}^{2+}(\text{aq}) + 2e^- \rightarrow \text{Fe(s)}$, | $E^0 = -0.44$ V; |
| $\text{Fe(s)} \rightarrow \text{Fe}^{2+}(\text{aq}) + 2e^-$, | $E^0 = +0.44$ V; |
| $\text{Fe(s)} + \text{Cu}^{2+}(\text{aq}) \rightarrow \text{Fe}^{2+}(\text{aq}) + \text{Cu(s)}$, | $E^0 = +0.78$ V. |

In this example, 2 moles of electrons are transferred in the reaction ($n = 2$). Calculate the reaction quotient:

$$Q = [\text{Fe}^{2+}/\text{Cu}^{2+}] = (0.2/1.2) = 0.17.$$

Determine the cell potential using the Nernst equation:

$$E_{cell} = E^0 - (0.06/2) \lg 0.17 = 0.803 \text{ V}.$$

**Control Exercise**   If $[\text{Cu}^{2+}] = 1.2$ M, what $[\text{Fe}^{2+}]$ is needed to have $E_{cell} = 0.75$ V?

## 3   CONDUCTIVITY

Convert the electric conductivity 100 µS/cm of a water sample into the concentration of TDS (ppm).

## 4   WATER QUALITY INDEX

National sanitation foundation (NSF) developed and proposed water quality index (WQI) to distinguish and compare rivers and lakes from different regions.

A reader can find and use this WQI on the site:

www.water-research.net/watrqualindex/waterqualityindex.htm

The following water quality parameters are included in this index:

1. DO (0.17);
2. Fecal coliform (0.16);
3. A pH (0.11);
4. Biochemical oxygen demand (0.11);
5. Temperature (0.10);
6. Total phosphates (0.10);
7. Nitrates (0.10);
8. Turbidity (0.08);
9. Total solids (0.07).

Here a weighting factor is given in brackets; it characterizes each quality parameter and indicates its importance in evaluating water quality.

The 100 point WQI can be divided into several criteria as follows:

90–100—Excellent; 70–90—Good; 50–70—Medium; 25–50—Bad; 0–25—Very bad.

**Control Exercise**   Calculate overall WQI for the following parameters:

1. DO = 20%sat;
2. Fecal coliform = 2 colonies/100 mL;
3. pH = 4;
4. Biochemical oxygen demand = 5 ppm;
5. Temperature = $-10°C$;
6. Total phosphates =1 ppm;
7. Nitrates = 10 ppm;
8. Turbidity =10 JTU;
9. Total solids = 50 ppm.

Estimate the quality of this water.

## QUESTIONS AND PROBLEMS

1. What is water quality?
2. Define potable water.

**3.** Name the main indicators of water quality.

**4.** What is the difference between distilled and sterile water?

**5.** What is demineralized water?

**6.** What is brackish water?

**7.** What is the difference between primary and secondary standards of water?

**8.** Name the main water quality parameters.

**9.** Using library sources name and explain the main laboratory methods of measuring water quality parameters.

**10.** Explain the principles of the purge-and-trap gas chromatography/MS.

**11.** Explain the principles of the MIMS.

## FURTHER READING

Ahuja, S. 2009. Overview. In: *Handbook of Water Purity and Quality*. Elsevier Inc.

Boyd, C.E. 1999. *Water Quality: An Introduction*. Kluwer Academic Publishers Group, The Netherlands.

DeZuane, J. 1997. *Handbook of Drinking Water Quality*, 2nd edition. John Wiley & Sons.

Ela, W.P. 2007. *Introduction to Environmental Engineering and Science*, 3rd edition. Prentice Hall.

U.S. Environmental Protection Agency (EPA). 1991. *Guidance for Water Quality-Based Decisions: The TMDL Process*. Doc. No. EPA 440/4-91-001.

UNICEF, WHO 2008. UNICEF and World Health Organization Joint Monitoring Programme for Water Supply and Sanitation. 2008. *Progress on Drinking Water and Sanitation: Special Focus on Sanitation*. UNICEF, New York and WHO, Geneva.

World Health Organization and United Nations Children's Fund. (WHO and UNICEF). 2000. Global Water Supply and Sanitation Assessment 2000 Report.

## ELECTRONIC REFERENCES

Brian, O. PG B.F. Environmental Consultants Inc. The Water quality index monitoring the quality of surfacewaters, www.water-research.net/watrqualindex/waterqualityindex.htm (accessed September 2, 2013).

Burden of disease and cost-effectiveness estimates, http://www.who.int/water_sanitation_health/diseases/burden/en/index.html (accessed September 3, 2013).

EU's drinking water standards, http://www.lenntech.com/applications/drinking/standards/eus-drinking-water-standards.htm (accessed September 4, 2013).

Gualtero, S.M. Pollution prevention measures for unwanted pharmaceuticals/industrial ecology, December 2005. http://www.seas.columbia.edu/earth/wtert/sofos/Gualtero_IETerm_.pdf (accessed September 8, 2013).

HORIBA – U-50 Series – Multiparameter water quality checker, http://www.environmental-expert.com/products/horiba-u-50-series-multiparameter-water-quality-checker-151247 (accessed September 11, 2013).

Helmenstine, A.M. Nernst equation. Electrochemistry calculations using the Nernst equation. http://chemistry.about.com/od/electrochemistry/a/nernstequation.htm?rd=1 (accessed September 14, 2013).

Helmenstine, T. Nernst equation example problem. Calculate cell potential in nonstandard conditions, http://chemistry.about.com/od/workedchemistryproblems/a/Nernst-Equation-Example-Problem.htm?rd=1 (accessed September 16, 2013).

Rizzardo, J. How to calculate the water quality index of an area, http://www.ehow.com/how_8613028_calculate-water-quality-index-area.html (accessed September 18, 2013).

Safe drinking water, http://www.nrdc.org/health/safe-drinking-water.asp (accessed September 20, 2013).

Secondary drinking water standards, www.epa.gov/safewater/ncl.html (accessed September 22, 2013)

Second UN World Water Development Report, WWDR 2006 "Water, a Shared Responsibility," http://www.unesco.org/new/en/natural-sciences/environment/water/wwap/wwdr/wwdr2-2006/ (accessed September 23, 2013).

Total dissolved solids, http://en.wikipedia.org/wiki/Total_dissolved_solids (accessed September 24, 2013).

The negative effects of water pollution to human health, http://www.whatiswaterpollution.org/the-negative-effects-of-water-pollution-to-human-health/ (accessed September 25, 2013).

USEPA. Drinking Water Standards and Health Advisories. U.S. Environmental Protection Agency. Office of Water 4304 EPA 822-B-00-001. Summer 2000, http://web.ics.purdue.edu/~peters/HTML/docs/drinking-water-standards.pdf (accessed September 26, 2013).

United States Environmental Protection Agency. Office of Water 4304. EPA 822-Z-99-001. April 1999. National Recommended Water Quality Criteria – Correction. http://water.epa.gov/scitech/swguidance/standards/upload/2008_03_11_criteria_wqctable_1999table.pdf (accessed September 27, 2013).

Water consumption Statistics – Worldometers, http://www.worldometers.info/water/ (accessed September 27, 2013).

Water quality facts and statistics, http://www.unwater.org/wwd10/downloads/WWD2010_Facts_web.pdf (accessed September 28, 2013).

WHO and UNICEF Joint Monitoring Programme for Water Supply and Sanitation, http://www.wssinfo.org/ (accessed September 29, 2013).

WHO/EU drinking water standards comparative table, http://www.lenntech.com/who-eu-water-standards.htm (accessed September 29, 2013).

WHO's drinking water standards 1993, http://www.lenntech.com/applications/drinking/standards/who-s-drinking-water-standards.htm (accessed September 29, 2013).

World Health Organization (WHO). 2002. World Health Report: Reducing Risks, Promoting Healthy Life. France. http://www.who.int/whr/2002/en/whr02_en.pdf (accessed September 30, 2013).

# PART IV

## EDAPHIC FACTORS

# 21

# SOIL QUALITY

## 21.1  SOIL AS A NATURAL BODY

Soil is not simply a physical non-living object. V. Dokuchaev was the first who created the doctrine of the soil as a special natural body with complex and multiform processes taking place within it. He established that the soil was formed as a result of the aggregate of a series of soil-formation factors: parent material, plants and animals, climate, relief, and age (Krasilnikov, 1961).

Really, soil is a complex mixture of minerals, organic materials, water, and air. It is possible to find in an acre of healthy topsoil about 900 pounds of earthworms, 2400 pounds of fungi, 1500 pounds of bacteria, 133 pounds of protozoa, and 890 pounds of arthropods and algae (Pimentel et al., 1995); 1 g of surface soil contains about $10^{8-9}$ of bacteria, $10^{5-8}$ of actinomycetes, $10^{5-6}$ of fungi, $10^{3-6}$ of micro-algae, $10^{3-5}$ of protozoa, $10^{3-5}$ of nematodes, and $10^{1-2}$, $10^{3-5}$ of other invertebrates (Dindal, 1990; Blaine Metting, 1993).

The most comprehensive definition of soil was published by the Soil Science Society of America: *soil* is the unconsolidated mineral or organic matter on the surface of the Earth that has been subjected to and shows effects of genetic and environmental factors of: climate (including water and temperature effects), and macro- and microorganisms, conditioned by relief, acting on parent material over a period of time (SSSA, 1997).

*Methods of Measuring Environmental Parameters*, First Edition. Yuriy Posudin.
© 2014 John Wiley & Sons, Inc. Published 2014 by John Wiley & Sons, Inc.

## 21.2   SOIL STRUCTURE AND COMPOSITION

*Soil* is a natural formation that consists of layers (soil horizons) of mineral components of variable thicknesses. Soil is formed due to the conversion of the surface layers of the lithosphere under the influence of biotic, abiotic, and anthropogenic factors and differs from the original materials in morphological, physical, chemical, and mineralogical characteristics.

Soil can be composed of solid, liquid, gaseous and living components.

The composition of the soil includes: *inorganic mineral parts*, which consist of aluminum, silicon, and other minerals whose sizes range from small particles of clay (0.002 mm) to large grains of sand, pebbles, and gravel; *organic residues* such as the remains of plants and animals which have undergone several stages of decomposition forming a stable substance (humus); *water* is a necessary component for the activity of microorganisms; gases, primarily nitrogen, oxygen, and carbon dioxide, which fill up the pores of the soil; and *biological systems* that include plant roots, small animals, and microorganisms.

The main sources of soil pollution are agricultural runoff, acid rain, industrial waste, fallout, and the remains of oil, resins, and agrochemical products.

The principal pollutants include organic soil substances such as hydrocarbons of petroleum products, polycyclic aromatic hydrocarbons (PAHs), polychlorinated biphenyls (PCBs), chlorinated aromatic compounds, detergents and pesticides; inorganic contaminants include nitrates, phosphates, heavy metals (cadmium, lead, chromium, copper, zinc, mercury, arsenic), inorganic acids, and radionuclides. Metals are conservative pollutants that do not decompose in the soil.

## 21.3   SOIL QUALITY

The definitions of *soil quality* were proposed by the European Commission's Joint Research Centre:

> "Soil quality is an account of the soil's ability to provide ecosystem and social services through its capacities to perform its functions under changing conditions" (Tóth et al., 2007)

and by the Soil Science Society of America:

> "The ability of a specific type of soil to function within natural or managed ecosystem boundaries, to sustain plant and animal productivity, maintain or improve air quality and water to support human health and livable" (Karlen et al., 1997; SSSA, 1997).

Soil quality can be used to evaluate the functions of the soil. It cannot be measured directly; indicators that are measurable properties of soil give information about how the soil performs all of its functions.

Soil quality is a combination of all available positive and negative characteristics and properties that determine its fertility; and is determined by the interaction of its physical, chemical, and microbiological properties.

Soils of high quality provide nutritional suitability, aeration, infiltration and detention of water, structural stability, and high biological activity.

## 21.4  SOIL QUALITY INDICATORS

*Soil quality indicators* are measurable physical, chemical, and biological properties, processes, or characteristics of soil that provide information on how well the soil functions. Soil quality indicators may be qualitative or quantitative. Some of the indicators of the soil quality are measured in the field, with the aid of a field kit (a set of tools or instruments which can be applied in the field), whereas other indicators are evaluated in the laboratory.

There are three principal categories of soil quality indicators: physical, chemical, and biological.

## REFERENCES

Blaine Metting, F. Jr., ed. 1993. *Soil Microbial Ecology.* Marcel Dekker, Inc.

Dindal, D.L., ed. 1990. *Soil Biology Guide.* Wiley-Interscience.

Karlen, D.L., Mausbach, M.J., Doran, J.W., Cline, R.G., Harris, R.F., and Schuman, G.E. 1997. Soil quality: A concept, definition, and framework for evaluation. *Soil Sci. Soc. Am. J.* 61:4–10.

Krasilnikov, N.A. 1961. *Soil Microorganisms and Higher Plants.* Israel Program for Science Translation, Jerusalem.

Pimentel, D., Harvey, C., Resosudarmo, P., Sinclair, K., Kurz, D., McNair, M., Crist, S., Shpritz, L., Fitton, L., Saffouri, R., and Blair, R. 1995. Environmental and economic costs of soil erosion and conservation benefits. *Science* 267(24):1117–1122.

Soil Science Society of America (SSSA). 1997. *Glossary of Soil Science Terms.* Soil Science Society of America Inc., Madison, WI.

Tóth, G., Stolbovoy, V., and Montanarella, L. 2007. Soil Quality and Sustainability Evaluation – An integrated approach to support soil-related policies of the European Union, EUR 22721 EN. p. 40. Office for Official Publications of the European Communities, Luxembourg.

# 22

# PHYSICAL INDICATORS

*Physical indicators* provide information concerning the arrangement of soil particles and pores. These indicators include aggregate stability, available water capacity, bulk density, infiltration, slaking, soil crusts, soil structure, and macropores.

## 22.1  AGGREGATE STABILITY

The United States Department of Agriculture (Natural Resource Conservation Service USDA-NRCS, 2005) proposed the following definition:

*Aggregate stability* refers to the ability of soil aggregates to resist disintegration when disruptive forces associated with tillage and water or wind erosion are applied.

Soil aggregates are the primary soil particles that are able to bind to each other more strongly than to other particles. Soil aggregates are formed of elementary soil particles due to their adhesion or bonding under the influence of physical, chemical, or biological processes. Aggregate stability demonstrates the ability of soil aggregates to resist disintegration processes that occur during tillage and wind or water erosion. The stability of wet aggregates indicates how well the soil resists rain drops or water erosion. Size distribution of dry soil aggregates allows to assess the resistance of soil abrasion and wind erosion. Changes in aggregation stability may serve as early indicators of soil degradation or recovery. Aggregate stability is also an indicator of organic matter content, biological activity, or circulation of nutrients in the soil.

*Methods of Measuring Environmental Parameters*, First Edition. Yuriy Posudin.
© 2014 John Wiley & Sons, Inc. Published 2014 by John Wiley & Sons, Inc.

Aggregate stability provides a larger pore space which is important from the point of view of aeration, transport of water and nutrients within the soil, and storage of organic carbon.

The greater the amount of stable aggregates, the better the quality of the soil. For example, water stable aggregation of 50–60% indicates weak structure and highly erodible soil, while water stable aggregation more than 80% corresponds to highly stable structure and small susceptibility of soil to erosion (Kinyangi, 2007).

## 22.2  MEASUREMENT OF AGGREGATE STABILITY

Traditional methods for measuring water stable aggregates (WSA) are based on the analysis of aggregates passing through a set of sieves of a particular mesh size. But these methods suffer disadvantages: only certain sieve-size fractions can be measured; the particle size distribution is not taken into account; the procedure of measurements is time-consuming and tedious. Therefore, we consider the modern methods of measurement of aggregate stability which makes it possible to overcome these limitations.

### 22.2.1  Ultrasound Dispersion

The main idea of the measurement of soil aggregate stability using ultrasonic dispersion tests is based on determining the mass fraction of macroaggregates after certain absorbed specific energies (Mentler et al., 2004). Ultrasonic dispersion of soil aggregates in soil water solution is accompanied with the processes of cavitation, stressing of soil aggregates, and breaking of aggregate bonds (Schomakers et al., 2011).

### 22.2.2  Laser Granulometer

A novel method for aggregate stability measurement using a laser granulometer was proposed by Rawlins et al. (2013).

The principle of operation of the laser granulometer is based on the analysis of the scattering and diffraction of the laser radiation on the soil aggregates. An instrument contains a 750 nm diode laser, system of filters, and 126 photodiode detectors.

The measurement procedure involves adding the WSA (1–2 mm) to the circulating water. The next step was the destruction of these aggregates with ultrasound (sonication). Then the particle size distributions of water-stable and disaggregated material were analyzed. Comparing the difference between the mean weight diameters (MWD) of both size distributions made it possible to estimate the number of aggregates which are resistant to mechanical failure. This difference was evaluated as disaggregation reduction DR ($\mu$m): more stable aggregates have larger values of DR.

*The advantages of a laser granulometer:*

- High speed of analysis.
- Does not require drying and weighing aggregates.
- High accuracy.

- Ease of operation.
- Large range (400 nm–2 mm) of detectable particle sizes.

## 22.3    AVAILABLE WATER CAPACITY

*Available water capacity* is the amount of water that a soil can store that is available for use by plants (USDA-NRCS, 1998).

Available water capacity is estimated as the amount of water available to plants that is held in soil between its *field capacity FC* (the maximum amount of water that a soil can hold before the water is drawn away by gravity) and *permanent wilting point (PWP)* (the minimal point of soil moisture the plant requires not to wilt).

Available water capacity is affected by such soil properties and factors as organic matter, bulk density, soil structure and texture, osmotic pressure, the rooting depth.

Available soil water capacity is calculated as follows:

$$\text{AWC}(\%) = \text{FC} - \text{PWP}, \tag{22.1}$$

where *FC* is the field capacity (%); *PWP* is the permanent wilting point (%).

Water capacity is usually expressed as a volume fraction ($in^3/in^3$) or percentage (%), or as a depth (in or cm). It is necessary to multiply volume fraction by 100 to get a percentage.

**Example**    The following data represent a volumetric soil moisture of soil sample (sand): soil moisture at field capacity 0.08 $in^3/in^3$; soil moisture at wilting point 0.03 $in^3/in^3$. Calculate available water capacity.

*Solution*    Available water capacity can be determine as follows:

$$\text{AWC}(\%) = \text{FC} - \text{PWP} = (0.08 - 0.03) \times 100 = 5\,\%.$$

**Control Exercise**    Calculate the available water capacity, if volumetric soil moisture of soil sample (loam) is 0.28 $in^3/in^3$ at field capacity and 0.11 $in^3/in^3$ at wilting point.

## 22.4    MEASUREMENT OF AVAILABLE WATER CAPACITY

The *pressure plate apparatus* consists of a pressure chamber, which encloses a porous plate (Figure 22.1). Water passes through the pores of the plate but not the air. The atmospheric pressure acts on the porous plate at the bottom, and the pressure of the gas (argon or air) is applied from the top surface. The soil sample is placed in a rubber ring in contact with the surface of the porous plate and is saturated with water. Then the gas pressure is applied to force water out of the soil and to pass through the porous plate. The flow of water continues until equilibrium is reached when the force due to the applied pressure, and the force with which the soil holds water, are equal.

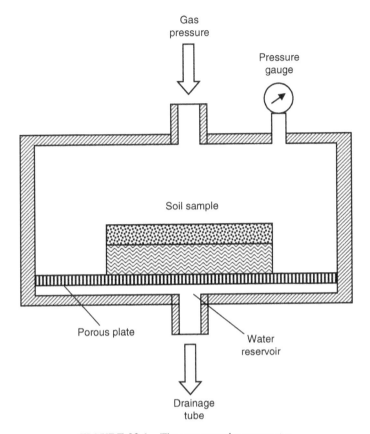

**FIGURE 22.1**    The pressure plate apparatus.

The soil water content is determined by the gravimetric method. This procedure is repeated for other required pressures. Then pF is determined; it corresponds to the force with which soil particles hold water. This force is expressed as a 10 logarithm (e.g., 10000 cm water column means pF4.0; 100 cm pF2.0; 10 cm is pF1.0).

The graph of pF against moisture content is plotted; the available soil water capacity can be determined using the formula in Equation 22.1.

**Water Content**    The *water content* is the liquid water present within a sample of soil; usually it is expressed in percentage by weight.

*Soil water content by weight,* $\theta_w$, is determined as (Don Scott, 2000)

$$\theta_w = \frac{m_w}{m_s},\qquad(22.2)$$

where $m_w$ is the mass of water and $m_s$ is the mass of solid.

The unit of measurement is kg(water)/kg(solids).

Soil water content by weight can be estimated in percentage:

$$\theta_w(\%) = \frac{m_{wet} - m_{dry}}{m_{dry}} \times 100 = \frac{\Delta m}{m_{dry}} \times 100, \qquad (22.3)$$

where $m_{wet}$ is the wet soil mass; $m_{dry}$ is the dried soil mass; $\Delta m = m_{wet} - m_{dry}$ is the water lost.

If we consider a container of mass $m_{cont}$, the latter relations are transformed as follows:

$$\theta_w(\%) = \frac{(m_{cont} + m_{wet}) - (m_{cont} + m_{dry})}{(m_{cont} + m_{dry}) - m_{cont}} \times 100. \qquad (22.4)$$

**Example**   Determine the percentage water by weight for the following parameters:

$$m_{cont} + m_{wet} = 56.49 \text{ g}; m_{cont} + m_{dry} = 55.23 \text{ g}; m_{cont} = 48.96 \text{ g}.$$

**Solution**   The percentage water by weight is:

$$\theta_w(\%) = \frac{(m_{cont} + m_{wet}) - (m_{cont} + m_{dry})}{(m_{cont} + m_{dry}) - m_{cont}} \times 100 = \frac{56.49 - 55.23}{55.23 - 48.96} \times 100 = 20.09\%.$$

Soil water content by volume, $\theta_v$, is defined as

$$\theta_v = V_w/V_t, \qquad (22.5)$$

where $V_w$ is the volume of water, $V_t$ is the total volume of soil (soil volume, $V_s$ + water volume, $V_w$ + air space, $V_a$).

The water volume ratio, $\theta_r$, is calculated as

$$\theta_r = V_w/V_s. \qquad (22.6)$$

## 22.5   BULK DENSITY

The *soil bulk or dry density*, $\rho_b$, is the ratio of the mass of the dried soil to its total volume (solid and pore volumes together) and is defined by formula:

$$\rho_b = M_s/V_t = M_s/(V_s + V_g + V_l), \qquad (22.7)$$

where $M_s$ is the mass of solids; $V_s$, $V_l$, and $V_g$ are the volume of solids, liquids (mostly water), and gas, respectively; $V_t$ is the total volume.

The SI unit for bulk density is kilogram per cubic meter ($kg/m^3$).

Since about 50% of soil belongs to the soil pores, dry soil bulk density is about half the density of solids ($2650 \text{ kg/m}^3$) and varies from 1000 to 1800 $kg/m^3$. Organic soils have lower bulk densities ranging from 800 to 1000 $kg/m^3$ (Don Scott, 2000).

The *soil wet bulk density* is defined as the mass of solids plus liquids divided by the total volume:

$$\rho_w = M_t/V_t = (M_s + M_l)/(V_s + V_g + V_l). \tag{22.8}$$

The bulk density as soil indicator determines the soil compaction and porosity; it affects water and solute movement, soil aeration.

Compacted soil layers are able to restrict root growth; for example, ideal bulk density for silt loams is less than $1300 \ kg/m^3$, whereas bulk density that restricts root growth is more than $1750 \ kg/m^3$.

*Soil porosity* is defined as the ratio of the total volume of all pores and spaces between the structural units to the total volume of the soil. The porosity of the soil is determined as follows:

$$f_a = V_p/V_t = (V_a + V_w)/(V_s + V_a + V_w), \tag{22.9}$$

where $V_p$, $V_a$, $V_w$, and $V_t$ are the volume of all pores between soil particles, air, water, and the total volume, respectively.

Soil porosity depends on the size, shape, and number of individual pores.

The *void ratio* determines the relationship between the volume that is occupied by solids and voids:

$$e = V_p/V_s = (V_a + V_w)/(V_t - V_p) = (V_t - V_s)/V_s. \tag{22.10}$$

The relation between soil porosity and void ratio is (Don Scott, 2000)

$$e = f/(1-f), \tag{22.11}$$

$$f = e/(1+e) = 1 - (\rho_b/\rho_s), \tag{22.12}$$

where $\rho_b$ and $\rho_s$ is the bulk density and particle density, respectively.

**Example**   Determine the bulk density of a cylindrical soil sample with diameter of 7 cm and height 3 cm, if the soil mass is 160 g.

**Solution**   Calculate the volume of the soil: $V_t = \frac{\pi D^2}{4}h = \frac{3.14 \times 7^2}{4} \times 3 = 115 \ m^3$.
Using the formula (22.7), find the soil bulk density:

$$\rho_b = M_s/V_t = 160 \ g/115 \ cm^3 = 1.391 \ g/cm^3 = 1391 \ kg/m^3.$$

**Control Exercise**   Determine the bulk density of dry soil that is shaped like a cube with sides of 10 cm and mass of 1300 g.

**Answer:**   $1300 \ kg/m^3$.

**Example**   Calculate the porosity of a soil sample and the void ratio if a bulk density is $1380 \ kg/m^3$ and the density of solids is $2650 \ kg/m^3$.

*Solution*    The soil porosity can be determined as

$$f = 1 - \rho_b/\rho_s = 1 - 1380 \, kg/m^3/2650 \, kg/m^3 = 1 - 0.521 = 0.479.$$

The void ratio is:

$$e = f/(1 - f) = 0.479/(1 - 0.479) = 0.92.$$

## 22.6  MEASUREMENT OF BULK DENSITY

### 22.6.1  Bulk Density Test

The procedure for measuring the bulk density requires the insertion of a 3-inch diameter ring to a depth of 3 inches. It is necessary to remove the ring with the soil sample; to push down the sample into a plastic bag; to weigh the soil bag with the sample, the empty bag without the soil, and to record these weights; to take 1/8-cup level scoop subsample in a paper cup; to weigh the paper cup with the soil and without it, and to record the weights; to place the paper cup in a microwave oven and to dry the sample; and to weigh the dry sample and to record the data.

Bulk density can be calculated by the mass of oven dry soil (g) divided by the total volume of soil (cm³).

*The advantages of a bulk density test:*

- Requires a relatively simple equipment.
- Undisturbed sample.

*The disadvantages of a bulk density test:*

- Small sampling area of the sample.
- Compression of soil inside the sample.
- Stones can distort the measurement results.

### 22.6.2  Clod Method

A clod (a large soil aggregate) is coated with paraffin or saran by immersing it into heated paraffin. The cooled sample is weighed in air, then in water to determine its volume. The weighted sample is removed from the water; it is blotted on filter paper and weighed to check the integrity of the shell. Soil density $\rho$ (g/cm³) is calculated by the formula:

$$\rho = \frac{m\rho_p\rho_w}{\rho_p(m_1 - m_2) - \rho_w(m_1 - m)}, \tag{22.13}$$

where $m$ is the mass of the sample before waxing (g); $m_1$ is the mass of waxed soil sample (g); $m_2$ is the result of weighing sample in the water—the difference between the mass of the sample and mass of the displaced water (g); $\rho_p$ is the density of the paraffin (0.900 g/cm$^3$); $\rho_w$ is the density of the water at the ambient temperature (g/cm$^3$).

*The advantage of a clod method:*

- Low cost.

*The disadvantages of a clod method:*

- This method is difficult and labor intensive.
- The clod is destroyed.

### 22.6.3   Three-Dimensional Laser Scanning

The main idea of this method is based on the action of laser radiation on the object (soil sample) under study. This radiation is scanned spatially, reflected from the object and detected by photodetectors that create a three-dimensional (3-D) image of the object.

An automated 3-D laser scanning to measure bulk density of soil clods and rock fragments was proposed by scientists at the University of California (Rossi et al., 2008). The 3-D image of the soil sample can be created as result of assembling the scanned images. Such a method of 3-D laser scanning can be applied to quantitative estimation of soil samples related to their structure, size, and grade.

### 22.7   INFILTRATION

*Infiltration* is a process which determines the amount of water from rainfall, irrigation, or snowmelt enters the soil and becomes runoff.

Infiltration as a soil indicator characterizes the ability of soil to pass water deep into the soil and thus to accumulate the water needed for the root system, plant growth, and activity of soil organisms.

The infiltration process is characterized by two parameters—the infiltration rate and cumulative infiltration.

*Infiltration rate* is the maximum rate at which soil in a given condition will absorb water (rainfall or irrigation). Infiltration rate is defined as the volume of water passing into the soil per unit of area per unit of time and has the dimensions of velocity (Johnson, 1963).

The SI unit of infiltration rate is $m^3/m^2 \cdot s = m/s$, but typical units for infiltration rates are inches per hour (in/h) or millimeters per hour (mm/h). Thus, the infiltration rate corresponds to the depth (in inches or mm) of the water layer, which is formed in the soil during 1 hour.

For example, the infiltration rate of less than 15 mm/hour is considered to be low, in the range of 15–50 mm/h it is medium; more than 50 mm/h is high infiltration rate (Brouwer et al., 1985).

*Cumulative infiltration* is the total amount of water infiltrated during a given period.

The SI unit of cumulative infiltration is $m^3/m^2 = m$, but such units as cm or mm also can be used.

The main factors that influence the infiltration are the soil texture, structure, compaction, tillage, hydrodynamic characteristics, soil coverage and vegetation, rainfall intensity, irrigation flow, water content in the soil, initial wetness, soil and water temperature, air entrapment, soil salinity topography and morphology of slopes, and parameters of irrigation channels.

## 22.8   MEASUREMENT OF INFILTRATION

### 22.8.1   Infiltration Test

This test provides the following operation: to immerse the 6-inch diameter ring to a depth of 3 inches; to cover the soil surface inside the ring with a sheet of plastic wrap; to pour 444 mL of distilled water into the ring with a plastic wrap; to remove the plastic wrap leaving the water in the ring; to record the interval of time that takes for the water to infiltrate the soil; to repeat infiltration test (The Soil Quality Test Kit Guide, Section II, Chapter 2. Infiltration. pp. 55–56).

### 22.8.2   Single-ring and Double-ring Infiltrometers

The *single ring infiltrometer* consists of a metal cylinder. Usually cylinders 20 inches high and of different diameters (12, 18, and 24 inches) are used. This ring infiltrometer is driven partially into the soil and filled with water (Bouwer, 1986).

The disadvantage of single-ring infiltrometers is the errors of measurement which are caused by not only the vertical flow of water beneath the cylinder, but also by the lateral movement of water. Such an infiltrometer gives much more infiltration rates than under real rainfall due to the so-called end effects. There are problems with the end effects. Another source of error is a poor connection between the walls of the ring and the soil. These inaccuracies can be avoided through the use of double-ring infiltrometers.

The *double-ring infiltrometer* contains two rings: an inner ring and outer one. Typical diameters of the double-ring infiltrometer are 12 and 18; 12 and 24; 18 and 24 inches.

There are two approaches to measuring the infiltration rate that are used with ring infiltrometers. The *constant head test* means that the water level in the inner ring is maintained at a fixed level and the volume of water required to maintain this level is measured simultaneously. Another approach, the *falling head test*, is based on measuring the time interval during which the water level is decreased in the ring due to infiltration (ASTM, 2003).

*The advantages of a ring infiltrometer:*

- Does not require expensive equipment.
- Simple in operation.

*The disadvantages of a ring infiltrometer:*

- Suffer with soil deformation and formation of cracks during the insertion of ring into the soil.
- Time-consuming.

### 22.8.3   Tension Infiltrometer

The *tension infiltrometer* is designed to measure the unsaturated flow of the water into soil. The unsaturated zone exists below the ground surface. The pore spaces in this zone are partially filled with water and partially with air.

The water infiltrating through the top soil will flow vertically through the unsaturated zone. The vertical gradient is the driving force for the flow of water into the soil; it is controlled mainly by gravity and capillary forces.

The tension infiltrometer was proposed at first by Perroux and White (1988). It consists of the bubble tower, the water reservoir, the 20-cm disk and the tube that connects the disk and the water tower (Figure 22.2). The bubble tower controls the tension at the soil surface. The water reservoir empties when the water flows into the soil. The water level in the water reservoir is determined due to the attached centimeter scale which indicates this level, or by measuring the pressure in the upper end of the water reservoir.

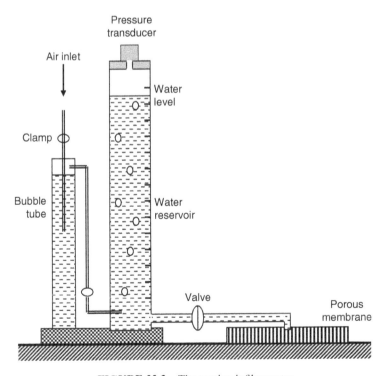

**FIGURE 22.2**   The tension infiltrometer.

As soon as the water tower empties the air pocket at the top of the water reservoir is formed. The tension (negative pressure) in this pocket depends on the height of water in the reservoir linearly.

The second pressure transducer is mounted on the infiltrometer disk; the combination of two pressure transducers makes it possible to eliminate the bubbling noise and to increase the precision of measurements.

The measurement procedure involves placing the disk on the soil surface so that the bottom of the bubble tower and the disk membrane are on the same level. The infiltration rate can be estimated by recording pressure changes which are measured with a pressure transducer.

*The advantages of a tension infiltrometer:*

- Separate infiltration disk provides greater stability (no wind effects).
- On-site determination of infiltration rate.
- Flow rates read directly from the water column or with a pressure transducer.
- Optional transducers and datalogger allow electronic data collection.
- Polycarbonate and plexiglass materials.
- Replaceable nylon mesh screen membrane.

***Infiltration Models***    Infiltration process may be modeled mathematically by a number of empirical equations. Consider the example of such a model.

***The Kostiakov Equation***    The infiltration rate of water can be expressed by the Kostiakov equation:

$$I = at^{-b}, \tag{22.14}$$

where $I$ is the infiltration rate (mm/h) which corresponds to the volume of water infiltrating a unit soil surface area per unit time $t$ (h); $a$ and $b$ are constants that depend on soil type.

**Example**    Calculate the infiltration rate of water after 10 minutes if the constants are: $a = 24$ mm/h and $b = 0.10$.

*Solution*    Substitute numeric values to Equation 22.14:

$$I = at^{-b} = 24 \times (10/60)^{-0.10} = 24 \times 0.167^{-0.10} = 24 \times 1.196 = 28.7\,\text{mm/h}.$$

***Constructive Test***    Find in scientific literature or in the internet empirical equations that describe the process of infiltration: modified form of the Kostiakov equation, Horton equation, Philip equation, Green–Ampt equation. Give a short description of these infiltration equations.

### 22.8.4   The Automatic Infiltration Meter

The *automatic infiltration meter (AIM)* is a stand-alone logging instrument for the measurement of infiltration rates in soil. It is equipped with two tension infiltrometers, 2 GB data-logging capacity and internal battery that provides the regime of several continuous days' field work. Communication is realized via a USB port or wireless connectivity. Wireless is compatible up to 250 m distance.

Both tension infiltrometers are installed in the field. A differential pressure transducer (ICTGT3-1D, ICT International) is installed near the bottom of each tension infiltrometer. This transducer records the pressure difference between the air pressure at the top of the water tower and water pressure at the bottom of the water tower. The pressure values are transferred to AIM where they are converted to metric values.

The AIM automatically records infiltration data from tension infiltrometers. These data are stored and downloaded as a file for further elaboration and analysis (Automatic Infiltration Meter, 2013).

*The advantages of an automatic infiltration meter:*

- Automatic conversion of pressure transducer measurements to cumulative infiltration.
- High accuracy ($\pm$ 1%) and resolution (0.01 kPa).
- Wireless communication with any information and communications technology (ICT) system.
- Portable, easy to use.
- Connects directly into any Windows-based computer via a USB cable.

### REFERENCES

ASTM. 2003. D3385-03 Standard test method for infiltration rate of soils in field using double-ring infiltrometer. *Annual Book of ASTM Standards 04.08*. West Conshohocken, PA.

Bouwer, H. 1986. Intake rate: Cylinder infiltrometer. In: *Methods of Soil Analysis* (ed. A. Klute). Part 1. 2nd edition. ASA and SSSA, Madison, WI, pp. 825–844.

Brouwer, C., Goffeau, A., Heibloom, M. 1985. *Irrigation Water Management: Training Manual No. 1 – Introduction to Irrigation*. Food and Agriculture Organization of the United Nations.

Don Scott, H. 2000. *Soil Physics. Agricultural and Environmental Applications*. Iowa State University Press, Ames.

Johnson, A.I. 1963. A field method for measurement of infiltration. *Geological Survey Water-Supply Paper 1544-F*. US GPO, Washington.

Mentler, A., Mayer, H., Strauß, P., and Blum, W.E.H. 2004. Characterisation of soil aggregate stability by ultrasound dispersion. *Int. Agrophysics*. 18:39–45.

Perroux, K.M. and White, I. 1988. Designs for Disc Permeameters. *Soil Sci. Soc. Am. J.* 52:1205–1215.

Rawlins, B.G, Wragg, J., and Lark, R.M. 2013. Application of a novel method for soil aggregate stability measurement by laser granulometry with sonication. *Eur. J. Soil Sci.* 64(1): 92–103.

Rossi, A.M., Hirmasb, D.R., Grahama, R.C., and Sternberga, P.D. 2008. Bulk density determination by automated three-dimensional laser scanning. *Soil Sci Soc Am J.* 72 (6): 1591.

# 23

# CHEMICAL INDICATORS

Soil contaminants can be analyzed by: high performance liquid chromatography (phenols, xylenols, cresols, naphthols), gas chromatography with flames ionization detector (polyaromatic hydrocarbons, polychlorinated biphenyls, volatile organic compounds, pesticides), a combination of gas chromatography with mass spectrometry GC/MS (radionuclides, volatile organic compounds, heavy metals).

We focus our attention on principal chemical indicators of soil quality such as pH and electrical conductivity, as well as on new instrumental approaches for assessing the composition and soil pollution.

## 23.1  pH OF SOIL

Chemical indicators include the results of measurements of pH, salinity, organic matter, phosphorus concentration, cycle nutrients, cation-exchange capacity, concentration of elements that can be potential contaminants (heavy metals, radionuclides, etc.) or that are required for plant growth and development. Soil pH is a measure of the acidity or alkalinity of the soil, which is defined as the negative logarithm of the molar concentration of dissolved hydrogen ions.

The pH scale is based on a logarithmic scale, that corresponds to an increase or decrease of acidity or alkalinity of the soil by a 10-fold.

The value of the pH of the soil varies from 0 to 14, with 7 being neutral. A pH below 7 indicates acidic soil, pH values greater than 7 correspond to basic soil. The range of pH 5.5–7.5 is optimal for plant growth.

*Methods of Measuring Environmental Parameters*, First Edition. Yuriy Posudin.
© 2014 John Wiley & Sons, Inc. Published 2014 by John Wiley & Sons, Inc.

The level of pH determines the solubility of soil minerals; soil pH affects the activity of soil microorganisms, plant growth, and consumption of nutrients. Thus, plants consume elements such as iron, manganese, copper, zinc, boron in highly acidic soils (pH <6), and phosphorus, calcium, magnesium, molybdenum in more alkaline soils (pH >7.5).

Brief description of the measurement of pH is presented in Section 20.1.2.

## 23.2   ELECTRICAL CONDUCTIVITY OF SOIL

Definition of electrical conductivity, the units and the basic principles of its measurement are presented in Sections 20.1.9–20.1.10. With regard to electrical conductivity of the soil, it should be noted that conductivity range 0–98 dS/m is almost negligible for crop; yields of most crops are restricted within the range 1.71–3.16 dS/m; only tolerant and very tolerant crops yield satisfactorily from 3.16–6.07 to more than 6.07 dS/m (Soil Survey Staff, 1993; Smith and Doran, 1996).

## 23.3   OPTICAL EMISSION SPECTROSCOPY WITH INDUCTIVELY COUPLED PLASMA

Optical emission spectroscopy with inductively coupled plasma (OES-ICP) can be also referred to as atomic emission spectroscopy with inductively coupled plasma (AES-ICP).

Inductively coupled plasma (ICP) is a type of gas discharge, excited by energy that is supplied by an electric current which is produced by electromagnetic induction during the application of a radio-frequency (1–100 MHz) magnetic field. This method is used to detect metals in trace quantities. The excitation of the plasma is accompanied by radiation at certain wavelengths, which characterize the elements of interest. The intensity of the radiation is proportional to the concentration of these elements (Hou and Jones, 2000).

The OES-ICP system consists of plasma source containing three concentric quartz tubes and coils, to which a radio-frequency field is applied. Argon gas is passed through the coils; this gas produces a torch under the influence of powerful radio-frequency field.

The ionizing process is induced by a discharge ark. A stable, high-temperature plasma (~7000 K) is created due to collisions between neutral atoms of argon and excited particles.

The sample is directly introduced into the plasma flame where it collides with charged particles and is broken down into charged ions. The process of losing electrons and their recombination with molecules in the plasma is accompanied with the emission of wavelengths characteristic of the element being studied.

The wavelength and intensity of spectral lines is measured using diffraction gratings and a photoelectronic multiplier combined with a means of recording the response and analyzing the data.

The OES-ICP method can be used for the analysis of soil contaminants (e.g., aluminum, barium, beryllium, boron, cadmium, calcium, chromium, cobalt, copper, iron, magnesium, manganese, molybdenum, silver, silicon, strontium, tin, vanadium, zinc) and the elementary constituents of fertilizer (Ca, Fe, K, Mg, Na, P, and S). The sensitivity of this method is 0.2–100 ppb.

*The advantages of an optical emission spectroscopy with inductively coupled plasma:*

- Relatively low sample size needed <10 mL.
- Automated analysis.
- Inexpensive to purchase and maintain.
- Easy to operate.
- Flexibility of wavelengths and elements.

*The disadvantages of an optical emission spectroscopy with inductively coupled plasma:*

- Interferences: plasma excites all atoms, so it is necessary to choose wavelengths for each element and analyze multiple elements at once.
- Standards have to be made for each element.

## 23.4   MASS SPECTROMETRY WITH INDUCTIVELY COUPLED PLASMA

This method utilizes ICP as the ion source and a mass spectrometer for separation and detection (Newberry et al., 1989).

Unlike OES-ICP, measurement of the wavelengths and intensities of the spectral lines is realized using a mass spectrometer. The sample is introduced into the central channel in the form of an aerosol, which is obtained by spraying a liquid sample. When the aerosol enters the central channel, it evaporates and breaks up into atoms. A significant part of the atoms are ionized due to the high temperature and pass into the input of the mass spectrometer. Here, the ions are separated according to their weight against the charge and the detector receives a signal proportional to the relative concentration of the particles.

Mass spectrometry with inductively coupled plasma (MS-ICP) is one type of mass spectrometry. It is characterized by high sensitivity and the ability to identify metals, and some non-metals, and for some at concentrations not exceeding $10^{-10}\%$ or one part for $10^{12}$ (trillion) parts.

The MS-ICP method allows simultaneously identifying and quantifying over 60 elements of the periodic table within 2 minutes at an accuracy of 0.1 mg/L.

The MS-ICP method is used for the analysis of soil contaminants (e.g., aluminum, antimony, arsenic, barium, beryllium, cadmium, chromium, cobalt, copper, lead, manganese, molybdenum, nickel, silver, thallium, uranium, vanadium, zinc).

*The advantages of a mass spectrometry with inductively coupled plasma:*

- Ability to identify and quantify all elements with the exception of argon.
- Is suitable for all concentrations at ppt levels.
- Low detection limit.
- A complete multi-element analysis can be undertaken in a very short time.
- Only small sample quantities required (about 3–5 mL).
- High sensitivity, good precision, and accuracy.
- Isotope ratio measurements are possible.

*The disadvantages of a mass spectrometry with inductively coupled plasma:*

- Destructive technique.
- Strong dependence of signal on plasma parameters.
- Cannot analyze very small (<3–5 mL) sample volumes for solution MS-ICP.

## 23.5    LASER-INDUCED BREAKDOWN SPECTROSCOPY

*Laser-induced breakdown spectroscopy (LIBS)* is a method of atomic emission spectroscopy which is based on the excitation of a matter by high energetic laser pulse. The Nd:YAG laser (wavelength of 1064 nm; power density 1 $GW/m^2$; pulse duration 10 ns) is used usually as the source of excitation.

Laser radiation is focused onto small area, where plasma is formed due to extremely high temperature. After the adiabatic expansion the plasma is cooled. This process is accompanied by the emission of characteristic spectral lines of the matter forming the plasma. The spectral lines of individual elements are selected with a spectrometer, which consists of the Czerny–Turner scanning monochromator with a photomultiplier or a polychromator with CCD (charge-coupled device) detector.

A typical design of laser-induced breakdown spectrometer is presented in Figure 23.1.

The practical application of this promising technique for the analysis of such chemical elements as As, Ba, Ca, Cd, Cr, Cu, Zn, Hg, Ni, Pb, Ti, Sr, Zn, or contaminants in the soil samples has demonstrated its efficiency (Bublitz et al., 2001; Yamamoto et al., 2005; Bousquet et al., 2007; Li et al., 2008; Yong et al., 2008; Yang, 2009; Madhavi et al., 2013).

Detection limits of *LIBS* are in the range from 1 to 100 ppm.

LIBS can be applied also for qualitative and quantitative analysis of such environmental objects as water and aerosols.

*The advantages of a laser-induced breakdown spectrometer:*

- Very fast and sensitive technique.
- A small amount of material is required.

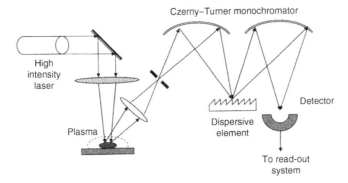

**FIGURE 23.1**    Typical design of laser-induced breakdown spectrometer.

- Sample preparation is typically minimized to homogenization.
- Possibility of contamination during chemical preparation steps is reduced.
- Effective penetration of laser radiation into the specimen with each shot.
- Can be applied for high volume analyses or online industrial monitoring.

*The disadvantages of a laser-induced breakdown spectrometer:*

- The laser spark and resultant plasma which often limits reproducibility.
- The detection limits vary from one element to the next depending on the specimen type and the experimental apparatus used.

## REFERENCES

Bousquet, B., Sirven, J.B., and Canioni, L. 2007. Towards quantitative laser-induced breakdown spectroscopy analysis of soil samples. *Spectrochimica Acta. Part B,* 62 (12):1582–1589.

Bublitz, J., Dölle, C., Schade, W., Hartmann, A., and Horn, R. 2001. Laser-induced breakdown spectroscopy for soil diagnostics. *Eur. J. Soil Sci.* 52(2):305–312.

Hou, X. and Jones, B.T. 2000. Inductively coupled plasma/optical emission spectrometry. In: *Encyclopedia of Analytical Chemistry* (ed. R.A. Meyers). John Wiley & Sons Ltd, Chichester.

Madhavi, Z.M., Mayes, M.A., Heal, K.R., Brice, D.J., and Wullschleger, S.D. 2013. Investigation of laser-induced breakdown spectroscopy and multivariate analysis for differentiating inorganic and organic c in a variety of soils. *Spectrochim. Acta. B.* 100–107.

Smith, J.L. and Doran, J.W. 1996. Measurement and use of pH and electrical conductivity for soil quality analysis. In: *Methods for Assessing Soil Quality.* (eds. J.W. Doran and A.J. Jones). *J Soil Sci. Soc. Am. Spec. Publ.* 49. SSSA, Madison, WI, pp. 169–185.

Soil Survey Staff. 1993. *Soil survey manual.* United States Department of Agriculture. Hnbk no. 18. U.S. Gov. Printing Office, Washington, DC.

Yamamoto, K.Y., Cremers, D.A., Foster, L.E., Davies, M.P., and Harris, R.D. 2005. Laser-induced breakdown spectroscopy analysis of solids using a long-pulse (150 ns) Q-switched Nd:YAG laser. *App Spectroscop.* 59(9):1082–1097.

Yang, N. 2009. *Elemental Analysis of Soils Using Laser-Induced Breakdown Spectroscopy (LIBS).* Master's Thesis, University of Tennessee. http://trace.tennessee.edu/utk_gradthes/89 (accessed October 3, 2013).

Yong, Li., Lu, J.-D., Lin, Z.-X., Xie, C.L., Li, J., and Li, P.-Y. 2008. Quantitative analysis of lead in soil by laser-induced breakdown spectroscopy. *J Appl Opt.* 29(5):789–792.

# 24

# BIOLOGICAL INDICATORS

The main biological indicators of soil include the total biomass of soil, microbial biomass, total number of bacteria and microscopic fungi, production of carbon dioxide ($CO_2$), soil respiration, enzymatic activity, etc.

Biological indicators include measurements of micro-and macroorganisms, estimation of their activity or formation of by-products. Populations of worms, nematodes, insects and pathogens, respiratory processes, or decomposition of organic matter—all these parameters can be used to assess soil quality.

## 24.1 EARTHWORMS AS SOIL BIOINDICATORS

Earthworms are recognized as effective biological indicators of soil quality which improve soil quality, including aeration, supply nutrients to plants, decomposition of organic matter, aggregate stability, water-holding capacity, pore size, and infiltration rate (Stockdill, 1982); they are sensitive to agrochemical compounds, particularly to pesticides and high concentrations of heavy metals (Cikutovic et al., 1999; Römbke et al., 2007), metal organic compounds (pentachlorophenol), and polychlorinated biphenyls (PCBs) (Bunn et al., 1996; Booth et al., 2000). Charles Darwin wrote: "It may be doubted whether there are many other animals which have played so important a part in the history of the world, as have these lowly organized creatures" (Darwin, 1881).

*Methods of Measuring Environmental Parameters*, First Edition. Yuriy Posudin.
© 2014 John Wiley & Sons, Inc. Published 2014 by John Wiley & Sons, Inc.

Population of worms varies from 10 to 1300 per 1 m$^2$ depending on the crop (Kladivko, 1993). A number of papers are devoted to the study of the effect of heavy metals, pesticides, and other hazardous soil contaminants on earthworms.

**Example**    The average number of earthworms in one sample of area 0.04 m$^2$ is 24. How many earthworms can be found in 1 hectare? What is the total mass of the earthworms in a hectare, if the mass of each worm is 0.2 g?

*Solution*

Number of sample areas in a hectare is (100 m $\times$ 100 m)/0.04 m$^2$ = 10,000 m$^2$/0.04 m$^2$ = 250,000.

A number total of earthworms in a hectare is 24 $\times$ 250,000 = 6,000,000.

A mass total of the worms in a hectare is 0.2 g $\times$ 6,000,000 = 1,200,000 g = 1200 kg = 1.2 tonnes.

## 24.2    ANALYSIS OF EARTHWORMS

The traditional method of measuring the earthworm populations is based on counting the number of earthworms/m$^2$ (Soil Quality Test Kit Guide, Section I, Chapter 10, pp. 22–23. See Section II, Chapter 9, pp. 73–75).

The quantitative parameters and physiological functions of earthworms as bioindicators of soil quality depend on the state of the soil and the presence of various contaminants in the soil.

Let us consider the method of analysis of soil contamination by lead with earthworms. The fact is that earthworms are characterized with the ability to excrete calcium carbonate granules of about 2 mm in diameter (Lee et al., 2008). The investigators (Fraser et al., 2011) conducted experiments to study the incorporation of lead into calcium carbonate granules secreted by the earthworm *Lumbricus terrestris*. The lead- and calcium-amended artificial soils were used in this experiment. The application of x-ray diffraction demonstrated that lead was incorporated into the calcite at the surface of granules.

Gago-Duport et al. (2008) used scanning and transmission electron microscopy (SEM and TEM), Fourier transform infrared (FTIR) spectroscopy and x-ray diffraction to study the properties of calcium carbonate granules.

The process of bioaccumulation of heavy metals (Cd, Cu, and Zn) by the earthworms *Aporrectodea caliginosa* and *Lumbricus rubellus* was studied (Hobbelen et al., 2006). The concentrations of heavy metals in soil extracts and worm digests were measured using a PerkinElmer atomic absorption spectrometer (AAS) (flame AAS and graphite furnace AAS). It was concluded that the earthworms contained elevated concentrations of Cu and Cd.

The heavy metals (Zn, Cd, Pb, and Cu) contents in earthworms (*A. caliginosa* and *L. rubellus*) tissue were determined by inductively coupled plasma–atomic emission spectroscopy (Dai et al., 2004).

A research group affiliated with the Karl Franzens Universität in Austria investigated the process of accumulation of the pentavalent form of arsenic, AsV by earthworm *L. terrestris* (Foun et al., 2006; Geiszinger et al., 1998). A correlation between soil arsenic concentrations and concentrations of arsenic in the tissues of the worms was established.

Mark Hodson, Professor of Reading University, informs about the results of application of UK's national synchrotron to study the process of accumulation of heavy metals by earthworms (Diamond Light Source, 2011).

Synchrotron is a particle accelerator that produces high intensity beams of focused radiation in the range from x-rays to infrared radiation.

The main objective of this study is to examine the earthworms, the mucus-rich linings of the burrows they create and the bulk soil around. Such information will make it possible to understand the movement and levels of metals in the environment. In addition, the participation of calcite granules in interacting with metals in the soil, their structure and stability will be investigated.

The scientists have used such techniques as x-ray fluorescence that allows the identification of relative metal concentrations, x-ray absorption near edge structure(XANES) or extended x-ray absorption fine structure (EXAFS) for analysis of the areas of interest reveals the exact form of the heavy metal.

As can be seen, the quantitative and qualitative analyses of earthworms as bioindicators of soil quality has experienced significant progress, rising from simple shovel to synchrotron.

## 24.3   A BIOTA-TO-SOIL ACCUMULATION FACTOR

Metal bioavailability to earthworms can be estimated due to a *Biota-to-Soil Accumulation Factor (BSAF)* (Cortet et al., 1999). This factor can be calculated by the following formula:

$$BSAF = \frac{M_{ew}}{M_s}, \qquad (24.1)$$

where $M_{ew}$ is the metal content in the tissues of earthworms (mg/kg); $M_s$ is the metal content in soil (mg/kg).

## 24.4   SOIL RESPIRATION

*Soil respiration* is the production of $CO_2$ by the living organisms and plant roots; this process is accompanied with the total $CO_2$ efflux at the soil surface.

Soil respiration rate depends on soil temperature and moisture, root nitrogen concentration, and soil texture. Human activity and global climate changes also affect the rate of global soil respiration.

$CO_2$ flux from soil is one of the principal components of the global carbon cycle.

It is necessary to note that the $CO_2$ concentration in the soil air space between soil particles is often an order of magnitude higher than in the atmosphere.

It should be noted that the concentration of $CO_2$ in the soil air space between the soil particles is about an order of magnitude greater than the concentration of $CO_2$ in the atmosphere. The transport of $CO_2$ from the soil to the atmosphere is caused by molecular diffusion.

*Diffusion* is the movement of substance molecules from an area where their concentration is high to an area that has low concentration. In soil, diffusion means the net motion of $CO_2$ molecules through pores.

Diffusion occurs as a result of the second law of thermodynamics which states that the entropy of any system must always increase with time.

The process of diffusion can be described by *Fick's first law*:

$$J_m = \frac{dm}{Adt} - D\frac{dc}{dx},\qquad(24.2)$$

or in generalized form:

$$J_m = L_m \times F_m,\qquad(24.3)$$

where $J_m = \frac{dm}{Sdt}$ is the diffusive flux (the rate at which mass is transported per unit area); $L_m = -D$ is the diffusion coefficient; $F_m = \frac{dc}{dx}$ is the concentration gradient; $m$ is the mass of the material that diffuses during the time interval $dt$ across the area $A$.

The concentration gradient occurs in the soil and across the soil surface due to the high resistance to gas transport in the soil; the $CO_2$ flux $F_c$ at the soil surface can be estimated as follows (Madsen et al., 2010):

$$F_c = \frac{pV}{RTA}\frac{dC_c}{dt},\qquad(24.4)$$

where $p$ is the atmospheric pressure (Pa); $V$ is the chamber volume ($m^3$); $T$ is the air temperature (K); $A$ is the soil surface area; $\frac{dC_c}{dt}$ is the $CO_2$ concentration increase inside the chamber ($\mu mol/mol/s$); $R$ is the gas constant (8.314 $Pa \cdot m^3/K \cdot mol$).

## 24.5    MEASUREMENT OF SOIL RESPIRATION

### 24.5.1    The Draeger Tubes

The rate of $CO_2$ release from the soil can be estimated by a simple field method which is based on the application of the Draeger tubes (The Soil Quality Test Kit Guide, Chapter 2, pp. 4–6. See Section II, Chapter 1, pp. 52–54). A Draeger tube is a glass cylindrical vessel equipped with a scale. The vessel is filled with a chemical reagent that can change color in the presence of $CO_2$. The length of the purple color change typically indicates the measured $CO_2$ concentration.

*The advantages of a Draeger tube:*

- Simplicity, can be applied in the field conditions.
- Low cost.

*The disadvantages of a Draeger tube:*

- Reagent required.
- Low accuracy.

### 24.5.2  Soil $CO_2$ Flux Chambers

There are two general chamber systems which can be used to measure soil $CO_2$ flux (Davidson et al., 2002; Xu et al., 2006): the open-chamber system (steady-state system) and the closed-chamber system (non-steady-state system).

In an open system, ambient air is pumped into the chamber, and $CO_2$ flux is estimated using the air flow rate and the difference in $CO_2$ concentrations between air entering and leaving the chamber after the air in the chamber has reached an equilibrium state. Commercial devices include LI-7500A (LI-COR Biosciences, Lincoln, NE, USA) and SRC-MV5 (Dynamax Inc., Houston, TX, USA).

The main disadvantage of the open systems is the effect of wind on the results of measurements.

The open system, particularly the LI-7500A Open Path $CO_2$/$H_2O$ Analyzer that is commonly used in Eddy covariance measurements to determine the vertical $CO_2$ flux, was considered in Section 14.2.

In a closed system, the air circulates from the chamber to an infrared gas analyzer and then returns to the chamber. Commercial devices can be presented by LI-8100 (LI-COR Biosciences, Lincoln, NE, USA)

There are several modifications of the chamber systems such as a static chamber containing NaOH base trap, open dynamic chamber methods, and head space analysis with gas chromatography (Knoepp and Vose, 2002, and references therein).

We shall discuss here the closed-chamber automated soil $CO_2$ system as an example of advanced technology.

### 24.5.3  The Automated Soil $CO_2$ Flux System

The automated soil $CO_2$ flux system is the closed-chamber system which consists of a chamber, an analyzer control unit (diaphragm pump, filter, and $CO_2$ gas analyzer), and a data-logging unit (Figure 24.1).

The camera is mounted on the surface of the soil surface. $CO_2$ flux enters from the soil to the chamber by diffusion. Air flow is generated by a diaphragm pump inside the analyzer control unit. Air that returns from the chamber passes through a filter before it enters the optical bench of the infrared gas analyzer.

The infrared gas analyzer is an absolute, non-dispersive device based upon a single path, dual wavelength, thermostatically controlled infrared detection system.

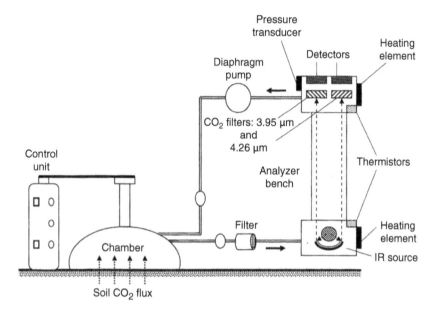

**FIGURE 24.1**    The automated soil $CO_2$ flux system.

This analyzer contains the source of infrared radiation with gold parabolic reflector. $CO_2$ measurements are made by estimating the relative intensities of selected pair of infrared wavelengths: one wavelength (4.26 μm) is strongly absorbed by $CO_2$ and the other (optical reference wavelength 3.95 μm) is not. The degree to which $CO_2$ is absorbed is directly related to the $CO_2$ concentration according to the Beer–Lambert law. The analyzer is equipped with a thermal equilibrium system and a pressure transducer.

The $CO_2$ flux is estimated as the rate of $CO_2$ concentration increase inside the chamber.

The commercial closed system such as LI-8100 (LI-COR Biosciences, Lincoln, NE, USA) is characterized by the following parameters:

- Measurement range: 0–20,000 ppm.
- Accuracy: 1.5% of reading.
- Drift at 0 ppm: <0.15 ppm/°C.
- Span drift: <0.03 ppm/°C.
- Total drift at 337 ppm: <0.4 ppm/°C.
- Signal noise at 370 ppm with 1 second signal averaging: <1 ppm.

*The advantages of an automated soil CO2 flux system:*

- Fast response and highly accurate measurements of the mole fraction of $CO_2$.
- High stability and accuracy.

- Low signal noise.
- Low zero and span drift.
- Low cost.
- Low power requirements.

*The disadvantages of an automated soil CO2 flux system:*

- The $CO_2$ concentration in the chamber headspace starts to change.
- A very small pressure front is generated in the soil immediately beneath the chamber, which may disturb the $CO_2$ concentration gradient in the soil by pushing ambient air into the soil pores.

## REFERENCES

Booth, L.H., Hodge, S., and O'Halloran, K. 2000. The use of enzyme biomarkers in *Aporrectodea caliginosa* (*Oligochaeta; Lumbricidae*) to detect organophosphate contamination: a comparison of laboratory tests, mesocosms and field studies. *Environ. Toxicol. Chem.* 19:417–422.

Bunn, K, Thompson, H., and Tarrant, K. 1996. Effects of agrochemicals on the immune systems of earthworms. *Bull. Environ. Contam. Toxicol.* 57:632–639.

Cikutovic, M.F., Fitzpatrik, L.C., Goven, A.J., Venables, B.J., and Giggleman, M.A. 1999. Wound healing in earthworms Lumbricus terrestris: a cellular-based biomarker for assessing sublethal chemical toxicity. *Bull. Environ. Contam. Toxicol.* 62:508–514.

Cortet, J., Vauflery, A.G.D., Balaguer, N.P., Gomot, L., Texier, Ch., and Cluzeau, D. 1999. The use of invertebrate soil fauna in monitoring pollutant effects. *Eur. J. Soil. Biol.* 35:115–134.

Dai, J., Becquer, T., Rouiller, J.H., Reversat, G., Bernhard-Reversat, F., Nahmani, J., and Lavelle, P. 2004. Heavy metal accumulation by two earthworm species and its relationship to total and DTPA-extractable metals in soils. *Soil Biol. Biochem.* 36:91–98.

Darwin, C. 1881. *The Formation of Vegetable Mould, Through the Action of Worms, With Observations on their Habits.* John Murray, London.

Davidson, E.A., Savage, K., Verchot, L.V., and Navarro, R. 2002. Minimizing artifacts and biases in chamber-based measurements of soil respiration. *Agr. Forest. Meteorol.* 113:21–37.

Fraser, A., Lambkin, D.C., Lee, M.R., Schofield, P.F., Mosselmans, J.F.W., and Hodson, M.E. 2011. Incorporation of lead into calcium carbonate granules secreted by earthworms living in lead contaminated soils. *Geochim. Cosmochim. Ac.* 75(9):2544–2556.

Gago-Duport, L., Briones, M.J.I., Rodriguez, J.B., and Covelo, B. 2008. Amorphous calcium carbonate biomineralisation in the earthworm's calciferous gland: Pathways to the formation of crystalline phases. *J. Struct. Biol.* 162:422–435.

Geiszinger, A, Goessler, W, Kuehnelt, D., Francesconi, K., and Kosmus, W. 1998. Determination of arsenic compounds in earthworms. *Environ. Sci. Technol.* 32:2238–2243.

Hobbelen, P.H., Koolhaas, J.E., and van Gestel, C.A. 2006. Bioaccumulation of heavy metals in the earthworms *Lumbricus rubellus* and *Aporrectodea caliginosa* in relation to total and available metal concentrations in field soils. *Environ. Pollut.* 144(2):639–646.

Kladivko, E.J. 1993. Earthworms and crop management. Agronomy 279. Purdue Univ. Extension Service. http://www.agcom.purdue.edu/AgCom/Pubs/AY/AY-279.html (accessed October 30, 2013).

Knoepp, J.D. and Vose, J.M. 2002. *Quantitative Comparison of* In Situ *Soil $CO_2$ Flux Measurement Methods.* U.S. Department of Agriculture, Forest Service, Southern Research Station. pp. 357–368.

Lee, M.R., Hodson, M.E., and Langworthy, G.N. 2008. Crystallization of calcite from amorphous calcium carbonate: earthworms show the way. *Mineral Mag.* 72(1):257–261.

Römbke, J., Jänsch, S., and Garcia, M. 2007. Earthworms as bioindicators (in particular for the influence of land use). In: *Minhocas na América Latina:biodiversidade e ecologia* (eds. G.G. Brown and C. Fragoso), Embrapa Soja, Londrina, pp. 455–466.

Stockdill, S.M.J. 1982. Effects of introduced earthworms on the productivity of New Zealand pastures. *Pedobiologia* 24:29e35.

Xu, L, Furtaw, M.D., Madsen, R.A., Garcia, R.L., Anderson, D.J., and McDermitt, D.K. 2006. On maintaining pressure equilibrium between a soil $CO_2$ flux chamber and ambient air. *J Geophys Res.* 111: D08S10. doi:10.1029/2005FD006435

# PRACTICAL EXERCISE **10**

# DETERMINATION OF THE SEDIMENTATION VELOCITY AND THE DENSITY OF SOLID PARTICLES

*Sedimentation* is the tendency for solid particles to settle out of a suspension to the bottom of the liquid.

## 1 DERIVATION OF THE SEDIMENTATION EQUATION

Let us consider a spherical solid particle of mass $m$, volume $V$, and density $\rho$, which is falling in a fluid of density $\rho_0$.

This particle reaches a terminal velocity when the retarding forces, viscosity ($F_s = 6\pi\eta R\upsilon$) and buoyancy ($F_A = \rho_0 Vg = \frac{4}{3}\pi R^3 \rho_0 g$) equal the weight ($mg = \frac{4}{3}\pi R^3 \rho g$) of the particle:

$$\frac{4}{3}\pi R^3 \rho g = \frac{4}{3}\pi R^3 \rho_0 g + 6\pi\eta RU, \tag{P10.1}$$

where $R = \frac{D}{2}$ is the radius of the sphere (m); $\rho$ and $\rho_0$ is the density of the sphere and fluid ($kg/m^3$); $\eta$ is the viscosity of the fluid (kg/m·s); $\upsilon$ is the velocity of the particle (m/s); $g$ is the gravitational acceleration ($m/s^2$).

*Methods of Measuring Environmental Parameters*, First Edition. Yuriy Posudin.

Hence the terminal velocity $v$ which is also called *sedimentation velocity* can be determined:

$$v = g\frac{\rho - \rho_0}{\eta}\frac{2R^2}{9}.$$ (P10.2)

Substituting the diameter for the radius gives the other form of the sedimentation equation:

$$U = g\frac{\rho - \rho_0}{\eta}\frac{d^2}{18}.$$ (P10.3)

This equation is valid either for settling particles ($\rho > \rho_0$), or for rising particles ($\rho < \rho_0$).

Assume that $v = l/t$, where $l$ is the path length and $t$ is the time needed for the particle to fall.

The diameter of a particle can be expressed as follows:

$$d = \left[\frac{18l\eta}{tg(\rho - \rho_0)}\right]^{1/2}.$$ (P10.4)

## 2   DETERMINATION OF THE SEDIMENTATION VELOCITY OF SOLID PARTICLES

1. Prepare three lead balls about the same (2–3 mm) diameter.
2. Measure the distance $l$ between the points $a$ and $b$ of a cylindrical vessel that is filled with glycerol.
3. Using a micrometer or slide caliper determine the diameter $d$ of each ball.
4. Throw each ball into the liquid and measure the fall time $t$ between the points $a$ and $b$.
5. Enter in the table the tabular values $\rho = 11.3 \times 10^3$ kg/m$^3$, $\rho_0 = 1.26 \times 10^3$ kg/m$^3$ and $g = 9.8$ m/s$^2$; and the results of measurements ($l$, $d_1$, $d_2$, $d_3$, and $t_1$, $t_2$, $t_3$).
6. Determine the mean values $\langle d \rangle$ and $\langle t \rangle$.
7. Substitute tabular values and values $l$, $\langle d \rangle$, and $\langle t \rangle$ into Equation P10.3 and determine sedimentation velocity $v$.
8. Calculate the error of indirect measurement by the following formulas:

$$\varepsilon_v = \varepsilon_g + \varepsilon_d.$$ (P10.5)

$$\varepsilon_\eta = \varepsilon_g + 2\varepsilon_d + \varepsilon_t + \varepsilon_l + \varepsilon_{\rho - \rho_0} = \frac{\Delta g}{g} + \frac{2\Delta d}{d} + \frac{\Delta t}{t} + \frac{\Delta l}{l} + \frac{\Delta\rho + \Delta\rho_0}{\rho - \rho_0}.$$ (P10.6)

9. Determine the confidence interval of total errors:

$$\Delta_v = \langle v_{cer} \rangle \cdot \varepsilon_v; \qquad \text{(P10.7)}$$

$$\Delta_\eta = \langle \eta \rangle \cdot \varepsilon_\eta \qquad \text{(P10.8)}$$

10. Enter in the table the values of random, systematic, and total errors:

**Tabular values and the results of direct and indirect measurements.**

| N | $\rho$, kg/m³ | $\rho_0$, kg/m³ | $g$, m/s² | $d$, m | $t$, s | $l$, m | $v$, m/s | $\eta$, Pa·s |
|---|---|---|---|---|---|---|---|---|
| | | Tabular Values | | | Results of Direct Measurements | | Results of Indirect Measurements | |
| 1 | — | — | — | | | | | |
| 2 | — | — | — | | | | | |
| 3 | — | — | — | | | | | |
| $\langle x \rangle$ | | | | | | | | |
| $\Delta_c\, P_c = 1$ | | | | | | | | |
| $P = 0.95$ | — | — | — | | | | | |
| $\Delta P \geq 0.95$ | | | | | | | | |
| $\varepsilon$ | | | | | | | | |

Note: The procedure of determination of Student's coefficient and errors of tabular values is described in "Theory of Errors" in Practical Exercise 1.

## 3   DETERMINATION OF THE DENSITY OF SOLID PARTICLES

1. Find a little (2–3 mm) stone, wash and measure its diameter with a micrometer.
2. Throw a stone into the liquid.
3. Measure the time $t$ that is needed for the stone to fall through a distance $l$.
4. Calculate the density $\rho$ of the stone if the ambient temperature is 20°C and viscosity of glycerol is $\rho_0 = 1.26 \times 10^3$ kg/m³ at this temperature.

**Constructive Test**   Particle densities of mineral soils vary between 2600 and 2700 kg/m³. Do the results of your measurements fall within this range?

**Example**   Calculate the sedimentation velocity of the soil particle with a diameter 0.2 mm. Assume that the density of the particle is 2600 kg/m³ and the density of water is 998.2 kg/m³. The temperature of water is 20°C, and viscosity at this temperature is $1.002 \times 10^{-3}$ Pa·s.

*Solution*    Substitute for variables into formula P10.3:

$$U = 9.8 \times \frac{2600 - 998.2}{1.002 \times 10^{-3}} \times \frac{2}{9}(1 \times 10^{-3})^2 = 355.25 \times 10^{-6} \text{ m/s} = 3.55$$
$$\times 10^{-4} \text{ m/s}.$$

**Control Exercise**    How long does a soil particle of diameter 0.4 mm take to fall 15 cm in water at 25°C? The density of the particle is 2650 kg/m$^3$.

## QUESTIONS AND PROBLEMS

1. Define the term "soil."

2. Describe the main components of soil.

3. Give the definition of soil quality.

4. What is the principle of optical emission spectroscopy with inductively coupled plasma?

5. What is the principle of mass spectrometry with inductively coupled plasma?

6. What forces act on the spherical particle, which is immersed in a liquid?

7. What the difference between settling and rising particles?

8. Write the Stokes terminal velocity equation. For which object it is valid?

## FURTHER READING

Amoozegar, A. Soil permeability and dispersion analysis. 2005. In: *Environmental Instrumentation and Analysis Handbook* (eds. R.G. Down and J.H. Lehr). Wiley-Interscience. pp. 625–677.

Arshad, M.A., Lowery, B., and Grossman, B. 1996. Physical tests for monitoring soil quality. In: *Methods for Assessing Soil Quality*. (eds. J.W. Doran and A.J. Jones). Madison, WI. pp. 123-141.

Artiola, J.F., Pepper, I.L., and Brusseau, M.L. 2004. *Environmental Monitoring and Characterization*. Elsevier Academic Press, San Diego, CA.

Doran, J.W., and Parkin, T.B. 1994. Defining and assessing soil quality. In: *Defining Soil Quality for a Sustainable Environment*. (eds. Doran J.W., D.C. Coleman, D.F. Bezdicek, and B.A. Stewart). SSSA Publ. No.35. Soil Science Society of America, 677 South Segoe Rd., Madison, WI, pp. 3–21.

Doran, J.W. and Parkin, T.B. 1996. Quantitative indicators of soil quality: a minimum data set. In: *Methods for Assessing Soil Quality*. (eds. J.W. Doran and A.J. Jones). SSSA, Inc., Madison, WI.

Hillel D. 1998. *Environmental Soil Physics*. Academic Press, San Diego-London-Boston, p. 771.

Kemper, W.D. and Rosenau, R.C. 1986. Aggregate Stability and Size Distribution. In: *Methods of Soil Analysis*. Part 1. Physical and mineralogical methods. (ed. A. Klute). Madison, WI, pp. 425–442.

Larson, W.E. and Pierce, F.J. 1991. Conservation and enhancement of soil quality. In: *Evaluation for Sustainable Land Management in the Developing World*. (eds. J. Dumanski, E. Pushparajah, M. Latham, and R. Myers). Vol. 2: Technical Papers. Proc. Int. Workshop, Chiang Rai, Thailand. September 15–21, 1991. Int. Board for Soil Res. and Management, Bangkok, Thailand.

Lili, M., Bralts, V.F., Yinghua, P., Han, L., and Tingwu, L. 2008. Methods of measuring soil infiltration: State of art. *Int. J. Agric. Biol. Eng.* 1(1):22–30.

Parkin, T.B., Doran, J.W., and Franco-Vizcaino, E. 1996. Field and laboratory tests of soil respiration. In: *Methods for Assessing Soil Quality*. (eds. J.W. Doran and A.J. Jones) SSSA Spec. Publ. # 49. Soil Science Society of America, Madison, WI, pp. 231–245.

Pirez, L.E. and Bacchi, O.O.S. 2006. Design, construction and performance of a pressure chamber for water retention curve determination through traditional and nuclear methods. *Nucleonika*. 51(4):225–230.

Radziemski, L.J. and Cremers, D.A. 2006. *Handbook of Laser-Induced Breakdown Spectroscopy*. John Wiley & Sons, New York.

Schechter, I., Miziolek, A.W., and Palleschi, V. 2006. *Laser-Induced Breakdown Spectroscopy (LIBS): Fundamentals and Applications*. Cambridge University Press, Cambridge.

## ELECTRONIC REFERENCES

Andrews, S., Andrews, S., Wander, M., and Kuykendall, H. Soil quality for environmental health, http://soilquality.org/home.html (accessed October 1, 2013).

Automatic infiltration meter, http://www.ictinternational.com/aa-shared/pdf/aim/aim.pdf (accessed October 3, 2013).

Capewell, M. The why and how to testing the electrical conductivity of soils, http://www.agriculturesolutions.com/resources/92-the-why-and-how-to-testing-the-electrical-conductivity-of-soils (accessed October 5, 2013).

Department of Sustainable Natural Resources. Soil survey standard test method. Available water capacity, http://www.environment.nsw.gov.au/resources/soils/testmethods/awc.pdf (accessed October 1, 2013).

Diamond light source: illuminating chemistry, http://www.rsc.org/Education/EiC/issues/2011May/DiamondLightSource.asp (accessed October 2, 2013).

Fortin, S. Infiltrometer tests, http://www.robertsongeoconsultants.com/hydromine/topics/Permeability_Testing/DoubleRing_Infiltrometer.html (accessed October 4, 2013).

Foun E., Shenk J., Kalinowski C., Grant K., Grabowski J., and Smith B. 2006. Arsenic accumulation in *Lumbricus terrestris*, http://courses.umass.edu/chemh01/Fall%202006/Proposals/Group%201%20proposal.htm (accessed October 5, 2013).

Inductively coupled plasma atomic emission spectroscopy, http://en.wikipedia.org/wiki/Inductively_coupled_plasma_atomic_emission_spectroscopy (accessed October 6, 2013).

Kinyangi, J. 2007. Soil health and soil quality: a review, http://worldaginfo.org/files/Soil%20Health%20Review.pdf (accessed October 7, 2013).

Kladivko, E.J. 1993. Earthworms and crop management. Agronomy 279. Purdue Univ. Extension Service, http://www.agcom.purdue.edu/AgCom/Pubs/AY/AY-279.html (accessed October 8, 2013).

Lecture 9. Chapter 4 Soil water properties, http://jan.ucc.nau.edu/~doetqp-p/courses/env320/lec9/Lec9.html (accessed October 9, 2013).

Madsen, R., Xu, L., and Medermitt, D. 2010. Considerations for making chamber-based soil CO2 flux measurements. 19th World Congress of Soil Science, Soil Solutions for a Changing World August1–6, 2010, Brisbane, Australia. Published on DVD, http://www.iuss.org/19th%20WCSS/Symposium/pdf/1272.pdf (accessed October 10, 2013).

Mendoza T.C. The living soil: The basis of ecologically sustainable agriculture. 2012. http://pt.scribd.com/doc/103113928/The-Living-Soil-The-Basis-of-Ecologically-Sustainable-Agriculture-by-Dr-Ted-C-Mendoza (accessed October 11, 2013).

Method 6020A. Inductively coupled plasma-mass spectrometry (ICP-MS), http://www.epa.gov/epawaste/hazard/testmethods/sw846/pdfs/6020a.pdf (accessed October 12, 2013).

Newberry, W.R., Butler, L.C., Hurd, M.L., Laing, G.A., Stapanian, M.A., Aleckson, K.A., Dobb, D.E., Rowan, J.T., and Garner, F.C. 1989. Final report of the multi-laboratory evaluation of method 6020 CLP-M inductively coupled plasma-mass spectrometry, http://www.epa.gov/osw/hazard/testmethods/sw846/pdfs/6020a.pdf (accessed October 14, 2013).

Operating instructions 09.09. Tension infiltrometer. Eijkelkamp Agriserch equipment. 2010. http://pkd.eijkelkamp.com/Portals/2/Eijkelkamp/Files/Manuals/M1-0909e%20Tension%20infiltrometer.pdf (accessed October 15, 2013).

pF values and measurements, http://pkd.eijkelkamp.com/Portals/2/Eijkelkamp/Presentations/pF%20measurements%20in%20the%20lab.pdf (accessed October 16, 2013).

Pumpanen, J., Laakso, M., Uusimaa, M. Measuring CO2 from our breathing soil, http://www.aweimagazine.com/article.php?article_id=567 (accessed October 18, 2013).

Schomakers, J., Mentler, A., Degischer, N., Blum, W.E.H., and Mayer, H. 2011. Measurement of soil aggregate stability using low intensity ultrasound vibration. *SJSS*. 1(1). http://sjss.universia.net/pdfs_revistas/articulo_234_1319128166703.pdf (accessed October 19, 2013).

Soil quality indicators. USDA Natural Resources Conservation Service. 2011. http://soils.usda.gov/sqi/assessment/files/aggregate_stability_sq_physical_indicator_sheet.pdf (accessed October 20, 2013).

Soil quality indicators: pH, http://soils.usda.gov/sqi/publications/files/indicate.pdf (accessed October 21, 2013).

Soil quality information sheet. Soil quality resource concerns: Available water capacity, http://soils.usda.gov/sqi/publications/files/avwater.pdf (accessed October 22, 2013).

Soil quality test kit guide, Section I, Chapter 10, p 22–23. See Section II, Chapter 9, pp. 73–75, http://soilquality.org/indicators/earthworms.html (accessed October 22, 2013).

Soil Science Glossary (Soil Science Society of America), https://www.soils.org/publications/soils-glossary/ (accessed October 23, 2013).

Soil survey standard test method bulk density of a soil: clod method, http://www.environment.nsw.gov.au/resources/soils/testmethods/bdsc.pdf (accessed October 24, 2013).

The soil quality test kit guide, Chapter 2, p 4–6. See Section II, Chapter 1.Respiration, p 52–54, http://soilquality.org/indicators/respiration.html (accessed October 28, 2013).

The soil quality test kit guide, Section I. Soil pH. Chapter 6, p. 15; Section II, Chapter 5, pp. 63–66, http://soilquality.org/indicators/soil_ph.html (accessed October 29, 2013).

The soil quality test kit guide, Section I. Aggregate Stability. Chapter 8, pp. 18–19; Section II, Chapter 7, pp. 69–71, http://www.soilquality.org/indicators/aggregate_stability.html (accessed October 29, 2013).

The soil quality test kit guide, Section II, Chapter 2. Infiltration. pp. 55–56, http://soilquality.org/indicators/infiltration.html (accessed October 30, 2013).

United States Department of Agriculture (USDA), Natural Resources Conservation Service, 2005. National Soil Survey Handbook, title 430-VI. Soil Properties and Qualities (Part 618), Available Water Capacity (618.05), http://soils.usda.gov/technical/handbook/ (accessed October 31, 2013).

USDA Natural Resources Conservation Service, January 1998. Soil Quality Information Sheet. Soil Quality Resource Concerns, Available Water Capacity, http://urbanext.illinois.edu/soil/sq_info/awc.pdf (accessed October 31, 2013).

# PART V

## VEGETATION FACTORS

# 25

# SPECTROSCOPIC ANALYSIS OF PLANTS AND VEGETATION

## 25.1 SPECTROSCOPIC APPROACH

*Spectroscopy* is a science that studies spectra of electromagnetic radiation for qualitative and quantitative analysis of the structure and properties of matter.

A *spectrum* is a dependence of the intensity of radiation absorbed on the frequency of electromagnetic radiation.

*Vegetation* is a complex concept that involves soil together with floral ensemble and the near-surface layer of the atmosphere. Supervision of such cover requires the development of new modern monitoring methods that enable to give accurate and comprehensive information on all stages of plants during development and under stress conditions.

Investigation of the dependence of the intensity of radiation absorbed, transmitted, reflected, scattered, or reemitted by a single leaf, plant, or vegetation as a whole on the wavelength or frequency of optical radiation provides qualitative and quantitative analysis of these plant objects during development and under stress conditions.

### 25.1.1 Optical Radiation

*Optical radiation* (*light* in wide meaning of the word) includes the electromagnetic waves that have a wavelength from about 300 nm (ultraviolet region), through the visible region (400–700 nm), and into the infrared region (from 760 to 2500 nm).

*Methods of Measuring Environmental Parameters*, First Edition. Yuriy Posudin.
© 2014 John Wiley & Sons, Inc. Published 2014 by John Wiley & Sons, Inc.

Optical radiation has a *dual nature* as it is found as both waves and particles. According to the classical interpretation, electromagnetic radiation consists of changing electric and magnetic fields that propagate through space forming an electromagnetic wave. This wave is characterized by the amplitude, wavelength, and polarization. In terms of the quantum theory, electromagnetic radiation consists of particles called *photons*—discrete packets (*quanta*) of electromagnetic energy that move at the speed of light. Each photon is characterized with its energy $E = h\nu$ and momentum $p = h/\lambda$, where $h$ is Planck's constant, $\nu$ is the frequency, and $\lambda$ is the wavelength of the electromagnetic wave. Dualism of light is presented by the formula $E = h\nu$, where the energy $E$ is related to the particle, and frequency $\nu$ is associated with the wave.

### 25.1.2 The Interaction of Light with Plant Objects

An optical radiation that interacts with plant object takes part in some processes: it can be absorbed by this object and transmitted by it, reflected by its surface, scattered by structural elements, and reemitted by internal pigments. If we shall neglect the participation of scattering and reemission, this interaction can be described by the following equation:

$$I_0 = I_r + I_a + I_t, \tag{25.1}$$

wwhere $I_0$ is the total intensity of radiation incident on the object; $I_r$, $I_a$, and $I_t$ describe the intensity of radiation reflected, absorbed, and transmitted by the object correspondingly.

### 25.1.3 Reflectance

A *reflectance* (or *coefficient of reflection*) $\rho$ describes the intensity of radiation $I_r$ reflected by the object relative to the total intensity $I_0$ of radiation incident on the object:

$$r = \frac{I_r}{I_0}. \tag{25.2}$$

*Specular reflection* is the reflection of light from a surface where at the point of reflection, an incident beam is reflected at (and only at) an angle equal to the angle of incidence (both taken with respect to the perpendicular at that point).

*Diffuse* (or *body*) *reflection* is the reflection of light from an uneven or granular surface such that an incident ray is seemingly reflected at a number of angles. Diffuse reflection provides a more rich in comparison with specular reflection information about the internal properties of the object due to numerous processes of random reflection, refraction, and scattering at interfaces of internal elements inside the sample (Fig. 25.1, b).

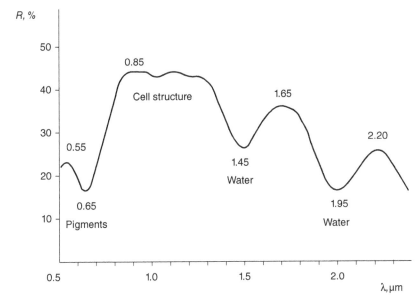

**FIGURE 25.1**   Reflectance spectrum of green leaf.

## 25.2   REFLECTANCE SPECTROSCOPY

*Reflectance spectroscopy* is the branch of spectroscopy which studies an optical radiation as a function of wavelength that has been reflected from an object.

The *reflectance spectrum* is the plot of the reflectance as a function of wavelength.

Reflectance spectrum of green leaf is shown in Figure 25.1. It includes three main parts: 500–750 nm, which characterizes the absorption of vegetable pigments, such as chlorophylls *a* and *b*, carotenoids, xanthophylls, and anthocyanins; 0.75–1.35 μm, which is characterized by high reflectance due to internal structures (e.g., cellulose) of the leaf; 1.35–2.50 μm is a spectral part where an intense water absorption with maxima at 1.45 and 1.95 μm takes place.

Thus, a single leaf reflectance spectrum demonstrates the maximum reflectance at 550 nm in the visible region of the spectrum, a broadband of reflectance at 0.75–1.35 μm, and the maxima of reflectance at 1.65 and 2.20 μm in the infrared spectrum.

## 25.3   METHODS OF REFLECTANCE SPECTROSCOPY

All methods of reflectance spectroscopy can be divided into laboratory methods, field methods (which are based on the application of portable reflectance instrumentation or near-field measurements that are carried out at a short distance from the plant site), and remote-sensing methods (which use the detection of natural radiation that is reflected by the object).

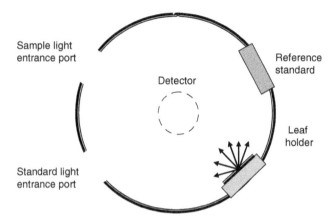

**FIGURE 25.2**    The integrating sphere.

### 25.3.1  Laboratory Methods

Laboratory methods include the study of reflectance spectra of green leaves using spectrophotometers equipped with integrating sphere (Figure 25.2), which collects all the radiation reflected from a sample and spreads this radiation over the entire surface area of the sphere with a high degree of homogeneity. The sphere is equipped with small holes for entrance and exit ports. A detector responds to a portion of this radiation and gives information about the average hemispherical reflectance of the sample. If the light is directed at the sample at an angle of 0°, specular reflected light is not detected as it exits the integrating sphere.

The integrating sphere can be covered by two types of reflective coatings such as Spectraflect® (300–2400 nm wavelength range) and Infragold® (0.7–20 μm wavelength range). Such factors of the integrating sphere as sphere efficiency, signal throughput, spectrum noise level, measurement accuracy, sphere port fraction, detector baffles, and sample beam size are considered by Taylor (2010).

The reflectance of a sample can be derived using the ratio between a known reference ($BaSO_4$) and unknown sample.

The combination of an integrating sphere with a UV-VIS spectrophotometer allows to widen its spectral range. For example, the spectral range of the Shimadzu UV-2600/2700 spectrophotometer (Shimadzu Scientific Instruments) can be extended from 185–900 nm to 200–1400 nm in combination with the optional dual detector integrating sphere ISR-2600Plus.

*The advantages of a spectrophotometer equipped with integrating sphere:*

- Wide spectral range.
- High sensitivity.
- Signal to noise in the laboratory is much higher.
- No single lamp can be used during the analysis; changing the lamp is a time-consuming process.

### 25.3.2  Portable Reflectance Instrumentation

The portable reflectance instruments can be used to measure the amount of chlorophyll content and nitrogen status in plant leaves that are perspective indicators of leaf physiology and plant health.

Typical portable reflectometer (UniSpec-SC, PP Systems) consists of the source of optical radiation (tungsten halogen lamp), miniature photodiode array detector, fiber optics, leaf clips, interface with integral computer and large, full color LCD, battery, and reference standard. A wavelength range is 310–1100 nm.

It can be applied for measurement of leaf or canopy reflectance. Spectral reflectance makes it possible to obtain information about leaf or canopy structure, nutrient status, and pigment content.

Dimensions of UniSpec-SC are 25 cm × 15 cm × 8.5 cm; weight is 1.7 kg (3.74 lb).

*The advantages of a portable reflectance instrument:*

- Non-invasive and non-destructive measurements.
- Compactness.
- Quick and easy measurements.
- High accuracy.

### 25.3.3  Near-Field Reflectance Instrumentation

Near-field methods are based on the measurements of vegetation reflectance spectrum in the field. A typical device for measuring near-field reflectance spectra of vegetation is shown in Figure 25.3.

A multi-angle spectrometer for automatic measurement of near-field spectral reflectance of plant canopies is described by Leuning et al. (2006). It contains a rotating periscope that has been installed on top of a 70-m tower. An optical periscope was connected via a fiber optic cable to a UniSpec-DC spectrometer and associated electronics.

The UniSpec-DC spectrometer (Spectronic Devices Ltd) is a dual channel, field portable instrument capable of simultaneous measurement of incident and reflected light. In addition to measuring, the device is capable of calculating a number of vegetation indices.

Such a system provides automatic measurements of spectral reflectance of a forest canopy at four azimuths at hourly intervals every day of the year. Wavelength range of the system is 300–1150 nm.

*The advantages of a near-field reflectance instrumentation:*

- Noninvasive and nondestructive measurements.
- Can be used for automated measurements.
- Is capable to calculate a number of vegetation indices.

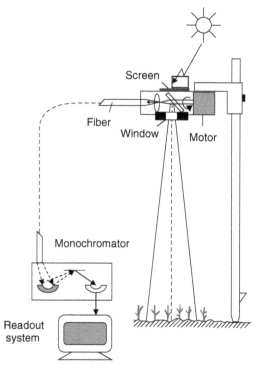

**FIGURE 25.3**   A typical device for measuring near-field reflectance spectra of vegetation.

### 25.3.4   Vegetation Indices

*Spectral vegetation indices* represent the sum, difference, or ratio of spectral parameters determined at some analytical wavelengths. Such vegetation indices make it possible to establish relationships between vegetation characteristics of plants under stress conditions and reflectance parameters of these plants.

Let us consider the most common vegetation indices.

*Ratio vegetation index* (RVI) is calculated using the following formula:

$$RVI = NIR/RED, \tag{25.3}$$

where NIR is the near-infrared (750–900 nm) and RED is the red (630–700 nm) parts of spectrum (Gorte, 2000).

The larger the RVI, the larger the percentage of healthy vegetation. The RVI ranges from 0 to infinity (Jordan, 1969).

*Normalized difference vegetation index* (NDVI) is calculated as follows:

$$NDVI = (NIR - RED)/(NIR + RED). \tag{25.4}$$

This index varies between −1 and 1. Healthy vegetation absorbs visible radiation and reflects near-infrared radiation. Thus, values of NDVI close to +1 indicate the highest possible biomass; a zero values mean no vegetation (soil, sand, and snow); negative values of NDVI (about −1) correspond to water. The advantage of this index is close to linear dependence of its value on the amount of plant biomass (Rouse et al., 1974).

NDVI can be used for identification of the health status of plants, quantitative and qualitative estimation of green biomass and crops, and to study the effect of agrochemical products on the plants.

### 25.3.5 Remote Sensing of Vegetation Reflectance

*Remote sensing* is the technique or method of obtaining information about an object being investigated without actual contact with the object.

There are two basic approaches to remote sensing.

*Passive Remote Sensing Systems* (or *Passive Sensors*) are based on the illumination of the object by natural radiation (e.g., sunlight), whereas *Active Remote Sensing Systems* (or *Active Sensors*) provide the exposure of object by artificial radiation (e.g., laser).

Remote sensing of vegetation reflectance is realized by a number of sensors mounted on air carriers or satellites. They differ in the sensor design, the spectral range, a number of spectral bands, and the spatial and temporal resolution. Let us mention as examples such systems as AVIRIS (Airborne Visible/Infrared Imaging Spectrometer, NASA Jet Propulsion Laboratory); CASI (Compact Airborne Spectrographic Imager, ITRES Research Limited); HyMap (HyVista Corporation, Integrated Spectronics Pty Ltd.); Landsat (NASA and the U.S. Geological Survey); MODIS (Moderate Resolution Imaging Spectrometer, NASA Jet Propulsion Laboratory); IKONOS (Lockheed Martin Corporation, USA); KOMPSAT-2 MSC (Korea Multi-Purpose Satellite).

### 25.3.6 Multispectral Scanning

One of the main methods of remote sensing is a multispectral scanning (MSS). The principle of operation of these systems is the registration of spectral reflectance of objects in certain spectral parts of visible and infrared spectrum (0.3–14 μm). MSS devices that are installed on satellites provide information with a resolution of about 10 m, while the scanning area is about 60–185 km.

The design of MSS systems can be realized as *across-track scanners* that employ a rotating or oscillating mirror to scan back and forth across the line of flight or *along-track scanners* that scan in parallel along the direction of flight (Sabins, 1987).

The reflected radiation flows from the vegetation surface through the atmosphere into the aperture of the scanner. Across-track scanners provide a series of lines which are oriented perpendicular to the direction of motion of the sensor carrier using a rotating mirror. Then, the scanned radiation enters the input of the monochromator and detector, respectively (Figure 25.4a).

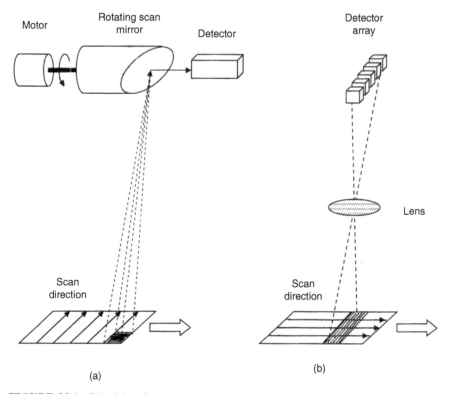

**FIGURE 25.4** Principle of operation of multispectral scanning: (a) across-track scanner; (b) along-track scanner.

Along-track scanners use the forward motion of the carrier; these scanners use a linear array of charge-coupled devices (CCDs) (dense arrays of photodiodes) that are located at the focal plane of the image formed by lens (Figure 25.4b). Each detector responds to a specific spectral region.

### 25.3.7 Spectral Bands MSS and TM

Let us consider an example of practical realization of remote sensing of vegetation such a classical satellite system as Landsat (Land Satellite). Two principal sensors were carried by Landsat: multispectral scanner (MSS) and thematic mapper (TM). The following spectral bands can be used as vegetation indices:

Band MSS4: 500–600 nm, green region. Associated with the absorption of chlorophyll and is responsible for the reflectance of healthy plants.

Band MSS5: 600–700 nm, red region. This band should be used to identify plants and estimate cultural features.

Band MSS6: 700–800 nm, near-infrared region. Used for estimation of vegetation biomass, crop identification, and delineation of the vegetation boundary between land and water.

Band MSS7: 800–1100 nm, near-infrared region. Applicable for inspection and assessment of vegetation.

Band TM1: 450–520 nm, blue region. Used for recognizing the boundaries between soil and vegetation, forest-type mapping, and cultural feature identification.

Band TM2: 520–600 nm, green region. Used for vegetation discrimination and vigor assessment. Also useful for cultural feature identification.

Band TM3: 630–690 nm, red area. Used for plant species differentiation and cultural feature identification.

Band TM4: 760–900 nm, near-infrared region. Used for determination of the vegetation biomass, crop identification, and delineation of the boundaries of the distribution of soil, plant, and water areas.

Band TM5: 1.55–1.74 μm, mid-infrared region. Useful for studying the impact of drought on crops and analysis of vegetation.

Band TM6: 10.40–12.50 μm, thermal infrared region. Used to assess the impact of stress on vegetation and crops, including thermal factors and insecticides.

Band TM7: 2.08–2.35 μm, mid-infrared region. This band is important for evaluation of vegetation moisture content.

### 25.3.8  Spectral Vegetation Indices that are used in the Remote Sensing

Currently, there are about 160 versions of vegetation indices. They are chosen by comparing the properties of vegetation with characteristics of its reflectance.

Let us consider principal vegetation indices that are based on the application of spectral bands of a series of MSS and TM:

*Vegetation Index, VI*

$$VI = \frac{MSS7 - MSS5}{MSS7 + MSS5}. \tag{25.5}$$

*Relative Vegetation Index, RVI*

$$RVI = \frac{TM4}{TM3}. \tag{25.6}$$

This index ranges from 0 to infinity. The values of RVI increase with green biomass (usually the values of RVI are 2–8) (Jordan, 1969).

*Normalized Differential Index, NDVI*

$$NDVI = \frac{TM4 - TM3}{TM4 + TM3}. \tag{25.7}$$

This index varies from $-1$ to $+1$. NDVI usually takes values from 0.2 (low foliar activity) to 0.8 (active level of foliar activity) (Rouse et al., 1974).

*Transformed Vegetation Index, TVI*

$$TVI = \sqrt{VI + 0.5}; \qquad (25.8)$$

here, 0.5 is added to avoid negative values of the square root. NDVI and TVI are functionally equivalent (Tucker, 1979).

*Normalized Difference Infrared Index, NDII*

$$NDII = \frac{TM4 - TM5}{TM4 + TM5}. \qquad (25.9)$$

The index can range from $-1$ to $+1$. NDII usually takes values from 0.02 to 0.6 for green vegetation (Hardisky et al., 1983).

*Moisture Stress Index, MSI*

$$MSI = \frac{TM5}{TM4}. \qquad (25.10)$$

NDII varies from 0.04 to 2.0 for green vegetation (Rock et al., 1985).

*Perpendicular Vegetation Index, PVI*

$$PVI = \sqrt{(MSS5_S - MSS5_V)^2 + (MSS7_S - MSS7_V)^2}, \qquad (25.11)$$

where subscripts S and V correspond to soil and vegetation, respectively. The use of this index makes it possible to avoid the influence of the soil background (Richardson and Wiegand, 1977).

*The advantage of a remote sensing of vegetation reflectance:*

- Relatively low power consumption (less than active sensors).

*The disadvantages of a remote sensing of vegetation reflectance:*

- Passive sensors can only be used to detect energy when the naturally occurring energy is available.
- Dependence of results of detection on the weather conditions.

## 25.4   EFFECT OF EXTERNAL FACTORS ON SINGLE LEAF AND CANOPY REFLECTANCE

Reflectance parameters of a single leaf are influenced by such factors as the level of pigmentation (e.g., high chlorophyll concentrations correspond to low values of the coefficient of reflection), the position of the leaf on a plant node (older leaves show greater reflectance), and side of the leaf (upper side of leaf contains more chlorophyll than the lower one). In addition, the reflectance of a leaf depends on the dehydration (lack of water), extreme temperatures, lack of nutrients, and excess ozone.

Reflectance properties of vegetation cover depend on its geometry (area size and orientation of the leaves, the number of leaf layers), type of plants, and forming the cover. In addition, a significant effect is caused by meteorological and climatic conditions, elevation of the sun, the presence of clouds, pollution, and aerosols in the atmosphere, the type and spectral properties of soil, and agricultural treatment of vegetation. Consequently, the reflectance spectrum of vegetation is characterized by a more contrasting reflectance band within 750–1350 nm range compared to a single leaf.

## 25.5  FLUORESCENCE SPECTROSCOPY

### 25.5.1  Photosynthesis and Chlorophyll Fluorescence

Green plants synthesize organic compounds (carbohydrates) as a fuel for their viability using inorganic raw materials. This process of converting solar energy into chemical energy that occurs in sunlight is called *photosynthesis.*

This process includes steps such as light absorption by pigment (chlorophyll) molecule, excitation energy transfer, and chemical reactions in photosystem II. De-excitation of the absorbed light energy is accompanied by heat release and light emission in the form of fluorescence.

*Fluorescence* is the process in which susceptible molecule (fluorophore) is able to absorb light at a certain wavelength and reemit light from electronically excited state at longer wavelength.

Fluorescence is the radiative process which is based on the transition between electronic states of the same multiplicity; usually it occurs from the ground vibrational state of the first electronic singlet state $S_1$ to various vibrational levels in ground singlet state $S_0$ and is accompanied by the emission of photon. The light absorbed by the accessory pigments (chlorophyll $b$ and carotenoids) is transferred to chlorophyll $a$. This is why the primary processes of photosynthesis are reflected by chlorophyll $a$ fluorescence.

It is found that about 2–5% of the absorbed light energy is converted into radiation energy by chlorophyll (Lichtenthaler and Rinderle, 1988). The relationship between chlorophyll fluorescence and the overall process of photosynthesis is quite complex, however, the registration of chlorophyll fluorescence of green leaves can be used for the analysis of plants under the influence of abiotic and anthropogenic factors in laboratory and field conditions.

*Fluorescence spectroscopy* is a branch of optical spectroscopy which analyzes fluorescence from a sample.

## 25.5.2  Fluorescence Properties of a Green Leaf

The fluorescence emission spectrum of a green leaf is characterized by a maxima at 735–740 nm and 685–690 nm (red part of spectrum) and at 440–450 nm (blue part); the leaves of some plants demonstrate a shoulder near 520–530 nm (green part). According to the modern ideas about the character of the fluorescence spectrum of a green leaf, the red maxima originate from chlorophyll *a*. The experimental investigations showed the presence of a number of native forms of chlorophyll. There are the fluorescence bands at room temperature near 672, 677, 682, 687, 693–695, 700, 720–725, 735, 750–760, 800–820 (the underlined figures correspond to the intense bands) (Kochubey, 1986). The appearance of these bands is related to the process of aggregation of pigment molecule and formation the chlorophyll–protein complexes.

It is not clear enough which compounds are responsible for blue and green part of fluorescence spectrum. There is an opinion (Lang et al., 1991; Lichtenthaler et al., 1991; Goulas et al., 1990) that various phenolic plant constituents in the vacuoles and cell walls of both the epidermal layer and the mesophyll cells such as the coumarines aesculetin and scopoletin, with a very high fluorescence yield, and cinnamic acids and derivatives, with a lower yield (such as caffeic and chlorogenic acid, sinapic acid and catechin) are responsible for blue-green fluorescence. Some investigators (Chappelle et al., 1984a, 1984b) consider that $\beta$-carotene, nicotinamide adenine diphosphate (NADPH), and vitamin $K_1$ participate in fluorescence at 440 nm. But purified $\beta$-carotene does not show any blue or green fluorescence (Lang et al., 1991); the yield of blue fluorescence of NAPDPH is extremely low (Lang et al., 1992); the phylloquinone, $K_1$, fluoresces only under decomposition after intense UV exposure (Interschick-Niebler and Lichtenthaler, 1981).

The red chlorophyll fluorescence with two maxima near 690 nm and 740 nm emanates from the chlorophyll *a* in the chloroplasts of the leaves' mesophyll cells (Stober and Lichtenthaler, 1993).

The shape and intensity of the main bands in fluorescence emission spectrum of green leaves depend on the pigment content which may vary due to such factors as the growth phase of the plants, natural stresses (high light, heat, water shortage, mineral deficiency), and anthropogenic stresses (agrochemical treatment, air pollution, ozone, acid rains, heavy metals, and ultraviolet irradiation).

## 25.5.3  Fluorescent Properties of Vegetation

Fluorescence of plant canopy in natural conditions differs from that observed at the level of a single leaf. First of all, heterogeneity of chlorophyll distribution among the plants should be taken into account. This heterogeneity is associated with various age of leaf and, respectively, different rate of photosynthesis, with pigment concentration, dependence of the physiological state of plants on natural light conditions, agrochemical treatment, water and temperature stress, time of day, and seasons.

## 25.6    LABORATORY METHODS OF FLUORESCENCE SPECTROSCOPY

### 25.6.1    Spectrofluorometry

Recording the fluorescence excitation spectra and the fluorescence emission spectra of the objects under investigation for the purpose of their analysis is called *spectrofluorometry*.

An *excitation spectrum* is the dependence of the emission intensity $I_{fl}$ on the excitation wavelength $\lambda_{exc}$ at a constant emission wavelength $\lambda_{em}$

$$I_{fl} = f(\lambda_{exc}); \lambda_{em} = \text{const.} \tag{25.12}$$

An *emission spectrum* is the dependence of the excitation intensity $I_{fl}$ on the emission wavelength $\lambda_{em}$ at a constant excitation wavelength $\lambda_{exc}$:

$$I_{fl} = f(\lambda_{em}); \lambda_{exc} = \text{const.} \tag{25.13}$$

Spectrofluorometry makes it possible to analyze the dependence of the shape and intensity of the excitation and emission spectra of chlorophyll fluorescence.

The typical design of spectrofluorometer is presented in Figure 25.5. It contains the source 1 of light, excitation and emission monochromators, a number of reflecting

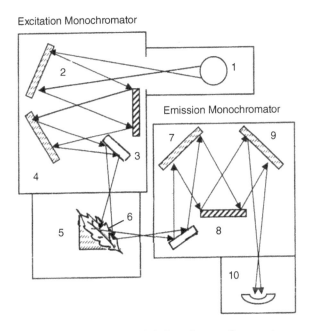

**FIGURE 25.5**   Typical design of spectrofluorometer.

mirrors 2, 4, 7, 9, diffraction gratings 3, 8, sample holder 5, sample 6, and detector 10. Excitation radiation from a source is directed to the monochromator of excitation where it successively passes optical elements (lenses and reflecting mirrors), dispersive element (diffraction grating), and reaches the sample (green leaf). Fluorescence emission of green leaf is directed to the emission monochromator and detector. Each monochromator is scanned in wavelength by rotating the dispersive element (diffraction grating) to record corresponding (excitation or emission) fluorescence spectrum.

*The advantages of a spectrofluorometer:*

- Nondestructive action.
- High sensitivity.
- Can be used for quantitation of fluorescent species.
- Easy and quick to perform analysis.
- Spectrofluorometry can be adapted to the microscopic level as microspectrofluorimetry.

*The disadvantages of a spectrofluorometer:*

- The long-term process of recording fluorescence spectrum during which the changes in the fluorescence intensity caused by the induction of chlorophyll fluorescence take place.
- It requires expensive and somewhat sophisticated equipment.
- Effect of optical components that provoke light scattering.
- Nonuniform spectral output of the light sources.
- Wavelength dependence of the monochromators.
- Wavelength and time dependence of detectors efficiency.
- Dependence of the efficiency of diffraction gratings on the light polarization.

### 25.6.2 Fluorescence Induction Kinetics

Photosynthetic activity of plant can be estimated through the measurement of *fluorescence induction kinetics* of dark-adapted green plant sample. The fact is that illumination of a pre-darkened leaf15–20 minutes induces induction kinetics of the chlorophyll fluorescence. This temporal behavior of the chlorophyll fluorescence intensity is known as "Kautsky effect" ( Kautsky and Hirsch, 1931). This induction kinetics consists of fast (100–500 ms) fluorescence rise to the maximum fluorescence level $f_m$ and slow (usually 4 minutes) fluorescence decrease to the steady-state level $f_s$. A typical fluorescence induction curve in minute range is shown in Figure 25.6. The fluorescence kinetics reflects the sum total of processes that are linked with photosynthesis activity of a plant object.

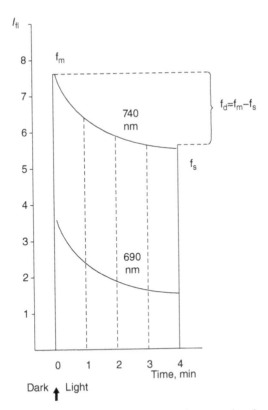

**FIGURE 25.6**   A typical fluorescence induction curve in minute range.

The parameters of the fluorescence induction kinetics have been measured in the laboratory and field conditions with a portable two-wavelength fluorometer (Posudin et al., 2007, 2008, 2010).

Fluorometer consists of light diode that is used as a source of fluorescence excitation; collimator and prism, beam splitter, sample (green leaf), interference filters with transmittance maxima at 690 and 740 nm, photodetectors, amplifier, and readout system. Two last units are connected with power supply (accumulator). The device is equipped with display, where fluorescence indices are indicated, and acoustic signalization that controls the 4-minute period of recording chlorophyll fluorescence.

*The advantages of a recording fluorescence induction kinetics:*

- Nondestructive action.
- High sensitivity.
- Can be used in the field.
- Easy and quick to perform analysis.

*The disadvantages of a recording fluorescence induction kinetics:*

- Dependence of the signal which is recorded by the intensity of radiation.
- Effects of surrounding light.
- Requirement of dark adaptation.

### 25.6.3   Optical Multichannel Analysis

*Optical multichannel analysis (OMA)* is based on the simultaneous detection of fluorescence in the blue, green, and red parts of the spectrum by polychromator with a diffraction grating and array detectors (their number can be up to 512). Chlorophyll fluorescence is excited by ultraviolet laser radiation. In fact, such a system does not record the fluorescence spectra but measures the fluorescence intensity at all wavelengths simultaneously. Thus, the induction of chlorophyll fluorescence can be neglected in this method. Schematic of optical multichannel analyzer and sum total of fluorescence emission spectra that are recorded simultaneously are shown in Figure 25.7.

*The advantages of an optical multichannel analyzer:*

- Avoids the temporal changes of chlorophyll fluorescence intensity.
- High sensitivity.
- Easy and quick to perform analysis.
- Spectrofluorometry can be adapted to the microscopic level as microspectrofluorimetry.

*The disadvantages of an optical multichannel analyzer:*

- It requires expensive and somewhat sophisticated equipment.
- Cannot be used in the field.

### 25.6.4   Pulse Amplitude Modulation Fluorometry

Pulse amplitude modulation (PAM) fluorometry method is based on the modulation of chlorophyll fluorescence by saturated light pulses (Schreiber et al., 1986).

The PAM measuring principle is based on the rapid switching on and off of the measuring light; it is not strong enough to stimulate photosynthesis but does promote a fluorescence signal. The fluorescence signal follows the on/off pattern (i.e., modulated) of the measuring beam and is measured with suitable light filters and electronics in the instrument. The fluorescence value obtained by the measuring beam is termed $F_0$ in dark-adapted sample and $F$ when the sample is illuminated by actinic light. Then the sample is illuminated with intense light pulse which saturates all reaction centers such that all reaction centers become closed. Closed reaction centers are reduced and unavailable, temporarily, to do photochemistry. At this point,

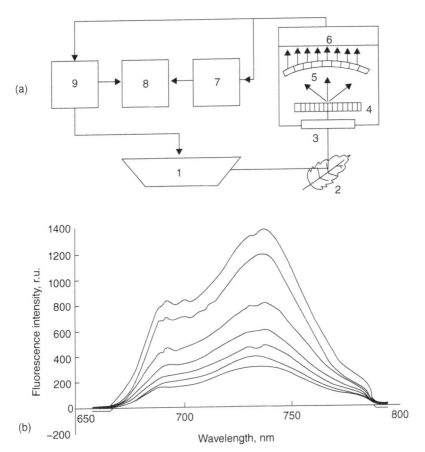

**FIGURE 25.7** Optical multichannel analyzer: (a) schematic representation of analyzer: 1—ultraviolet laser, 2—leaf, 3—window, 4—diffraction grating, 5—array of photodiodes, 6—optical multichannel system, 7—amplifier, 8—readout system, 9—trigger; (b) sum total of fluorescence emission spectra that are recorded simultaneously at different wavelengths. Here the time of recording spectra (in ms) is 0 (upper curve); 200; 5000; 10,000; 30,000; 60,000; 120,000; 300,000 (lower curve).

fluorescence becomes maximal, and the value is noted as $F_m$ in dark-adapted samples or as $F_m'$ in samples under actinic light.

With a very strong light pulse, the electron transfer chain between PSII and PSI is quickly interrupted, photochemical quenching becomes zero, and any remaining quenching must be nonphotochemical.

Company Heinz Walz GmbH (Germany) is one of the world's well-known producers of modern photosynthesis measuring systems, which are based on PAM fluorometry: monitoring PAM, phyto-PAM, diving-PAM, water PAM, PAM-2500, dual-PAM-100.

*The advantages of a PAM fluorometer:*

- Estimates optimal quantum yield, coefficients of photochemical quenching qP and nonphotochemical quenching qN and NPQ, photosynthetic efficiency $Y_{eff}$, and photosynthetic capacity electron transport rate (ETR).
- High sensitivity.
- Can be used in the field.

*The disadvantages of PAM fluorometer:*

- It requires expensive and somewhat sophisticated equipment.

### 25.6.5   Fluorescence Indices

*Spectrofluorometry.* When spectrofluorometry method is used, the ratio $F(690)/F(740)$ of the two chlorophyll maxima of green leaves can serve as fluorescent index.

*Fluorescence induction kinetics.* The control of chlorophyll fluorescence kinetics of dark-adapted leaves can be realized by the following fluorescence indices: vitality index $Rfd = f_d/f_s$ ($Rfd'$ at 690 nm and $Rfd''$ at 740 nm), and stress adaptation index $A_p = 1 - [Rfd(740) + 1]/[Rfd(690) + 1]$.

*PAM fluorometry.* Photosynthetic activity of green leaves is estimated using PAM fluorometry by measuring optimal quantum yield:

$$Y_{opt} = F_v/F_m = (F_m - F_o)/F_m, \qquad (25.14)$$

where $F_v/F_m$—the ratio of variable fluorescence; $F_v = F_m - F_o$ to maximal fluorescence $F_m$ of dark-adapted samples; here, $F_o$ is the initial fluorescence when all PSII reaction centers are opened and $F_m$ is maximal fluorescence when PSII centers are closed. Dark adaptation was provided by covering the sample with a black paper. The same sample was divided into two parts—one was kept in darkness, the another, under high irradiation. All the samples were dark adapted 15 minutes before the measurements.

Besides, the fluorescence parameters of dark-adapted samples such as coefficients of photochemical quenching qP and nonphotochemical quenching qN and NPQ were measured:

$$qP = (F_m' - F)/(F_m' - F_0); \qquad (25.15)$$

$$qN = (F_m - F_m')/(F_m - F_0); \qquad (25.16)$$

$$NPQ = (F_m - F_m')/F_m'. \qquad (25.17)$$

The useful information can be obtained without previous dark adaptation of the samples. The maximal fluorescence $F_m$ corresponds to the situation when PSII centers are closed; the more reaction centers are closed, the less photosynthetic efficiency of photosynthesis.

The fluorescence parameters of light-adapted samples such as photosynthetic efficiency, $Y_{eff}$, and photosynthetic capacity, ETR, were measured:

$$Y_{eff} = (F'_m - F)/F'_m = \Delta F/F'_m; \qquad (25.18)$$

$$ETR = \Delta F/F'_m \cdot PAR, \qquad (25.19)$$

where PAR is photon fluence rate of photosynthetic active radiation which is measured in $\mu mol/m^2 \cdot c^{-1}$. Light curves were defined as relative ETR of photosynthesis versus irradiance (samples should be well adapted to a moderate light intensity, which is close to the light intensity experienced by the alga in its natural environment); the photosynthetic efficiency $Y_{eff}$ was estimated as linear part of light curve $ETR = f(I)$ (where $I$ is light intensity); photosynthetic capacity was defined as ETRmax when the light curve is saturated.

Fluorescence is related to photochemical and heat processes: the measurement of fluorescence parameters such as effective and optimal quantum yields, photosynthetic efficiency, and capacity makes it possible to estimate the effects of stress factors—for example, such as high-intensity irradiation or ultraviolet radiation, on photosynthetic organisms (Posudin et al., 2004a, 2004b).

**Exercise**    Calculate and compare the vitality index ($Rfd'(690)$ and $Rfd''(740)$) and stress adaptation index $A_p$ for dark-green and light-green leaves (Figure 25.8).

## 25.7    REMOTE SENSING OF VEGETATION FLUORESCENCE

### 25.7.1    Laser-Induced Fluorescence Spectroscopy for *In Vivo* Remote Sensing of Vegetation

Unique properties of laser radiation such as monochromaticity, coherence, directionality, and brightness provide the application of efficient nondestructive, fast, and precise methods of *in vivo* remote sensing of plants.

There are a number of convincing examples of practical realization of remote sensing of agricultural and natural vegetation on the basis of laser-induced fluorescence spectroscopy (Chappelle et al., 1984a,b; Emmett et al., 1985; Svanberg, 1995; Hilton, 2000; Romanovskii et al., 2000; Lefsky et al., 2002; Saito et al., 2005; Posudin, 2007).

### 25.7.2    Laser Spectrofluorometer

A laser spectrofluorometer is a combination of laser as the source of high-intensity radiation and conventional fluorometer which makes it possible to record fluorescence

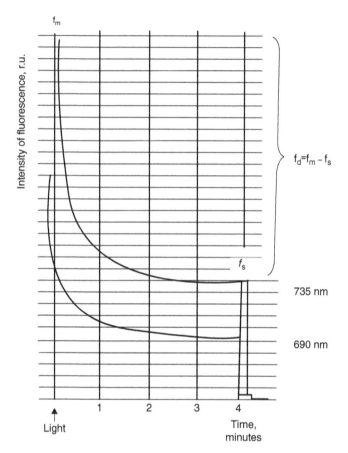

**FIGURE 25.8**   Laser-induced chlorophyll-fluorescence induction kinetics in dark-green and light-green leaves.

emission spectra of the samples. Such a system can be used for remote sensing of vegetation.

A laser spectrofluorometer, which was worked out by the author of this textbook (Posudin, 2007), consists of nitrogen laser as the source of excitation of chlorophyll fluorescence, optical system (semitransparent plate, concave mirror, and filter), sample holder, monochromator, detector, and readout system.

The laser radiation is directed through the semitransparent glass plate to the sample (a single intact leaf) and to the fluorescent standard (a solution of fluorescein). Either the sample or the cell with fluorescent standard are oriented at the angle 45° to the laser beam in order to escape nondesirable absorption of laser radiation by the sample (or standard) volume.

The fluorescence emission of the leaf (or the standard) is collected by concave mirror, passes through the cutoff filter, and focused on the entrance slit of monochromator. The dispersive element (prism) of this monochromator is linked with the electrical motor which provides the rotation of the prism and selection of the wavelength. The intensity of fluorescence is detected by photomultiplier; the electrical signal of it is amplified and recorded by the read-out system (amplifier and recorder).

The main characteristics of this $N_2$ laser are wavelength of generation 337 nm, frequency of pulse repetition 50 Hz, an average power output 3 mW, duration of pulse about 10 ns, divergence of beam $3 \times 10^{-3}$ radians.

## 25.8   THE EFFECT OF VARIOUS FACTORS ON THE CHLOROPHYLL FLUORESCENCE

The parameters of chlorophyll fluorescence at a single-leaf level depend on the chlorophyll content, leaf development phase, side and segment of the leaf, illumination, mechanical injury of leaves, dehydration, and extreme temperatures.

Methods of fluorescence spectroscopy of plant canopies make it possible to investigate the effects of various natural and anthropogenic stresses on vegetation in the field conditions such as solar high-intensity and ultraviolet radiation, agrochemical treatment (nitrogen deficiency/excess, salt imbalance, and herbicides), dehydration, mechanical damage, temperature, environmental pollution, and meteorological conditions.

## REFERENCES

Chappelle, E.W., Wood, F.M., McMurtrey, J.E., and Newcomb, W.W. 1984a. Laser-induced fluorescence of green plants. 1. A technique for remote detection of plant stress and species differentiation. *Appl. Optics.* 23:134–138.

Chappelle, E.W., Wood, F.M., McMurtrey, J.E., and Newcomb, W.W. 1984b. Laser-induced fluorescence of green plants. 2. L1F caused by nutrient deficiencies in corn. *Appl. Optics.* 23:139–142.

Emmett, W., Chappelle, E.W., Frank, M.W. Jr., Newcomb, W.W., and McMurtrey, J.E. 1985. Laser-induced fluorescence of green plants. 3. LIF spectral signatures of five major plant types. *Appl. Optics.* 24:74–80.

Gorte, B.G.H. 2000. Land-use and catchment characteristics. In: *Remote Sensing in Hydrology and Water Management* (eds G.A. Shhultz and E.T. Engman). Springer, pp. 133–156.

Goulas, Y., Moya, I., and Schmuck, G. 1990. Time-resolved spectroscopy of tine blue fluorescence of spinach leaves. *Photosynth. Res.* 25:299–307.

Hardisky, M.A., Klemas, V., and Smart, R.M. 1983. The influences of soil salinity, growth form, and leaf moisture on the spectral reflectance of *Spartina alterniflora* canopies. *Photogramm. Eng. Rem. Sens.* 49:77–83.

Hilton, P.J. 2000. Laser-induced fluorescence for discrimination of crops and weeds. Proc. SPIE 4124, High-Resolution Wavefront Control: Methods, Devices, and Applications II, p. 223 (November 22, 2000).

Interschick-Niebler, E. and Lichtenthaler, H.K. 1981. Partition of phyllo quinone K-1 between digitonin particles and chlorophyll proteins of chloroplast membranes from Nicotiana-tabacum. *Zeitschrift fur Naturforschung. Section C J. Biosci.* 36:276-283.

Jordan, C.F. 1969. Derivation of leaf area index from quality of light on the forest floor. *Ecology.* 50:663–666.

Kautsky, H. and Hirsch, A. 1931. Neue Versuche zur Kohlenstoffassimilation. *Naturwissenschaften.* 19:964.

Kochubey, S.M. 1986. *Organization of Pigments of Photosynthetic Membranes as the Basis of Energy Supply of Photosynthesis.* Naukova Dumka, Kiev (In Russian).

Lang, M., Stober, F., and Lichtenthaler, H.K. 1991. Fluorescence emission spectra of plant leaves and plant constituents. *Padiat. Environ. Biophys.* 30:333–347.

Lang, M., Siffel, P., Braunova, Z., and Lichtenthaler, H.K., 1992. Investigation of the blue-green fluorescence emission of plant leaves. *Rot. Acta.* 105:435–440.

Lefsky, M.A., Cohen, W.B., Parker, G.G., and Harding, D.J. 2002. Lidar remote sensing for ecosystem studies. *BioScience.* 52(1):19–30.

Leuning, R., Hughes, D., Daniel, P., Coops, N.C., and Newnham, G. 2006. A multi-angle spectrometer for automatic measurement of plant canopy reflectance spectra. *Rem. Sens. Environ.* 103(3):236–245.

Lichtenthaler, H.K. and Rinderle, U. 1988. The role of chlorophyll fluorescence in the detection of stress conditions in plants. *CRC Cr. Rev. Anal. Chem.* 19:29–85.

Lichtenthaler, H.K., Lang, M., and Stober, F. 1991. Laser-induced blue and red chlorophyll fluorescence signatures of differently pigmented leaves. In: *Proc. 5th Intern. Collotj. Pln/s. Measurements and Signatures in Remote Sensing.* Courchevel, pp. 727–730.

Posudin, Y.I. 2007. *Practical Spectroscopy in Agriculture and Food Science.* Science Publishers, Enfield.

Posudin, Yu.I., Murakami, A., Kamiya, M., and Kawai, H. 2004a. Effect of light of different intensity on chlorophyll fluorescence of *Ulva pertusa* Kjellman (*Chlorophyta*). *Int. J. Algae.* 6(3):235–250.

Posudin, Y.I., Hanelt D., and Wiencke C. 2004b. Effect of ultraviolet radiation on green alga *Ulva lactuca. Sci. Herald Nat. Agr. Univ.* 72:13–31.

Posudin Y.I., Gural, T.I., Milutenko, V.M., and Sobolev, O.V., 2007. Device for registration of fluorescence induction. Patent of Ukraine N 20668.

Posudin, Y.I., Melnychuk, M., Kozhem'yako, Y., Zaloilo, I., and Godlevska, O. 2008. Portable fluorometer for fluorescence analysis of agronomic plants under stress conditions. Pittsburgh Conference on Analytical Chemistry and Applied Spectroscopy, March 2–7, 2008. New Orleans, USA.

Posudin, Y.I., Godlevska, O.O., Zaloilo, I.A., and Kozhem'yako, Ya.V. 2010. Application of portable fluorometer for estimation of plant tolerance to abiotic factors. *Int. Agrophysics.* 24(4):363–368.

Richardson A.J. and Wiegand, C.L. 1977. Distinguishing vegetation from soil background information. *Photogramm. Eng. Rem. Sens.* 43:1541–1552.

Rock, B.N., Williams, D.L., and Vogelmann, J.E. 1985. Field and airborne spectral characterization of suspected acid deposition damage in red spruce (*Picea rubens*) from Vermont. Proc. Symp. on Machine Processing of Remotely Sensed Data, Purdue University, West Lafayette, IN, pp. 71–81.

Romanovskii, O.A., Matvienko, G.G., Grishin, A.I., and Kharchenko, O.V. 2000. Fluorescence lidar system for remote monitoring of the state of vegetative cover for the purpose of its prediction. Proc. SPIE 4035, Laser Radar Technology and Applications V, p. 143 (September 5, 2000).

Rouse, J.W., Haas, R.H., Schell, J.A., and Deering, D.W. 1974. Monitoring vegetation systems in the Great Plains with ERTS. Proc. Third ERTS-1 Symposium, NASA Goddard, NASA SP-351, pp. 309–317.

Sabins, F.F. 2007. *Remote Sensing: Principles and Interpretation*, 3rd edition. Waveland Pr Inc.

Saito, K. 2012. Plant and vegetation monitoring using laser-induced fluorescence spectroscopy. In: *Industrial Applications of Laser Remote Sensing* (eds T. Fukuchi and T. Shiina). Bentham Science Publishers, pp. 99–114.

Saito, Y., Matsubara, T., Kobayashi, F., Kawahara, T., and Nomura, A. 2005. Laser-induced fluorescence spectroscopy for *in-vivo* monitoring of plant activities information and technology for sustainable fruit and vegetable production. FRUTIC 05, 12 (September 16, 2005). Montpellier, France.

Schreiber, U., Schliva, U., and Bilger, W. 1986. Continuous recording of photochemical and non-photochemical chlorophyll fluorescence quenching with a new type of modulation fluorometer. *Photosyn. Res.* 10:51–62.

Stober, F. and Lichtenthaler, H.K. 1993. Studies of the localization and spectral characteristics of the fluorescence emission of differently pigmented wheat leaves. *Bot. Acta.* 106:365–370.

Svanberg, S. 1995. Fluorescence lidar monitoring of vegetation status. *Phys. Scripta.* T58(79).

Tucker, C.J. 1979. Red and photographic infrared linear combinations for monitoring vegetation. *Rem. Sens. Environ.* 8(2):127–150.

# PRACTICAL EXERCISE 11

## DETERMINATION OF PERPENDICULAR VEGETATION INDEX

**Example** Calculate perpendicular vegetation index (PVI) of green vegetation using the hypothetical reflection spectra of soil and vegetation (Figure P11.1).

*Solution*

1. To restore the vertical lines on the graphs $R = f(\lambda)$ (where $R$ is coefficient of reflectance; $\lambda$ is the wavelength) at the wavelengths corresponding to the spectral bands MSS5 $= \dfrac{600 + 700}{2} = 650$ nm and MSS7 $= \dfrac{800 + 1100}{2} = 950$ nm.
2. To enter the numerical data obtained from the intersection of the vertical lines with the graphs in the table.

1. Values of $\mathrm{MSS5_S}$, $\mathrm{MSS5_V}$, $\mathrm{MSS7_S}$, $\mathrm{MSS7_V}$, and level of vegetation color

| No. | $\mathrm{MSS5_S}$ (RED$_S$) | $\mathrm{MSS7_S}$ (NIRS) | $\mathrm{MSS5_V}$ (REDV) | $\mathrm{MSS7_V}$ (NIR$_V$) | Level of vegetation color |
|---|---|---|---|---|---|
| 1 | 28 | 38 | 11 | 39 | Green leaves |
| 2 | 25 | 33 | 33 | 45 | Yellow-green leaves |
| 3 | 21 | 28 | – | – | |
| 4 | 16 | 24 | – | – | – |
| 5 | 13 | 20 | – | – | – |

*Methods of Measuring Environmental Parameters*, First Edition. Yuriy Posudin.
© 2014 John Wiley & Sons, Inc. Published 2014 by John Wiley & Sons, Inc.

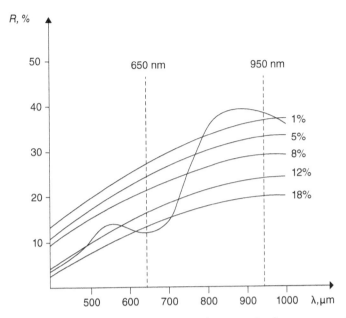

**FIGURE P11.1**    Dependence of soil and vegetation spectral reflectance on moisture.

3. To build a graph of dependence of $MSS5_S$ on $MSS7_S$ for soil in the coordinates MSS7 and MSS5 (Figure P11.2, line AB).
4. Find a point E on the graph with coordinates ($NIR_V = 39$; $RED_V = 11$) corresponding to green leaves.

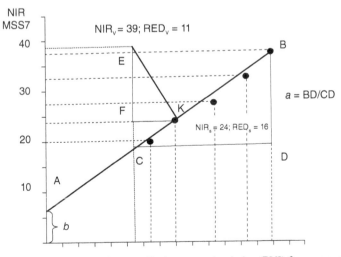

**FIGURE P11.2**    Calculation of perpendicular vegetation index (PVI) for green vegetation.

5. Drop a perpendicular from point E on the line AB.
6. Then, using the formula 25.11, we obtain

$$PVI = [(RED_S - RED_V)^2 + (NIR_V - NIR_S)^2]^{1/2}$$
$$= [(16 - 11)^2 + (39 - 24)^2]^{1/2} = (25 + 225)^{1/2} = 250^{1/2} = 15.8.$$

7. PVI can also be calculated as follows:

$$PVI = \frac{NIR_V - aRED_V - b}{\sqrt{1 + a^2}}, \qquad \text{(P11.1)}$$

where $a$ and $b$ are constants; $a$ is defined as the slope of the dependence $NIR_S = f(RED_S)$ and $b$ is the intersection point of this relationship with the axis of ordinates (Figure P11.2).

The main idea of PVI is based on the exclusion of the effect of soil reflectance. This concept can be presented graphically as rotation of line AB at certain angle in such a way that this line becomes the $x$-axis.

Thus, the PVI allows to avoid the influence of soil background.

**Control Exercise**    To repeat the operations 4–6 for green-yellow leaves and to find corresponding values of PVI, if $NIR_V = 45$; $RED_V = 33$. To analyze the effect of the senescing of the leaves on the PVI value.

## QUESTIONS AND PROBLEMS

1. What are the main factors that affect the spectral reflectance of a single leaf plant canopy?

2. Explain the mechanisms of specular and diffuse reflectance.

3. Explain the nature of the spectral reflectance curve of green leaf.

4. Name the basic vegetation indexes.

5. What is the role of the vegetation index PVI? and why is it useful?

6. Define photosynthesis.

7. What are the pigments responsible for the fluorescence in the red part of the spectrum?

8. Explain how chlorophyll fluorescence associated with photosynthesis.

9. Name the basic laboratory techniques of fluorescence spectroscopy.

10. Provide a comparative analysis of spectrofluorometry and recording fluorescence induction.

11. Explain the principle of PAM fluorometry.

12. Name the basic fluorescent indexes.

13. What is remote sensing?

14. What is the principal difference between passive and active sensors?


## FURTHER READING

Clark, R.N. 1995. *Reflectance Spectra*. AGU Handbook of Physical Constants.

Kortüm, G. and Lohr, J.E. 2012. *Reflectance Spectroscopy: Principles, Methods, Applications.* Springer.

Posudin, Y.I. 1998. *Lasers in Agriculture*. Science Publishers, Ltd, New York.

Wooley, J.T. 1971. Reflectance and transmittance of light by leaves. *Plant Physiol*. 47:656–662.

## ELECTRONIC REFERENCES

Instrumentation for Fluorescence Spectroscopy, http://www.springer.com/cda/content/document/cda_downloaddocument/9780387312781-c2.pdf?SGWID=0-0-45-351714-p134266323 (accessed November 4, 2013).

Matvienko, G.G., Grishin, A.I., Kharchenko, O.V., and Romanovskii, O.A. 2005. Application of laser-induced fluorescence for remote sensing of vegetation. *Opt. Eng.* 45(5), 056201 (May 10, 2006). doi:10.1117/1.2202366, http://opticalengineering.spiedigitallibrary.org/article.aspx?articleid=1076664 (accessed November 6, 2013).

Taylor, J.L. 2010. Design Considerations of Small (60 mm) vs. Large (150 mm) Integrating Spheres. Diffuse Reflectance, http://pe.taylorjl.net/PE_Blog/?p=303 (accessed November 12, 2013).

UniSpec-SC Spectral Analysis System, http://www.wittich.nl/EN/PDF/photosynthesis/leaf-reflectance/UniSpec-SC.pdf (accessed November 24, 2013).

UniSpec-DC Spectral Analysis System, http://www.wittich.nl/EN/PDF/photosynthesis/leaf-reflectance/UniSpec-DC.pdf (accessed November 30, 2013).

# PART VI

# PHYSICAL TYPES OF POLLUTION

Today we are surrounded by about 20,000 hazardous pollutants, among which about 3000 natural and anthropogenic substances are toxic. Chemical contaminants are discussed in detail in the scientific literature; that is why we pay attention to physical pollution that surrounds us in everyday life.

*Physical pollution* is the action of harmful physical factors that influence on human health and environment and lead to short- and long-term disruption of quality of life and ecological balance.

Basic physical pollutions are realized through vibration, noise, electromagnetic radiation, and undesirable light distribution.

# 26

# MECHANICAL VIBRATION

## 26.1  PARAMETERS OF VIBRATION

*Vibration* (from Latin *vibratio*—oscillation) is a periodic motion of the material system (particles or elastic body) with high frequency and small amplitude. This motion is accompanied with the displacement of a system in opposite directions from the position of equilibrium when that equilibrium has been disturbed.

Any vibration is described by the following parameters:

- *Displacement* is a distance moved by vibrating particle or body from the state of equilibrium at the given moment of time.
- *Amplitude of vibration* is the maximum displacement of a vibrating particle or body from the position of equilibrium.
- *Phase of vibration* is the fraction of a period (the time of one complete vibration) that a particle or a body completes after last passing through the reference, or zero.
- *Frequency of vibration* is the number of oscillations the particle or body makes per unit time. The unit of frequency in SI system is hertz (Hz).
- *Velocity* is the rate of change of displacement. The unit of velocity in SI system is m/s.
- *Acceleration* is the rate of change of velocity. The unit of acceleration in SI system is m/s$^2$.

*Methods of Measuring Environmental Parameters*, First Edition. Yuriy Posudin.
© 2014 John Wiley & Sons, Inc. Published 2014 by John Wiley & Sons, Inc.

## 26.2    VIBRATION LEVEL

Vibration is estimated by a *vibration level* which is expressed in decibel as

$$L_V = 20 \log(A_i/A_0) \, \text{dB}, \qquad (26.1)$$

where $A_i$ is the measured amplitude of vibrational velocity or vibrational acceleration and $A_0$ is the reference amplitude ($10^{-3}$ m/s for velocity and $10^{-5}$ m/s$^2$ for acceleration).

## 26.3    SOURCES OF VIBRATION

The main sources of mechanical vibration are jackhammers, concrete vibrators, power drill, electric motors, transport means (jet planes, automobile engines, subway trains), construction techniques, and industrial units.

Typical values of vibration velocity for various sources of vibration are presented in Table 26.1.

**TABLE 26.1    Principal Sources of Vibration**

| Source | Vibrational Velocity, mm/s |
| --- | --- |
| Rail transport | 0.3–160 |
| Industrial units | 0.05–5 |
| Construction techniques | 0.002–1.6 |
| Vehicles | 0.005–0.07 |
| Day city background | 0.006–0.02 |
| Night city background | 0.003–0.01 |

## 26.4    EFFECT OF VIBRATION ON HUMAN HEALTH

The human body is a complex system that has a large number of mobile elements. Each element is characterized by a certain degree of freedom and the natural frequency of oscillation. In general, the human body is sensitive to various parameters of vibration such as intensity, frequency, and duration of exposure. For example, the thorax–abdomen system is characterized by a resonance frequency in the range of 3–6 Hz; eyeball responds to the resonance frequency in the range of 60–90 Hz; the head–neck shoulder system is responsible to vibrations in the range of 20–30 Hz (Balachandran and Magrab, 2008).

The human body is able to perceive the effect of rhythmic oscillations generated by a variety of mechanisms and tools with which the body is in contact. Vibration should be regarded as undesirable movements that lead to *vibration disease*, which provokes disturbances of the vestibular apparatus, respiratory and cardiovascular systems, visual and auditory analyzers, and degenerative changes in bone tissue.

There are two main types of vibration disease:

*Whole-body vibration (WBV)* is a mechanical vibration which is transmitted into the body, when seated or standing, through the supporting surface, during work activity or exposure in rest facilities supervised by the employer (HSPS02 April 2008 revised).

This disease is caused by frequent exposure of drivers of buses, tractors, scrapers, and bulldozers that are on the surface that vibrates. The WBV in the range of 1–80 Hz can cause significant changes in the central nervous system, dystonia (change of tone) of vessels, fatigue, insomnia, stomach problems, and headache. Vibration at frequencies below 1 Hz provokes *kinetosis* which induces the disturbance of vestibular apparatus and cardiovascular and neuroendocrine system diseases.

*Hand–Arm Vibration (HAV)* is vibration, which reaches the hands and arms when working with hand-held power tools, hand-guided machinery, or when holding materials, which are being processed by machinery (HSPS02 April 2008 revised).

This disease is caused by permanent vibrating tools: hammers, circular saws, electric drills, grinders, and impact wrenches. These instruments transmit vibration into the operator's hands and arms. The typical frequency range of these tools is 8–100 Hz. The main symptoms are limb vascular spasms, impaired circulation of the blood vessels, finger blanching, or "white finger disease." Prolonged exposure to local vibration can lead to more serious disorders such as gangrene. Children who have been playing computer games experience a symptom of "shaking hands."

The impact of vibration can be reduced by the use of anti-vibration gloves with viscoelastic layer; limitation of the time during which an employee is on a vibrating surface; mechanical isolation of vibration sources; the establishment of vibration absorbing seats; use of insulation materials, spring isolators, and vibration dampers.

The Control of Vibration at Work Regulations requires the following standards (HSPS02 April 2008 revised):

For hand–arm vibration:

(a) The daily exposure limit value standardized to an 8-hour reference period shall be 5 m/s$^2$.

For whole-body vibration:

(a) The daily exposure limit value standardized to an 8-hour reference period shall be 1.15 m/s$^2$.

# 27

# MEASUREMENT OF VIBRATION

There are such types of devices as vibrometer, velometers, accelerometers, and frequency analyzers depending on the parameters measured (displacement, velocity, acceleration, phase, and frequency).

A *vibration transducer* is a device that converts a vibration parameter into electrical signal for a practical purpose.

## 27.1 RESISTIVE TRANSDUCERS

The *variable resistance transducer* can be used for measuring vibrations. The principle of operation of this transducer is based on the fact that the resistance of the conductor is directly proportional to the length of the conductor and inversely proportional to the area of the conductor:

$$R = \rho L / A, \qquad (27.1)$$

where $L$ is the length of the conductor (in m); $A$ is its area (in m$^2$); and $\rho$ is the resistivity of the material (ohm/m).

When the tension is applied to the electrical conductor, its length increases, the cross-section area decreases, and the resistance of the conductor changes.

*Methods of Measuring Environmental Parameters*, First Edition. Yuriy Posudin.
© 2014 John Wiley & Sons, Inc. Published 2014 by John Wiley & Sons, Inc.

Conductor requires a mechanical stress, which is associated with the resistance value:

$$\varepsilon = \frac{\Delta l}{l} = \frac{\Delta R}{kR}, \tag{27.2}$$

where $k$ is a factor that takes into account the type of conductor.

The change of the conductor's resistance is measured by means of an electric bridge; the signal that is measured by the bridge is proportional to the mechanical stress $\varepsilon$.

## 27.2  ELECTROMAGNETIC TRANSDUCERS

*Electromagnetic transducers* operate on principle of estimation of voltage at the ends of the conductor in the form of a coil during its motion in a magnetic field. The magnitude of this voltage is determined by the expression:

$$U = Blv, \tag{27.3}$$

where $B$ is the magnetic induction; $l$ is the length of the conductor; $v$ is the velocity of its motion.

## 27.3  CAPACITIVE TRANSDUCERS

*Capacitive transducer* is a non-contact device capable of high-resolution position or displacement measurement. Its principle of operation is explained in Section 2.5. Vibration measurement can be performed up to 20–100 KHz frequency bandwidth.

## 27.4  PIEZOELECTRIC TRANSDUCERS

*Piezoelectric transducers* contain a piezoelectric element (quartz, barium titanate, lead niobate, and tourmaline) that produces electrical signals when subjected to mechanical deformation (see Section 2.6).

Note that vibration can be regarded as the pressure of variable intensity. That is why the output signals of the capacitive and piezoelectric transducers can be amplified and the sensitivity of readout systems can be increased in such a way.

Photoelectric transducer consists of an electric bridge, one arm of which is a resistor that is placed on a body that vibrates and is illuminated through the diaphragm. The periodic light intensity changes under the applied vibration affect the value of the resistance of the shoulder of the bridge and its imbalance.

## 27.5   LASER DOPPLER VIBROMETER

*Laser Doppler vibrometer* is based on the detection of the Doppler shift of coherent laser radiation that is scattered from a small area of the vibrating object.

As the laser radiation has a very high frequency (about $4.74 \times 10^{14}$ Hz), a direct demodulation of this radiation is impossible. An optical interferometer is therefore used to mix the scattered beam coherently with a reference beam (see Section 16.5.2, Figure 16.10). Indeed, if one of the mirrors mounted on the vibrating surface of Michelson interferometer, the frequency of modulated light will combine with the reference light. As a result, the superposition of the two signals arises and oscillation beats occur. The photodetector measures the intensity of the mixed light whose beat frequency is equal to the difference in frequency between the reference and the measuring beam.

*The advantages of a laser Doppler vibrometer:*

- The possibility of remote sensing of vibration (variable working distance from 0.2 m up to 30 m).
- High sensitivity (about 50 nm).
- Can be directed at targets that are difficult to access, or that may be too small or too hot.

*The disadvantages of a laser Doppler vibrometer:*

- Limited portability.
- Potential setup difficulties.
- Measurement limitations if the surface is not optically visible.

# 28

---

# NOISE

## 28.1 MAIN DEFINTIONS OF NOISE

There are several definitions of noise. First definition: noise is the sound vibrations, which intensity and frequency are changing irregularly and aperiodically.

According to the second definition, noise is a sound that is superimposed on other sound, interacts with it, and in such a way is undesired to our hearing sound. Imagine three musicians who play on three different instruments simultaneously, loudly and in one place three different melodies; such an ensemble of course creates the noise.

The third version of noise definition describes it as any sound that interferes with a person. For example, the music is pleasant for a musician, but is an irritating noise for the person who wants to sleep.

An expert on noise, K. D. Kryter (1996), defined noise as "acoustic signals which can negatively affect the physiological or psychological well-being of an individual."

## 28.2 SOURCES OF NOISE

The most significant sources of noise are urban traffic (subway, automobiles) and highway traffic (trucks and buses), airplanes, and domestic activity (vacuum cleaners, drills, and toys). Noise occurs during entertainment, children's games, action of alarm systems and emergency sirens, collection of waste, construction and repair work, running the model airplanes, go-karting, sports cars, operation of audio systems, and rock concerts.

*Methods of Measuring Environmental Parameters*, First Edition. Yuriy Posudin.
© 2014 John Wiley & Sons, Inc. Published 2014 by John Wiley & Sons, Inc.

## 28.3   PARAMETERS OF NOISE

Range of sound intensities around us is very large. The minimum intensity (e.g., bees buzzing, rustling leaves) is $I_{min} = 10^{-12}$ W/m², while the maximum (e.g., roaring of jet engine) is $I_{max} = 10$ W/m². Thus, the range of possible sound intensities is 13 orders of magnitude.

The logarithm of the ratio $I/I_0$ gives us a value for the relationship, the unit being the *bel* after Alexander Graham Bell, the inventor of the telephone:

$$B = \log(I/I_0).\qquad(28.1)$$

A still more practical unit of *sound intensity level L*, that is used more widely today, is the 10th part of a bel, which is estimated as follows:

$$L = 10\lg\left(\frac{I}{I_0}\right),\qquad(28.2)$$

where $I$ is the sound intensity; $I_0$ is minimum intensity ($10^{-12}$ W/m² in air). Here, $L$ is measured in *decibels (dB)*.

The level of sound intensity can be written also as

$$L = 20\lg\left(\frac{p}{p_0}\right),\qquad(28.3)$$

where $p$ is sound pressure; $p_0 = 20 \times 10^{-6}$ Pa.

**Example**   The intensity of noise is 10 W/m². Express this in decibels.

**Solution**   Using Equation 28.1, we shall find:

$$dB = 10 \cdot \lg\frac{I}{I_0} = 10 \cdot \lg\frac{10}{10^{-12}} = 10 \times 13 = 130\,dB.$$

**Control Exercise**   Determine how many decibels correspond to the scale of sound intensity levels $I_{min} = 10^{-12}$ W/m²; $I_{max} = 10^2$ W/m².

## 28.4   EQUIVALENT SOUND LEVEL

*Equivalent sound level*, $L_{eq}$, quantifies the noise environment as a single value of sound level for any desired duration. It can be expressed as

$$L_{eq} = 10\lg\left[(1/T)\int_0^T 10^{L/10}dt\right],\qquad(28.4)$$

where $L$ is instantaneous sound level; $T$ is averaging time (usually, 1 hour, 8 hours, 24 hours).

**Example**     Consider four 15-minute intervals, during which corresponding sound levels are 90, 83, 75, and 95 dB. Determine equivalent sound level for this hour.

**Solution**     Using formula 28.4 equivalent sound level can be calculated:

$$L_{eq} = 10\lg(1/4)(10^{90/10} + 10^{83/10} + 10^{75/10} + 10^{95/10})$$
$$= 10\lg(1/4)(33816228) = 10 \times 6.927 = 69.27\,\text{dB}.$$

## 28.5   INTEGRATING SOUND LEVEL

If there are two or more uncorrelated sources of noise occur, the integrating sound level is described by the expression:

$$L_{int} = 10\lg \sum_{i=1}^{N} 10^{L_i/10}, \tag{28.5}$$

where $L_i$ is the sound-level intensity of each source; $N$ is the number of the sources.

**Example**     Separate contributions into noise of five sources are 84, 89, 79, 71, and 90 dB. Determine integrating sound level.

**Solution**     Using Equation 28.5 of integrating sound level, it is possible to obtain

$$L_{int} = 10\lg(10^{84/10} + 10^{89/10} + 10^{79/10} + 10^{71/10} + 10^{90/10})$$
$$= 10\lg 2052589254 = 93\,\text{dB}.$$

## 28.6   SPECTRAL DENSITY OF NOISE

Spectral density (power distribution in the frequency spectrum) is such a property that can be used to distinguish different types of noise. The power spectrum of a noise signal is usually described by the concept of "color."

The classification of different types of noise by color is borrowed from optics, where the long-wavelength part of the spectrum belongs to the red color, and a short-wavelength part to the blue one.

Spectral density of noise of different color is described by the expression $1/f^{\beta}$ where $f$ is a frequency and $\beta$ is a coefficient.

For instance, *white noise* is characterized by a flat constant power spectral density $(\beta = 0)$.

*Pink noise* or $1/f$ noise is a signal with a frequency spectrum such that the power spectral density is inversely proportional to the frequency $(\beta = 1)$; *red* (or *Brown*) *noise* $(\beta = 2)$ is usually refer to a power density which decreases with increasing frequency; *blue noise* $(\beta = 1)$ and *violet noise* $(\beta = 2)$ have power density that increases with increasing frequency.

## 28.7    EFFECT OF NOISE ON HUMAN HEALTH

The action of noise on human organism is characterized by either its intensity or the duration of action.

From the point of view of intensity, noise can be classified as painful (150–120 dB), extremely loud (110–90 dB), very loud (80–70 dB), moderate (60–40 dB), and (30 dB).

The duration of action is also important factor: for example, hearing loss is induced by noise level 90 dB during 8 hours or by 100 dB during 2 hours.

Noise causes a variety of disruptions of the human body such as annoyance, aggression, mental stress and illness, cardiovascular disease, upset stomach or ulcer, and sleep disturbance. The children living in noisy areas such as near airports or highways have problems with learning and reading abilities.

According to the World Health Organization (WHO), in European Union countries alone, 20% of the population or approximately 80 million people, including children, suffer from stress and sleep disorders that have a considerable negative effect on health. About 57 million people are annoyed by road traffic noise, 42% of them seriously. A preliminary analysis shows that each year over 245,000 people in European Union are affected by cardiovascular diseases that can be traced to traffic noise. About 20% of these people (almost 50,000) suffer a lethal heart attack, thereby dying prematurely (den Boer and Schroten, 2007).

According to the National Institute on Deafness and Other Communication Disorders (NIDCD), more than 30 million Americans are exposed to hazardous sound levels on a regular basis. Of the 28 million Americans who have some degree of hearing loss, over one-third have been affected, at least in part, by noise (Noise is Difficult to Define—Part 1).

## 28.8    MECHANISMS OF NOISE ACTION

The noise level of 90 dB and more, provokes various physiological disorders. The maximum limit for humans is 140 dB; the noise level of 160–170 dB causes the destruction of the eardrum of the human ear. In addition, noise can cause destruction of Corti organ. The hair cells of the inner ear are most sensitive to noise elements of the auditory analyzer. After termination of the noise level below 30 dB, hearing can be restored within 16–24 hours. The action of high levels causes irreversible damage to the hair cells. If the hair cells undergo serious damage, they are not able to restore their function and cannot be replaced by other cells. The result may be partial or complete loss of hearing.

To quantify the effects of noise on hearing, such a parameter that characterizes the change in auditory sensitivity as *noise-induced threshold shift (NITS)* is used. It is determined by measuring the threshold of auditory sensitivity before and after exposure to noise. This shift may be temporary or permanent—depending on the noise parameters (intensity, duration, and frequency composition).

## 28.9   HOW TO PROTECT YOURSELF FROM NOISE

In order to avoid the harmful effects of noise, it is necessary to wear hearing protectors, to use the noise insulation, to limit periods of exposure to noise, to inspect your children's toys, to check your hearing by an audiologist.

As for the impact of noise on the environment, it is necessary to inspect the types of noise sources and to map their location, to take into account the complaints of residents, to study weather conditions (wind, temperature, and humidity), and to use the acoustical barriers and reflectors.

Environmental noise regulations usually specify levels of 45 dB that are associated with indoor residential areas, hospitals, and schools, whereas 55 dB are identified for certain outdoor areas where human activity takes place; the level of 70 dB is identified for all areas in order to prevent hearing loss (EPA identifies noise levels affecting health and welfare).

## 28.10   EFFECT OF NOISE POLLUTION ON ECOSYSTEM

The environmental noise affects animals, having a serious influence on their ability to mate, live, and hunt. Birds and frogs with lower frequency calls avoid noisy areas where it is difficult to communicate or to find the mates. Owls that are characterized by their acute sense of hearing to detect the slightest movement of prey suffer from noise which limits their ability to survive and thrive.

In addition, intense noise can increase heart rate, blood pressure, and provoke stress in animals.

Sometimes the noise has the opposite effect on the living world. For example, the hummingbirds actually prefer noisy sites; such a behavior can be explained by the fact that their predators, Western scrub jays that eat the hummingbirds' nestlings, leave the noise areas. If the noise-sensitive scrub jays that pick up the seeds disappear from the noise sites, a mice, which is more inclined to eat the seeds, appears more frequently there. A hundred times more intense artificial sounds can alter the behavior of marine animals.

An active sonar is used by navy to detect submarines at a long distance. The sound level of such a sonar reaches 215–235 dB; even 300 miles from the source, these sonic waves can retain an intensity of 140 dB. Such a noise level is dangerous for marine animals. Whales and other marine mammals use ultrasound for orientation and communication, in search of food and partners.

Thus, the noise affects the representatives of the living world and causes changes in biodiversity.

# 29

# MEASUREMENT OF NOISE

## 29.1  SOUND LEVEL METERS

*Sound level meter* measures sound level pressure (intensity of sound) at one point and at given moment and evaluates quantitatively any noise of industrial, environmental, and *aircraft* origin.

The procedure of estimation of sound level requires a number of measurements at different times during the workday.

*Loudness meter* is a device that measures the loudness levels that listeners will perceive. The response of this device is modeled on the human ear, which is not uniform over the entire frequency range; human ear perceives loudness in logarithmic scale.

A common version of the sound level meter is a *noise dosimeter* which can be used for "personal" noise monitoring in a given environment. This device is a sound level meter that can accumulate the results of sound level measurements, integrate them over time, and produce results in the form of average noise exposure reading for a given period of time (e.g., during an 8-hour workday). This compact noise dosimeter is fixed to the worker's clothing; it does not interfere with the worker and does not affect the sound field.

In fact, a typical sound level meter consists of a microphone that is connected to a voltmeter calibrated in decibels. Since the output electric signal from the microphone is proportional to the original sound signal, the change of the sound pressure acting on the microphone membrane is converted into electrical signals of voltmeter.

*Methods of Measuring Environmental Parameters*, First Edition. Yuriy Posudin.
© 2014 John Wiley & Sons, Inc. Published 2014 by John Wiley & Sons, Inc.

## 29.2   TYPES OF MICROPHONES

*Condenser microphone* (*capacitor microphone*) contains a diaphragm that acts as one plate of a capacitor. The external sound pressure produces change in the distance between the plates and corresponding change in the capacitance of the plates which is inversely proportional to the distance between them for a parallel-plate capacitor (see Section 2.5, formula 2.2).

*Piezomicrophone* uses the phenomenon of *piezoelectricity* (see Section 2.6)—the ability of some materials to produce a voltage when subjected to sound pressure and to convert mechanical vibrations into an electrical signal.

*Electret microphone* is based on the application of electret—a ferroelectric material that has been permanently electrically charged or polarized. This term "electret" comes from *electro*static and magn*et*; it indicates the analogy to the formation of a magnet by aligning the magnetic domains in a piece of iron.

## 29.3   NOISE FREQUENCY ANALYZERS

Noise analyzers make it possible to estimate noise signal at each frequency simultaneously. The results are demonstrated at the display as dependence of sound intensity level on the frequency.

## 29.4   SOUND INTENSITY MEASUREMENT

*Vector sound intensity* is the net rate of sound energy. The measurement of this parameter is useful for determination of noise source power, measurement of a moving sound source, and for location of noise sources. The system of measurement provides the installation of several microphones at various locations.

# PRACTICAL EXERCISE 12

## SOUND INSULATION AND REVERBERATION TIME

### 1 SOUND INSULATION

If the source of an unwanted sound is in the next room, the direct transmission of sound through the wall depends on the insulating properties of the latter.

Sound insulation of the wall $R$ is defined by the expression:

$$R = 10 \lg I_i / I_t,  \tag{P12.1}$$

where $I_i$ is the intensity of unwanted sound; $I_t$ is the intensity of the sound which passes through the wall.

If the sound passes through the wall diffusely, a level of sound insulation can be expressed as

$$R = L_e - L_0 + 10 \lg \left( \frac{A}{S_{cr}} \right),  \tag{P12.2}$$

where $L_e$ and $L_0$ is the sound pressure level in the room where the sound source is located, and in the room, where the sound is perceived, respectively.

**Example**  My room is separated from my neighbors' room of a student by a concrete wall of area 15 m$^2$. The student is a big fan of rock music. Sound pressure level in the student's room where the sound source is located is $L_e = 110$ dB. Determine the

*Methods of Measuring Environmental Parameters*, First Edition. Yuriy Posudin.
© 2014 John Wiley & Sons, Inc. Published 2014 by John Wiley & Sons, Inc.

sound pressure level in my room, if the sound insulation of the wall is 40 dB and the absorption of sound in my room is $A = 2.96$ m$^2$.

**Solution**    Using Equation P12.2, we can find:

$$L_0 = L_e - R + 10\lg\left(\frac{A}{S}\right) = 110 \text{ dB} - 40 \text{ dB} + 10\lg\left(\frac{2.96}{15}\right)$$

$$= 70 - 10 \times 0.7 = 63 \text{ dB}.$$

This sound level can be considered as moderate noise.

## 2   REVERBERATION TIME

*Reverberation time* is defined as the time required, in seconds, for the average sound in a room to attenuate to a level 60 dB after a source stops producing sound.

Reverberation time is proportional to the volume $V$ of a room and inversely proportional to the absorption $A = \sum S_i\alpha_i$:

$$T(s) = 0.16V/\sum S_i\alpha_i, \tag{P12.3}$$

where $S_i$ is area of each surface, through which the sound passes; $\alpha_i$ is the absorption coefficient of each surface.

Since the absorption coefficient depends on the frequency, in practice the average frequency reverberation time $\langle T \rangle$ is measured:

$$\langle T \rangle = \frac{T(500 \text{ Hz}) + T(1000 \text{ Hz}) + T(2000 \text{ Hz})}{3}. \tag{P12.4}$$

**Example**    Specifications require during construction of a room the optimum value of reverberation time $T(s) = 0.8 - 1.1$ for speech. Dimensions of the room are 3.5 m × 5 m × 3 m. Room has a door area of 2 m$^2$ and a window area of 6 m$^2$, located on the wall 3.5 m × 5 m. Find real reverberation time.

**Solution**    Select the materials for equipment of the room.

**Materials for equipment of the room**

| Surface | Cover | Area, m$^2$ |
| --- | --- | --- |
| Floor | Linoleum | 17.5 |
| Door | Wood | 2 |
| Walls (excluding door and window) | Concrete | 15.75 |
| Ceiling | Concrete | 17.5 |
| Window | Glass | 5.25 |

Determine the coefficients of absorption for various materials.

**Values of absorption coefficients for various materials depending on sound frequency**

| Material | Absorption coefficient, $\alpha$ | | | | |
|---|---|---|---|---|---|
| | 250 Hz | 500 Hz | 1000 Hz | 2000 Hz | 4000 Hz |
| Concrete | 0.05 | 0.06 | 0.07 | 0.09 | 0.08 |
| Glass | 0.08 | 0.05 | 0.04 | 0.03 | 0.02 |
| Wood | 0.10 | 0.08 | 0.08 | 0.08 | 0.08 |
| Linoleum | 0.03 | 0.04 | 0.05 | 0.05 | 0.06 |

Calculate the total absorption of the sound in the room, where it is presented as the effective area $S$, through which all the sounds pass, multiplied by the absorption coefficient $\alpha$ for the frequencies of 500, 1000, and 2000 Hz

$$\text{Floor:} \quad A = S_{Fl}[\alpha(500 \text{ Hz})] = 17.5 \times 0.04 = 0.7 \text{ m}^2;$$
$$A = S_{Fl}[\alpha(1000 \text{ Hz})] = 17.5 \times 0.05 = 0.875 \text{ m}^2;$$
$$A = S_{Fl}[\alpha(2000 \text{ Hz})] = 17.5 \times 0.05 = 0.875 \text{ m}^2.$$

$$\text{Door:} \quad A = S_{D}[\alpha(500 \text{ Hz})] = 2 \times 0.08 = 0.16 \text{ m}^2;$$
$$A = S_{D}[\alpha(1000 \text{ Hz})] = 2 \times 0.08 = 0.16 \text{ m}^2;$$
$$A = S_{D}[\alpha(2000 \text{ Hz})] = 2 \times 0.08 = 0.16 \text{ m}^2.$$

$$\text{Ceiling:} \quad A = S_{Cl}[\alpha(500 \text{ Hz})] = 17.5 \times 0.06 = 1.05 \text{ m}^2;$$
$$A = S_{Cl}[\alpha(1000 \text{ Hz})] = 17.5 \times 0.07 = 1.225 \text{ m}^2;$$
$$A = S_{Cl}[\alpha(2000 \text{ Hz})] = 17.5 \times 0.09 = 1.575 \text{ m}^2.$$

$$\text{Walls:} \quad A = S_{Wl}[\alpha(500 \text{ Hz})] = 15.75 \times 0.10 = 1.575 \text{ m}^2;$$
$$A = S_{Wl}[\alpha(1000 \text{ Hz})] = 15.75 \times 0.05 = 0.7875 \text{ m}^2;$$
$$A = S_{Wl}[\alpha(2000 \text{ Hz})] = 15.75 \times 0.05 = 0.7875 \text{ m}^2.$$

$$\text{Window:} \quad A = S_{Wn}[\alpha(500 \text{ Hz})] = 5.25 \times 0.05 = 0.2625 \text{ m}^2;$$
$$A = S_{Wn}[\alpha(1000 \text{ Hz})] = 5.25 \times 0.04 = 0.21 \text{ m}^2;$$
$$A = S_{Wn}[\alpha(2000 \text{ Hz})] = 5.25 \times 0.03 = 0.1575 \text{ m}^2.$$

Determine the total amount of absorption for frequencies of 500, 1000, and 2000 Hz:

$$A(500 \text{ Hz}) = 3.7475 \text{ m}^2;$$
$$A(1000 \text{ Hz}) = 3.2675 \text{ m}^2;$$
$$A(2000 \text{ Hz}) = 3.555 \text{ m}^2.$$

Calculate the reverberation time for the frequencies 500, 1000, and 2000 Hz (here the volume of the room is $V = 52.5 \text{ m}^3$) using formula P12.2:

$$T(500 \text{ Hz}) = 0.16 \times 52.5/3.7475 = 2.24 \text{ s};$$
$$T(1000 \text{ Hz}) = 0.16 \times 52.5/3.2675 = 2.57 \text{ s};$$
$$T(2000 \text{ Hz}) = 0.16 \times 52.5/3.555 = 2.36 \text{ s}.$$

Find the average frequency reverberation time:

$$\langle T \rangle = (2.24 \text{ s} + 2.57 \text{ s} + 2.36 \text{ s})/3 = 2.39 \text{ s}.$$

This value exceeds the optimum values (1.2–1.4) that are predicted by specifications for an acoustic environment.

**Constructive Test**   Find in internet or scientific literature and explain, what does it mean "Environmental quality standards for noise"?
   What is A-weighting?

# 30

# THERMAL POLLUTION

## 30.1  SOURCES OF THERMAL POLLUTION

*Thermal pollution* is the discharge of heated water into bodies of water (rivers or lakes).

The main contributors to thermal heat pollution are thermal or nuclear power plants; industrial effluents such as petroleum refineries, pulp and paper mills, chemical plants, steel mills and smelters; sewage effluents; and biochemical activity. Over 60% of the original energy is dissipated into the environment as heated water and hot gases during the production of electricity and heat.

The use of water as a coolant is the principal cause of thermal pollution, while deforestation and soil erosion are other causal factors.

## 30.2  THE EFFECT OF THERMAL POLLUTION ON LIVING ORGANISMS

The ambient water temperature is the most vital requirements for survival of aquatic fauna and flora.

The effects of thermal pollution include decrease the amount of dissolved oxygen in the water, which aquatic life requires, damage to larvae and eggs of fish in rivers, killing off some species of fish and macroinvertibrates that have a limited tolerance for temperature change, and migration of living entities from their environment. The

*Methods of Measuring Environmental Parameters*, First Edition. Yuriy Posudin.
© 2014 John Wiley & Sons, Inc. Published 2014 by John Wiley & Sons, Inc.

reduction in the level of dissolved oxygen in the water can lead to the death of fish, insects, plants, amphibians, and marine or freshwater crustaceans.

The sudden opening or shutting down of power plants and respective abrupt change in water temperature known as "thermal shock" causes the death of aquatic organisms.

Temperature changes can cause significant changes in organism metabolism and other adverse cellular effects.

# 31

# MEASUREMENT OF THERMAL POLLUTION

## 31.1 THERMAL DISCHARGE INDEX

A *Thermal Discharge Index (TDI)* estimates the thermal efficiency of any power plant. It can be calculated as follows:

$$TDI = \frac{TP}{EP},$$

(31.1)

where TP is the thermal power discharged to the environment in $MW_{th}$ and EP is electrical power output in $MW_e$.

The typical values of TDI are

- fossil plant: TDI = 1.5 (thermal efficiency is 40%);
- nuclear plant: TDI = 2 (thermal efficiency is 33%);
- geothermal plant: TDI = 4 (thermal efficiency is 20%).

## 31.2 INDIRECT MEASUREMENT OF THERMAL POLLUTION

Indirect methods of thermal pollution measurement include determination of physical parameters such as turbidity, pH, conductance (see Chapter 20); electronic temperature meters (see Chapter 5); and dissolved oxygen meters (see Chapter 20).

*Methods of Measuring Environmental Parameters*, First Edition. Yuriy Posudin.
© 2014 John Wiley & Sons, Inc. Published 2014 by John Wiley & Sons, Inc.

# 32

# LIGHT POLLUTION

*Light pollution* is a periodic or continuous excessive or obtrusive artificial light that causes adverse effects such as sky glow, glare, light trespass, light clutter, decreased visibility at night, and energy waste, that obscures the night sky and interferes with astronomical observations and affects human light and ecosystems.

## 32.1   THE SOURCES OF LIGHT POLLUTION

The main sources of light pollution include streetlights, illumination of airports, city advertising, building exterior and interior lighting, commercial and industrial enterprises, offices, and sporting venues. Light pollution is inherent first of all to highly industrialized and densely populated areas. Inefficient design of many lighting systems leads to the waste of energy and illumination of the sky. This effect is enhanced by air aerosols, which refract, reflect, and scatter the emitted light.

## 32.2   TYPES OF LIGHT POLLUTION

Let us consider specific categories of light pollution.

*Methods of Measuring Environmental Parameters*, First Edition. Yuriy Posudin.
© 2014 John Wiley & Sons, Inc. Published 2014 by John Wiley & Sons, Inc.

### 32.2.1   Light Trespass

Light trespass means light violation of one's property or penetration of light into a foreign territory. For example, you experience sleep disturbances as a result of intense external light that penetrates into your window.

### 32.2.2   Over-Illumination

This type of light pollution is related to the excessive use of light. It is caused by improper lightning design, erroneous calculations of workplace illumination, the lack of timer-controlled lighting regime, improper installation, and orientation of the light sources.

### 32.2.3   Glare

This phenomenon is induced by excessive contrast between bright and dark areas in the field of view. Bright lights around roads can partially blind drivers or pedestrians and lead to accidents. The following types of glare are distinguished as *blinding glare* (caused by staring into the Sun), *disability glare* (provoked by oncoming cars lights), *discomfort glare* (that irritates, annoys, and causes fatigue during prolonged action).

### 32.2.4   Clutter

Clutter refers to excessive groupings of lights which can be caused by badly designed lights on roads, array of various commercial lightning and advertising, which can distract drivers or pilots and contribute to accidents.

### 32.2.5   Sky Glow

This type of light pollution is a result of a combination of light scattered or reflected from the densely populated areas. Sky glow is caused by the ionization of the atmosphere layers that are located above the mesosphere, followed by collisions between ions and neutral particles, their recombination, and emission of photons. In addition, sunlight is reflected and backscattered by interplanetary dust particles. As soon as sky glow reduces contrast in the night sky we cannot see the brightest stars.

## 32.3   EFFECTS OF LIGHT POLLUTION ON HUMAN HEALTH

There is evidence that excess light at night disturbs the body's internal clock and circadian rhythm (Schulmeister, 2002). Impaired biological cycle leads to hormonal imbalance. The presence of night illumination leads to reduced levels of hormone melatonin in the blood streams (even a 39-minute low-energy light exposure causes the suppression of melatonin levels by 50%) (Revell et al., 2006).

Excess light can trigger symptoms such as headaches, sexual dysfunction, fatigue in the workplace, increase in blood pressure, insomnia, migraine, and depression.

Children who sleep in the illuminated rooms demonstrate increased anxiety and stress responses.

## 32.4    EFFECTS OF LIGHT POLLUTION ON WILDLIFE

Artificial light at night may disrupt the capacity for orientation and navigation in nocturnal insects, fish and animals, migration of birds, food chains in insects and microorganisms, and periodicity of flowering in plants.

The female sea turtles lay their eggs in the sand at night and return to the sea, using the orientation of the moon light. Artificial lighting at the coast breaks the orientation of either adult animals or their descendants, who need to find their way to the ocean.

Birds that are guided by the bright stars during the migration process are confused by the night lights of the city. In addition, intense artificial light leads to a collision of birds with buildings and their death. According to the US Fish and Wildlife Service, about 4–5 million of birds are killed per year after being attracted to tall towers (Ecological light pollution).

Many nocturnal insects are attracted by city lights and are killed at collisions with lamps. In addition, they become victims of such predators as bats.

Illumination of watersheds through sky glow forces to reduce the height of staying zooplankton, which is the food for the fish and at the same time feeds on phytoplankton. Reduction in zooplankton predation causes the growth of phytoplankton (surface algae) that leads to algal blooms and reduces the quality of water.

## REFERENCES

Revell, V.L., Burgess, H.J., Gazda, D.J., Smith, M.R., Fogg, L.F., Charmane, I., and Eastman, C.I. 2006. Advancing human circadian rhythms with afternoon melatonin and morning intermittent bright light. *Endocrine Care,* 91 (1): 54.

Schulmeister, K., Weber, M., Bogner, W., et al. 2002. Application of melatonin action spectra on practical lighting issues. Final Report. The Fifth International LRO Lighting Research Symposium, Light and Human Health.

# 33

# MEASUREMENT OF LIGHT POLLUTION

## 33.1 DIGITAL PHOTOGRAPHY

*Digital Photography* is a photographic technology that is based on the conversion of light by sensitive matrix (array of electronic photodetectors) to capture the image which is then digitized and stored as a computer file for further processing and printing.

Commercial digital cameras Fuji S5000 and EOS D60 (Digital Camera Resource Page) were used for the estimation of circadian effective luminance (Hollan, 2004). These cameras were calibrated by solar spectrum. The spectral sensitivity of the cameras is in good agreement with the spectrum of action of the photosensitive hormone melatonin. Obviously, the method of digital photography can be used to quantify light pollution acting on the physiology of living organisms.

*The advantages of a digital camera:*

- Immediate image review and deletion is possible.
- High volume of images.
- Faster workflow.
- A digital camera can easily store up to 10,000 photos.
- Digital manipulation.

*Methods of Measuring Environmental Parameters*, First Edition. Yuriy Posudin.
© 2014 John Wiley & Sons, Inc. Published 2014 by John Wiley & Sons, Inc.

*The disadvantages of a digital camera:*

- The battery consumption of a digital camera is quite high.
- Digital cameras are sensitive to weather changes.
- Digital cameras are sensitive to breaking after being dropped or hit.
- A computer crash can delete all stored images.
- A good digital camera can cost a lot.
- Digital cameras can have disappointingly poor focusing ability in low-light situations.
- The complexity of downloading and editing images.

## 33.2   PORTABLE SPECTROPHOTOMETERS

Principles of spectrophotometry are discussed in Section 16.5.1. A *portable spectrophotometer* for the measurement of light pollution is proposed by Cinzano (2004). It consists of a cooled CCD camera and a small spectrographic head which is equipped with a De Amici prism composed of two external crown prisms and an inner Flint prism. This spectrographic head can be compared with a small "telescope" with external size of less than 17 cm of length and 5.5 cm of diameter; that is why this device was named "small spectrographic head." The camera lenses focus the image of the dark sky on the slit of telescope. The light is collected from a solid angle 0.2° × 3.8° when slit width is 100 μm. Such a photometric system allows hyperspectral mapping of the night sky at sites or across the territory in any photometrical band for light pollution evaluation in the visual range 420–950 nm.

*The advantages of a portable spectrophotometer:*

- Lightness, compactness, portability.
- Possibility of automatic mapping of the entire sky.
- Automatic registration of position, elevation, date, time, and space coordinates.
- Automatic data reduction.
- Low cost.

*The disadvantage of a portable spectrophotometer:*

- It requires calibrations with a standard lamp.

## 33.3   SKY QUALITY METER

*Sky quality meter (SQM)* is a portable photometer for measuring sky brightness and for light pollution monitoring (Teikari, 2007). This device collects the light from a wide solid angle (1.532 steradians or a cone with 89 degrees angle). It measures the sky brightness in such astronomical units as magnitude per square arc second. These units can be converted into candelas per square meters. Size of device is 3.8 in × 2.4 in × 1 in.

*The advantages of a sky quality meter:*

- Lightness, compactness, and portability.
- Infrared blocking filter restricts measurement to visual bandpass.
- Precision readings at even the darkest sites.
- Low cost.

*The disadvantages of a sky quality meter:*

- The presence of water vapor, clouds, and Milky Way in the sky affects a general sky brightening.

## 33.4   THE BORTLE SCALE

John E. Bortle (W. R. Brooks Observatory) proposed in 2001 a system of assessment of light pollution on the basis of its effect on the observability of celestial bodies (stars, star clusters, and planets) in the sky. The Bortle scale consists of nine classes. Let us discuss some of them.

Class 1: The Milky Way is visible distinctly which clearly stand out stars and clusters.

Class 5: The Milky Way is very weak or invisible near the horizon and looks washed out overhead.

Class 7: The Milky Way is invisible.

Class 9: The only object to observe the Moon, the planets, and a few of the brightest star clusters such as the Pleiades.

In a city where the light pollution is high, the sky illumination is estimated as 9 class; in the countryside, the impact of light pollution is negligible and the sky illumination is estimated by 1 class.

## REFERENCES

Bortle, J.E. 2001. The Bortle dark-sky scale. In: *Sky & Telescope*. Sky Publishing Corporation.

Cinzano, P. A portable spectrophotometer for light pollution measurements. *Mem. S.A.It.* (Suppl. 5): 395–398.

# 34

# ELECTROMAGNETIC POLLUTION

## 34.1   PRINCIPAL TERMINOLOGY AND UNITS

*Electromagnetic radiation (EMR)* is a form of energy produced by electrically charged objects; it consists of oscillating electric and magnetic fields at right angles to each other and to the direction of propagation.

Electromagnetic radiation is measured in such units as frequency (Hz, MHz, GHz) and wavelength (m, mm, nm).

The range of all types of electromagnetic radiation is called the *electromagnetic spectrum*. It extends from the short wavelengths (high frequencies) to long wavelengths (low frequency): gamma-rays, X-rays, ultraviolet, visible, infrared, microwaves, and radio waves.

The *electromagnetic field* is the coupled electric and magnetic fields that are generated by time-varying currents and accelerated charges.

An *electric field* is generated by electrically charged particles and time-varying magnetic fields.

The electric field is characterized by an *electric field strength $\vec{E}$* equal to the force the field would induce on a unit electric charge at a given point in space. Also called electric intensity, electric field intensity.

The SI unit of the electric field strength is *newtons per coulomb (N/C)* or *volts per meter (V/m)*, which in terms of SI base unit is $kg/m \cdot s^3 \cdot A^1$.

A *magnetic field* is a region of space near a magnet, electric current, or moving charged particle in which a magnetic force acts on any other magnet, electric current, or moving charged particle.

*Methods of Measuring Environmental Parameters*, First Edition. Yuriy Posudin.
© 2014 John Wiley & Sons, Inc. Published 2014 by John Wiley & Sons, Inc.

The magnetic field is characterized by a *magnetic field vector* $\vec{B}$—the amount of magnetic flux through a unit area taken perpendicular to the direction of the magnetic flux. Also called *magnetic flux density* or *magnetic induction*.

The SI unit of the magnetic field is the *weber per square meter (Wb/m²)* that is called *tesla (T)*. The SI unit of tesla is equivalent to

$$T = \frac{Wb}{m^2} = \frac{N}{C \cdot m/s} = \frac{N}{A \cdot m}.$$

The cgs unit for magnetic field is the *gauss (G)*: $1\,T = 10^4\,G$.

$1\,T = 1\,kg/A \cdot s^2 = 10^4\,G = 10^9$ gamma; $1\,G = 10^{-4}\,T = 10^5$ gamma; $1$ gamma $= 10^{-5}\,G = 1\,nT$.

Another characteristic of the magnetic field is *magnetic field strength* $\vec{H}$, which indicates the ability of a magnetic field to exert a force on moving electric charges. It is equal to the magnetic flux density divided by the magnetic permeability of the space where the field exists.

The SI unit of magnetic field strength is *amperes per meter (A/m)*, and *oersteds (Oe)* in cgs system of units.

$1\,A/M = 4\pi \times 10^{-3}\,Oe$; $1\,Oe = (1/4\pi) \times 10^3\,A/M = 79.5775\,A/M$.

These two quantities are related to the expression:

$$\vec{B} = \mu\vec{H}, \tag{34.1}$$

where $\mu$ is the magnetic permeability ($\mu = 4\pi \times 10^{-7}\,H/m$).

## 34.2  ELECTROMAGNETIC POLLUTION

*Electromagnetic pollution* is the sum total of all the man-made electromagnetic fields of various frequencies, which fill our homes, workplaces, and public spaces and are harmful to human.

The main sources of electromagnetic pollution are high-voltage electrical transmission lines, radar sites, radio and TV transmitters, cell phone masts, power tools, electric stoves, heaters, boilers, freezers, information networks, extension cords, house wiring, cell (and other mobile) phones, radios, video display systems, television sets, lamps, typewriters, photocopiers, fluorescent light fixtures, computers and related equipment, microwave ovens, toaster, coffee pot, hair dryer, and refrigerator.

Our special attention will be paid to *extremely low frequency (ELF)* waves (3–300 Hz), to *radiofrequency (RF) radiation* (3 kHz–300 MHz), and to *microwave (MW) radiation* (300 MHz–300 GHz) that are harmful to human health.

The majority of national standards for exposure to electromagnetic fields draw on the guidelines set by the International Commission on Non-Ionizing Radiation Protection (ICNIRP) (ICNIRP, 2003; ICNIRP Guidelines, 1998, 2010). The Institute of Electrical and Electronics Engineers (IEEE) published also the recommendations for preventing harmful effects from exposure to ELF fields (IEEE C95.6 Standard).

Let us give some examples demonstrating the use of reference levels for general public and occupational exposure to time-varying electric and magnetic fields (ICNIRP Guidelines, 2010).

Public exposure limits (frequency 25–50 Hz): electric field strength $E = 5$ kV/m; magnetic field strength $H = 1.6 \times 10^2$ A/m; magnetic flux density $B = 2 \times 10^{-4}$ T.

Occupational exposure limits (frequency 25–50 Hz): Electric field strength 10 kV/m; magnetic field strength $H = 8 \times 10^2$ A/m; magnetic flux density $B = 10^{-3}$ T.

Magnetic fields that surround us in everyday life depend on the type of appliance (electric shaver, vacuum cleaner, microwave oven, refrigerator, coffee maker, hair dryer, television set, and computer) and the distance from it. Typical values of magnetic field strength (at distance 1 m) vary from 0.01 to 0.6 μT (Federal Office for Radiation Safety, Germany 1999).

Microwave range of electromagnetic waves that surround us is created by a system of mobile communication, which is characterized by the following parameters:

The frequency range of mobile communication in most countries is 900–1800 MHz.

Mobile phone base station power density: public exposure limits 4.5 W/m$^2$ (900 MHz) and 9 W/m$^2$ (1800 MHz); occupational exposure limits 22.5 W/m$^2$ (900 MHz) and 45 W/m$^2$ (1800 MHz) (ICNIRP, 1998).

Maximum power of electromagnetic radiation emitted by mobile cell is 1–2 W (Yakymenko and Sydorik, 2010).

The electric field generated by a mobile phone depends on the category (U0, U1, U2, or U3) and varies from 46–51 to 36 V/m (Hoolihan, ANSI C63.19).

The magnetic field strength transmitted from the antenna of the mobile phones at the distance 7.5–10 cm from the antenna, is well over 2 μT; if a mobile cell is held to the user's head, the magnetic field strength reaches 1–6 μT (The Physics Factbook, G. Elert, ed.).

A personal computer that fills our life both at work and at home is the source of the electric and magnetic fields. Typical maximum public exposure to these fields according to WHO information is as follows:

Electric field at operator place $E = 10$ V/m; magnetic flux density $B = 0.7$ μT.

## 34.3 EFFECT OF ELECROMAGNETIC POLLUTION ON HUMAN HEALTH

### 34.3.1 Extremely Low Fields

In 2005, the World Health Organization, particularly a Task Group of scientific experts, studied the effects of ELF on human health and assesses the risks to health that might exist from exposure to these fields (WHO, 2007). It is known that electric current carriers produce both electric and magnetic fields. Typical values of electric field in the home are up to several volts per meter; the average values of magnetic field are 0.07 μT in Europe and 0.11 μT in North America.

Electric fields of electrical appliances (30 cm away) measured in V/m are electric range 4, broiler 40, vaporizer 60, color TV 30, toaster 40, coffee pot 16, hair dryer 40. stereo 90, refrigerator 30, iron 60, and electric blanket 250. The electric field strength underneath the electric power lines is about several thousand volts per meter and magnetic induction is about 20 μT.

Levels of risk to adverse electric and magnetic fields that produce the biological effect to humans are normal $E = 0$–5.9 V/m; $B = 0$–64 nT; threshold $E = 6.0$ V/m; $B = 65$ nT; dangerous $E = 6.1$–8.9 V/n; $B = 66$–99 nT; very dangerous $E = 9.0$–13.9 V/m; $B = 100$–249 nT (Michrowski, 1991).

Acute exposure at high levels (above 100 μT) causes nerve and muscle stimulation and disturbances of central nervous system.

Long-term exposure to strong electromagnetic fields of about 50–60 Hz may cause lack of energy or fatigue, irritability, aggression, hyperactivity, sleep disorders, and emotional instability.

Long-term exposure to ELF magnetic fields can cause the childhood leukemia: the Task Group has established that a twofold increase in childhood leukemia can be associated with average exposure to magnetic field above 0.3–0.4 μT.

The conclusions concerning possible biological effects of ELF are very controversial. Some investigators point to possible carcinogenic, reproductive, and neurological effects; others mention cardiovascular, brain and behavior, hormonal, and immune system changes.

The detailed summary of biological effects of electric and magnetic fields of ELF is given in the ICNIRP Guidelines (1998).

### 34.3.2  Estimation of Health Effects of EMF Through the Questionnaires

The effect of ELF electromagnetic fields exposure on sleep quality of workers who had been exposed to electromagnetic fields in high-voltage substations (132, 230, and 400 kV) in Kerman city, Iran, was investigated by group of researchers (Barsam et al., 2012). The average values of the electric field intensity and magnetic flux density in a shift work and sleep quality of workers were compared to assess the level of correlation. Sleep quality of the case and control groups of workers was evaluated by the Pittsburgh Sleep Quality Index (PSQI) questionnaire.

The PSQI was developed by Buysse et al. (1989). The questionnaire included several categories: subjective sleep quality, sleep latency, sleep duration, habitual sleep efficiency, sleep disturbances, use of sleeping medication, and daytime dysfunction. A global PSQI score had a range of 0–21; higher scores indicated worse sleep quality.

It was shown that occupational exposure to power frequency electric and magnetic fields might have a detrimental influence on night sleep.

### 34.3.3    Radiofrequency and Microwave Fields

Radiofrequency (RF) exposure from various devices provokes diseases such as cancer, neurological disease, reproductive disorders, immune dysfunction, and electromagnetic hypersensitivity.

It should be noted that the effect of electromagnetic fields on living organism depends on several factors, such as frequency, electromagnetic field intensity, duration of action, and individual properties of organism. Often, information about the possible effects is controversial, indicating a need for further in-depth investigations (Health Effects from Radiofrequency Electromagnetic Fields, 2012).

### 34.3.4    Effect of Mobile Phones on Human Health

The number of mobile subscribers in the world is close to 4 billion. Of course, the problem of the effect of mobile phone radiation on human health is very actual.

It is believed that the level of the magnetic field strength 0.2 µT is safe for the human health, while 0.7 µT is dangerous.

The debate as to whether the mobile phones are dangerous to health or not continues for a long time. Numerous studies on epidemiological level with laboratory animals and *in vitro* have not yet given a clear answer.

Point is that the brain absorbs a significant portion of the electromagnetic energy that is radiated by phone during its interaction with base stations. That is why people who often use a mobile phone are at risk. After all, most people talking on a cell for several hours a day and at any time of the day. And worst of all, it concerns our children (Yakymenko and Sydorik, 2010). So additional research into the long-term use of mobile phones needs to be conducted.

### 34.3.5    Effect of Computer on Human Health

With regard to the effects of electromagnetic fields generated by the computer, there is no consensus. The "Encyclopedia Britannica" reports that "prolonged exposure to computer radiation leads to an increased risk of tumors, cancers, miscarriage, blood disorders, insomnia, headaches, anxiety, and skin disorders" (Negative Effects of Computer Radiation).

At the same time, other researchers have argued that the electromagnetic field of the computer is safe for human health. For example, the American Academy of Ophthalmology (AAO) claims that "there is no convincing scientific evidence that computer video display terminals (VDTs) are harmful to the eyes" (Computer/VDT Screens).

Thus, the information concerning the computer radiation hazards is contradictory. In any case, we must remember that long-term use of the computer is undesirable.

## REFERENCES

Barsam, T., Monazzam, M.R., Haghdoost, A.A., Ghotbi, M.R., and Dehghan, S.F. 2012. Effect of extremely low frequency electromagnetic field exposure on sleep quality in high voltage substations Iranian. *J. Environ. Health Sci. Eng.* 9:15.

Buysse, D.J., Reynolds, C.F., Monk, T.H., Berman, S.R., and Kupfer, D.J. 1989. The Pittsburgh Sleep Quality Index (PSQI): a new instrument for psychiatric research and practice. *Psychiatry Res.* 28(2):193–213.

Health Effects from Radiofrequency Electromagnetic Fields. 2012. Report of the Independent Advisory Group on Non-ionizing Radiation. Document of the Health Protection Agency. Radiation, Chemical and Environmental Hazards.

ICNIRP. 1998. Guidelines for limiting exposure to time-varying electric, magnetic and electromagnetic fields (up to 300 GHz). *Health Phys.* 74(4):494–533.

ICNIRP. 2010. Guidelines for limiting exposure to time-varying electric and magnetic fields (1 Hz–100 KHz). *Health Phys.* 99(6):818–836.

ICNIRP—International Commission on Non-Ionizing Radiation Protection. 2003. *Exposure to Static and Low Frequency Electromagnetic Fields, Biological Effects and Health Consequences (0–100 kHz)* (eds R. Matthes, A.F. Mckinlay, J.H. Bernhardt, P. Vecchai, and B. Veyret). International Commission on Non-ionizing Radiation Protection (ICNIRP 13/2003), Oberschleissheim.

Michrowski, A. 1991. Electromagnetic pollution. *Consum. Health*, 14 (3).

WHO. 2007. Extremely low frequency fields. Environmental Health Criteria. Volume 238. World Health Organization, Geneva.

Yakymenko, I.L. and Sydorik, E.P. 2010. *Mobile Phone and Human Health.* Kiev, Znannya.

# 35

# MEASUREMENT OF ELECTROMAGNETIC POLLUTION

## 35.1 EMF METER

An *EMF meter* is a device which is intended for measuring the electromagnetic radiation and displaying the results of measurement on a dial or a digital display. The EMF meter can be called also as AC gaussmeter, an electromagnetic field meter, a field strength meter, an electrosmog meter, or an EMF detection meter. If the device provides an audible alarm or visual alert, but not actual measurement, it is called a *detector*.

The principle of operation of EMF meter is based on the conversion of the magnetic field oscillations into electric voltage oscillations, frequency filtering, and amplification of these oscillations, followed by digitization and numerical analysis of the results.

The device consists of an electric field sensor, magnetic flux density sensor, power operational amplifiers, power processor, liquid crystal display, battery, and charging unit. Electric sensor contains an electrically short dipole antenna and diode detector connected to an instrumentation amplifier. Magnetic flux density sensor consists of the pick up (some kind of loop or Hall effect sensor) and the detector which converts the signal to a form that can be registered on a readout system.

## 35.2 TYPES OF EMF METERS

All EMF meters can be classified as *low-frequency EMF meters* (measuring house wiring, domestic appliances, TVs, computers, office equipment, and vehicles) and

*Methods of Measuring Environmental Parameters*, First Edition. Yuriy Posudin.
© 2014 John Wiley & Sons, Inc. Published 2014 by John Wiley & Sons, Inc.

*radio frequency (RF) EMF meters* that are suitable for measuring radio waves and microwaves from cell phone towers, cell phones, wireless baby monitors, wireless networks, wireless modems, microwave ovens, TV, and radio broadcasting antennae (EMF Meters and Detectors).

EMF meters can be isotropic, mono-axial, or triaxial, active and passive.

The meters can display results of measuring the electric field strength in V/m or mV/m, magnetic field strength in A/m, mA/m, $\mu$A/m, magnetic flux density in mT, $\mu$T, nT, G, mG. In addition, power density values can be displayed in W/m$^2$, mW/m$^2$, $\mu$W/m$^2$, mW/cm$^2$, $\mu$W/cm$^2$.

We give as examples the EMF meters, which represent both types of devices.

*Lutron 822-A Digital EMF Meter* (Manufacturer: Lutron) belongs to a class low-frequency EMF meters. It is characterized by the following parameters:

Measurement range 0.1–199.9 mG; 0.01–19.99 $\mu$T.

Accuracy 4% ± 3% at 50/60 Hz.

Frequency range 30–400 Hz.

Requires one 9-V battery.

*The advantages of a Lutron 822-A digital EMF meter:*

- Lightweight, compact.
- Very easy to operate.
- Does not cost much.
- High resolution.

*The disadvantage of a Lutron 822-A digital EMF meter:*

- The only compromise with this meter is the fact that it is single axis: it is necessary to take readings in three planes manually.

*Extech 480836 RF/EMF Meter* (Manufacturer: Extech Instruments) represents radio frequency EMF meters. It has the following parameters.

Frequency range 50 MHz–3.5GHz (measurement optimized for 900 MHz, 1800 MHz, and 2.7 GHz);

Measuring Ranges
20 mV/m to 108.0 V/m
53 $\mu$A/m to 286.4 mA/m
1 $\mu$W/m$^2$ to 30.93 mW/m$^2$
0 $\mu$W/cm$^2$ to 3.093 mW/cm$^2$
Resolution
0.1 mVm, 0.1 mA/m, 0.1 mW/m$^2$, 0.001 mW/cm$^2$

*The advantages of an Extech 480836 RF/EMF meter:*

- Non-directional (isotropic) measurement with three-channel (triaxial) measurement probe.
- High-frequency range.

*The disadvantages of an Extech 480836 RF/EMF meter:*

- It seems to slowly run down the 9-V battery, even when the meter is switched off.
- This meter certainly is not cheap.

# 36

# RADIOACTIVE POLLUTION

## 36.1 PRINCIPAL DEFINITIONS

The *radioactive pollution* is defined as the physical pollution of living organisms and their environment (atmosphere, hydrosphere, and lithosphere) as a result of release of radioactive substances into the environment during nuclear explosions and testing of nuclear weapons, nuclear weapon production and decommissioning, mining of radioactive ores, handling and disposal of radioactive waste, and accidents at nuclear power plants.

In 1896, French physicist Becquerel discovered the phenomenon of *radioactivity*—spontaneous transformation of atoms of one element into atoms of other elements, accompanied by the emission of particles and electromagnetic radiation. The three types of radioactive emission include alpha particles (two protons and two neutrons bound together into a particle identical to a helium nucleus), beta particles (high-energy, high-speed electrons or positrons), and $\gamma$-radiation (electromagnetic radiation of high frequency and therefore high energy per photon). These emissions constitute *ionizing radiation*.

The nuclei of elements exhibiting radioactivity are unstable; the atoms with unstable nuclei are called *radionuclides*. There are natural and artificial radionuclides. Artificial radionuclides can penetrate into the environment and present a real danger to ecosystems and human health.

From the point of view of ecology, strontium $^{90}$Sr and cesium $^{137}$Cs are the most dangerous radionuclides because of their long period of half-decay: 28 years for $^{90}$Sr and 30 years for $^{137}$Cs.

*Methods of Measuring Environmental Parameters*, First Edition. Yuriy Posudin.

## 36.2 UNITS OF RADIOACTIVITY

The unit of radioactivity is *curie (Ci)* defined as

$$1 \text{ Ci} = 3.7 \times 10^{10} \text{ decays/s.}$$

Curie is large unit (it corresponds to the approximate activity of 1 g of radium), so common fractions of the curie are 1 mCi $= 10^{-3}$ Ci; 1 $\mu$Ci $= 10^{-6}$ Ci; 1 nCi $= 10^{-9}$ Ci; 1 pCi $= 10^{-12}$ Ci.

Curie is a non-SI unit of radioactivity. The SI unit of radioactivity is *becquerel (Bq)*

$$1 \text{ Bq} = 1 \text{ decay/s.}$$

One Bq corresponds to the activity of a quantity of radioactive material in which one nucleus decays per second.

As soon as 1 Bq is a small unit for practical application, the prefixes are used: 37 GBq $= 37 \times 10^{10}$ Bq $= 1$ Ci; 1 MBq $= 1 \times 10^6$ Bq $= 37 \times 10^3$ Ci; 1 GBq $= 1 \times 10^9$ Bq $= 37$ Ci; 1 TBq $= 1 \times 10^{12}$ Bq $= 0.037$ Ci;1 PBq $= 10^{15}$ Bq $= 3.7 \times 10^{-5}$ Ci.

Surface contamination is usually expressed in units of radioactivity per unit of area, for example, becquerels per square meter (or $Bq/m^2$), picoCuries per 100 $cm^2$ ($pCi/100 \text{ cm}^2$), $Ci/km^2$ and $kBq/km^2$, disintegrations per minute per square centimeter ($1 \text{ dpm/cm}^2 = 167 \text{ Bq/m}^2$). Contamination of water is expressed in Bq or Ci per liter.

## 36.3 NUCLEAR EXPLOSIONS AND TESTING OF NUCLEAR WEAPONS

A *nuclear weapon* is an explosive device that derives its destructive force from nuclear reactions. Its damaging factors are the blast wave, thermal radiation, prompt ionizing radiation, radioactive pollution, and electromagnetic pulse.

*Nuclear tests* are carried out to determine the effectiveness, yield, and explosive capability of nuclear weapons.

The nuclear powers have conducted at least 2119 nuclear test explosions in the atmosphere, under water or in space, and the rest underground. Of these 1032 tests belong to the United States, 727 to the Soviet Union, 88 to the United Kingdom, 217 to France, 47 to China, 3 to India, 2 to Pakistan, and 3 to North Korea.

The first nuclear test was approbated by the United States at the Trinity site in 1945, with a yield approximately equivalent to 20 kilotons.

The Soviet "Tsar Bomba" test was the largest nuclear explosion that was conducted at Novaya Zemlya in 1961. The yield of this explosion was estimated around 50 megatons (3800 times more powerful than the Hiroshima bomb).

The proportion of radioactive pollution is 15% of the total energy of the explosion. Radioactive pollution of water, water sources, and air space is the result of radioactive fallout from the cloud of a nuclear explosion. Radionuclides are the main sources of pollution; they emit beta particles and gamma rays, radioactive substances. Radioactive pollution, as well as ionizing radiation, does not cause damage to buildings, facilities, equipment, but affects living organisms, agricultural crops, and livestock facilities.

## 36.4 ACCIDENTS AT NUCLEAR POWER PLANTS

### 36.4.1 Three Mile Island Accident

The accident occurred on March 28, 1979, with failures in the non-nuclear secondary system of a cooling system, caused by malfunction of the pressurized relief valve in the primary system, followed by the termination of the water supply in both the steam generator of the primary system and partial meltdown of the unit TMI-2. As a result, there was a leak of large amounts of nuclear reactor coolant and the release of an amount of radioactivity, estimated at 43,000 Ci (1.59 PBq) of radioactive krypton-85 gas, but less than 20 Ci (740 GBq) of the especially hazardous iodine-131, into the surrounding environment.

According to the U.S. Nuclear Regulatory Commission (NRC), the accident did not provoke any deaths or injuries to plant workers or nearby inhabitants.

### 36.4.2 Kyshtym Accident

The Kyshtym accident happened on September 29, 1957, in a nuclear plant "Mayak" in Russia (previously the Soviet Union). In fact the disaster occurred in Ozersk, top-secret town (also known as Chelyabinsk-40 and Chelyabinsk-65) that was not marked on the map. Therefore, this event was named after Kyshtym, the nearest known city.

The cause of the disaster was damage of the cooling system in a tank, which contained about 70–80 tons of radioactive waste. The result was a nonnuclear burst and release of dried waste into the air. As soon as the level of radioactivity was 20 MCi (800 PBq), about 10,000 people were evacuated. The fallout of radioactive cloud provoked a long-term contamination by radionuclides (cesium-137 and strontium-90) which spread out over a wide area (Kabakchi and Putilov, 1995).

### 36.4.3 Chernobyl Accident

The Chernobyl disaster is the destruction of April 26, 1986, of the fourth unit of the Chernobyl nuclear power plant, located on the territory of the Ukrainian SSR (now Ukraine), about 130 km north of Kiev. The reactor was completely destroyed as a result of the explosion and the large quantities of radioactive substances were released to the environment.

There is no consensus about the causes of the accident: some experts believe that the personnel of the station is responsible, while others argue that the poor design of the reactor led to the accident.

The resulting steam explosion and fires released the radioactive substances, including isotopes of uranium, plutonium, iodine-131, cesium-134, cesium-137, and strontium-90.

Dmytro Grodzinsky noted that the total amount of radioactivity released exceeded significantly 50 Mega Curies; on the roof of the destroyed reactor building, dose rate in May–June 1986 reached a very high level of 100,000 Roentgens per hour. Grigory Medvedev wrote: "I must emphasize that the radioactivity of the ejected fuel reached 15,000–20,000 roentgens per hour ..." (Medvedev, 1991).

Over 200,000 km$^2$ of territories were polluted. The radionuclide-contaminated areas of Ukraine were divided into the following zones: *zone of exclusion* (the 30 km zone around CNPS); *zone of obligatory resettlement* ($^{137}$Cs more than 15 Ci/km$^2$ or more than 555 kBq/m$^2$); *zone of guaranteed voluntary resettlement* ($^{137}$Cs 5–15 Ci/km$^2$ or 185–555 kBq/m$^2$); *zone of enhanced radiolecological control* ($^{137}$Cs 1–5 Ci/km$^2$ or 37–185 kBq/m$^2$).

About 350,000 people were evacuated from contaminated zones to other places after accident.

### 36.4.4    Fukushima Accident

The Fukushima Dai-ichi Nuclear Power Plant (FDNPP) accident occurred on March 11, 2011, as result of the Tōhoku earthquake and tsunami which caused the failure of the power lines and backup generators followed by nuclear meltdowns and releases of radioactive materials (Krolicki, 2012; Matsunagi, 2012; Lipsy et al., 2013).

Maximum level of radiation was estimated as 72,900 mSv/h (inside reactor 2) and 400 mSv/h between reactor units 3 and 4, respectively.

Air-borne radioactivity release into the atmosphere due to the Fukushima Daiichi Nuclear Disaster from March 12 to 31, 2011 was 11,000 PBq of Xe-133; 160 PBq of I-131; 18 PBq of Cs-134; and 15 PBq of Cs-137.

According to TEPCO information, the cesium-134 and cesium-137 levels in the soil (500 m from the reactors) were between 7.1 and 530 kBq per kilo of undried.

The evacuation zone was designated as 40 km (25 mile) to a northwest direction and 20 km (12 mile) to the remaining direction from the FDNPP. About 300,000 people were relocated.

### 36.4.5    Effect of Radioactive Pollution

Radioactive pollution is accompanied with the *ionizing radiation*—emission of high-energy particles or gamma radiation that has a high frequency and therefore high energy.

Particles or high-energy electromagnetic radiation penetrate into the human body and cause ionization of molecules present in the body. This ionizing radiation is a serious threat to human health. Thus formed free radicals react with the components of the living organism, causing the destruction of proteins, membranes, and nucleic acids. Depending on the intensity and duration of exposure to ionizing radiation, its effects on a living organism may be accompanied by mild, moderate, or even fatal consequences.

Low-level exposure may induce only superficial effect and mild skin irritation.

The short-range exposure which occurs few days provokes the loss of hair or nails, subcutaneous bleeding, and impairment of cells.

The long but low-intensity exposure leads to nausea, vomiting, diarrhea, and bruises.

The acute exposure to ionizing radiation is characterized by the damage of DNA cells that results in cancer, genetic defects, and even death.

## REFERENCES

Kabakchi, S.A. and Putilov, A.V. 1995. *Data Analysis and Physicochemical Modeling of the Radiation Accident in the Southern Urals in 1957.* Atomnaya Energiya, Moscow (1):46–50.

Krolicki, K. (2012). Fukushima radiation higher than first estimated. *Reuters*, May 24.

Lipscy, P., Kushida, K., and Incerti, T. 2013. The Fukushima disaster and Japan's nuclear plant vulnerability in comparative perspective. *Environ. Sci. Technol.* 47(May): 6082–6088.

Matsutani, M. 2012. Reactor 2 radiation too high for access. *Japan Times*, March 29.

Medvedev, G. (Author), Sakharov, A. (Foreword). 1991. *The Truth About Chernobyl.* Basic Books; First US edition.

# 37

# MEASUREMENT OF IONIZING RADIATION

## 37.1  DOSES OF IONIZING RADIATION

*Radiation dosimetry* is the measurement and calculation of the amount, rate, and distribution of radiation or radioactivity from a source of ionizing radiation during interaction of the ionizing radiation with the environment.

*Dose* is the energy of ionizing radiation absorbed by a matter that is irradiated. The dose depends on the type of radiation, its intensity, duration of exposure, and the composition of matter that is irradiated. Let us consider the principal doses.

*Absorbed dose* is measured as the energy absorbed per unit mass of matter that is irradiated:

$$D_a = \frac{dE}{dm},$$ 

(37.1)

where $dE$ is the mean absorbed energy; $dm$ is the mass of the matter.

The unit of absorbed dose is the *gray (Gy)*:

$$1\ Gy = 1\ J/kg = 100\ rad.$$
$$1\ rad = 0.01\ Gy = 0.01\ J/kg.$$

*Absorbed dose rate* is the quantity of radiation absorbed per unit time:

$$P_a = \frac{dD_a}{dt}.$$

(37.2)

The unit of absorbed dose rate is the *gray per second (Gy/s)* or *rad/s*.

*Methods of Measuring Environmental Parameters*, First Edition. Yuriy Posudin.
© 2014 John Wiley & Sons, Inc. Published 2014 by John Wiley & Sons, Inc.

*Exposure dose* is defined as the ratio of the total charge of all ions of the same sign produced in the elementary volume of air by secondary particles (electrons and positrons) divided by the mass that would completely stop the radiation:

$$D_e = \frac{dQ}{dm}, \tag{37.3}$$

where $dQ$ is the total charge of all ions of the same sign; $dm$ is the air mass.

The SI unit for exposure is C/kg. The old unit is *roentgen (R)*:

$$1\ R = 2.58 \times 10^{-4}\ C/kg;$$
$$1\ C/kg = 3.88 \times 10^3\ R.$$

The *exposure dose rate:*

$$P_e = \frac{dD_e}{dt}. \tag{37.4}$$

The unit of exposure dose rate is C/kg·s.

*Equivalent dose* quantifies the biological effectiveness of an absorbed dose of ionizing radiation. The fact is that biological effects depend not only on the dose but also on the type of ionizing radiation. It is necessary to compare the biological effects caused by any ionization radiation with biological effects that occur under the influence of X-rays with the utmost energy of 250 keV.

The equivalent dose $H$ is the product of the absorbed dose $D_a$, radiation weighting factor $W_R$ appropriate to the type and energy of radiation, and tissue weighting factor $W_T$ which depends on the part of the irradiated body:

$$H = D_a W_R W_T. \tag{37.5}$$

So, for X-ray and $\gamma$-radiation $W_R = 1,0$; $\alpha$-particles—20; $\beta$-particles and electrons—1,0; neutrons—5 to 20; protons—5; heavy ion—20. A weighting factor $W_T$ takes into account the influence of body area or its volume, the duration of exposure, type of living organism. Thus, for the bones, breast, colon, lung, stomach factor $W_T = 0,12$; for bladder, gonads, liver, thyroid—0.05; for brain, kidneys, skin, salivary gland—0.01; for remaining tissues—0.1.

Equivalent dose is measured in sieverts or rems ("rem" is an abbreviation of "the roentgen equivalent in man"):

$$1\ Sv = 1\ J/kg = 100\ rem;$$
$$1\ rem = 0.01\ Sv = 0.01\ J/kg.$$

## 37.2   GAS-FILLED DETECTORS

The principle of operation of these detectors is based on the application of voltage to the spatially separated electrodes that are placed in a chamber filled with gas. As ionizing radiation passes through the gas, positive ions and electrons are generated

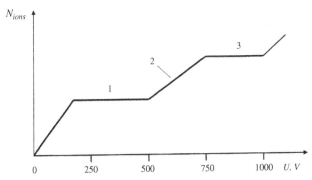

**FIGURE 37.1** Dependence of the collected charges on the applied voltage in gas-filled detectors: 1—ionization chamber; 2—proportional counter; 3—Geiger–Müller counter.

in the chamber where they are collected by the electrodes. These charges lead to an electric current or pulses in the electric circuit containing the detector.

The main types of gas-filled detectors are the ionization chamber, proportional counters, and Geiger–Müller counter. The difference between these detectors is explained in Figure 37.1, which shows the dependence of ions collected by the electrodes on the applied voltage.

This voltage causes a competition between loss of ion pairs by recombination and their collection at the electrodes. Increasing the voltage (region 1) increases the velocity of the ions, reducing the time required for recombination; the charge on the electrodes is proportional to the number of ion pairs formed in the inter-electrode space. Operation of ionization chamber is based on this principle. Further increase of voltage is accompanied by the additional ionization of which takes part in the primary ionization processes. The charge on the electrodes in this situation is directly proportional to the applied voltage (region 2). This dependence is used in the proportional counter. Continued increase of voltage results in situation when the charge on the electrodes is independent of the applied voltage (region 3). This situation is realized in the Geiger–Müller counter.

### 37.2.1  Ionization Chamber

This detector is based on the ability of charges to cause ionization within the gas through the application of an electric field. It makes it possible to obtain direct information on exposure or absorbed dose. Indeed, since roentgen is a unit of exposure dose and corresponds to the number of charges that is formed by ionizing radiation of 1 cm$^3$ of air under normal conditions, namely ionization chamber provides an opportunity to assess ionizing radiation in these units.

Applying voltage about several hundred volts enables to collect all the electrons and positive ions on the electrodes.

This chamber can be applied for the measurement of $\gamma$-radiation and X-ray (if special window is provided) and assessment of radiological conditions near the respective sources.

*The advantages of an ionization chamber:*

- High precision.
- Directly measures exposure rate (mR/h).
- Little-to-no dead time.
- Can measure very high exposure rates.

*The disadvantages of an ionization chamber:*

- Slow response time.
- Sensitive to temperature, pressure, and humidity.
- Large physical size with low spatial resolution.
- Cannot measure energy of radiation.

### 37.2.2  Proportional Counter

This counter produces the output signal that is proportional to the radiation energy. The design of the counter implies the existence of a central electrode, which bound the electrons released by ionization.

The field strength $E$, which is formed at a distance $r$ from the electrode, is equal to:

$$E = U/[r\ln(d_1/d_2)],   (37.6)$$

where $U$ is the applied voltage (V); $d_1$ is the diameter of counter; and $d_2$ is the diameter of central electrode.

The electric field near the electrodes is so large that the primary electrons acquire energy sufficient for secondary ionization. As a result, an avalanche of electrons comes to the central electrode. The ratio of the total number of electrons collected at the electrode to their original number is called the *coefficient of gas amplification*; its value can reach $10^2$–$10^4$.

The counter of this type can be used for $\alpha$- and $\beta$-contamination surveys.

*The advantages of a proportional counter:*

- Can discriminate between alpha and beta particles.
- Counting gas is readily obtainable.
- Insensitive to humidity.
- Can be used in the presence of high-ambient gamma fields.
- Alpha and beta counting efficiency is about 40%.
- Can measure energy of radiation and provide spectrographic information.
- Large area detectors can be constructed.

*The disadvantages of a proportional counter:*

- Relatively heavy.
- Counting gas is flammable.
- Anode wires delicate and can lose efficiency in gas flow detectors due to deposition.
- Efficiency and operation affected by ingress of oxygen into fill gas.
- Measurement windows easily damaged in large area detectors.

### 37.2.3  Geiger–Müller Counter

This gas-discharge detector operates during passage through its volume of charged particles. The construction of the counter is shown in Figure 37.2.

This counter differs from an ionization chamber in being filled with an argon or neon rather than air. Voltage applied to the electrode reaches several hundred volts. Free electrons are created in a gas as a result of the passage of ionizing particles; these electrons are attracted to the central electrode which is held at positive potential. The electric field increases near the electrodes, thus the electrons are accelerated to such an extent that they begin to ionize the gas. Electric pulses are produced in an external electrical circuit. Number of pulses per unit of time equals to the number of ionizing particles.

The Geiger–Müller counter can be used for measuring $\alpha$-, $\beta$-, and $\gamma$-radiation during personnel and contamination surveys.

*The advantages of a Geiger–Müller counter:*

- Cheap, robust detector with a large variety of sizes and applications.
- Large output signal from tube requiring minimal electronic processing for simple counting.
- Can measure overall gamma dose when using energy compensated tube.

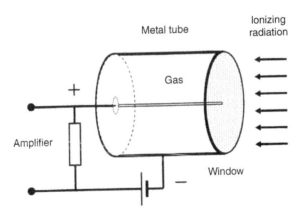

**FIGURE 37.2**  Geiger–Müller counter.

*The disadvantages of a Geiger–Müller counter:*

- Cannot measure energy of radiation—no spectrographic information.
- Will not measure high-radiation rates due to dead time.
- Sustained high-radiation levels will degrade fill gas.

## 37.3  SCINTILLATION COUNTER

*Scintillation counter* is a detector which consists of a substance ("scintillator") that fluoresces in response to incident radiation, a photomultiplier which converts the light to an electrical signal, and the electronics to process the photomultiplier signal.

Charged particle passes through scintillator, causing not only the ionization of atoms and molecules, but their excitement. The transition of atoms and molecules from the excited state to the ground one is accompanied by a quantum of visible or ultraviolet radiation. Each such a light flash is recorded by photomultiplier, electrical pulses from the output of which are processed by the output system. Crystals of ZnS (Ag), NaI (Tl), CsI (Tl) (specified in parentheses activator, causing scintillations in the crystal) are typical materials which are most commonly used in scintillation counters. Scintillator is in contact with the cathode of a photomultiplier tube, which is described in Section 12.8.

The scintillation counter is used for the measurement of $\gamma$-radiation during surveys of low-level ambient $\gamma$-radiation environments; it can be used for X-ray measurement also.

The advantage of this meter is high sensitivity (due to the high density of the working medium), especially $\gamma$-radiation, performance and the ability to determine the energy of a particle or photon radiation.

*The advantages of a scintillation counter:*

- Very sensitive to the presence of gamma radiation.
- Rapid response time.

*The disadvantages of a scintillation counter:*

- Detects only $\gamma$-radiation, X-rays.
- Relatively expensive.
- High background.
- Fragile.

## 37.4  SEMICONDUCTOR DIODE DETECTOR

Consider the contact of two semiconductors: $p$-type and $n$-type. The electron density is high in the $n$-region, while the concentration of holes is high in the $p$-region. Due to the diffusion the holes are moving to the *n*-region and electrons to the *p*-region. As a result, the so-called *p–n* junction is formed in the contact area. If the external electric field is applied to such a junction, the width and the resistance of *p–n* junction will depend on the direction of electric field. In such a way, a *p–n* junction acts as a diode.

If an ionizing radiation is directed through the *p–n* junction, the diode produces free electrons and holes. The electrons are excited into the conduction zone, and the holes are formed respectively in the valence band. Number of pairs of "electron–hole" is proportional to the energy of radiation. The electrons move toward the positive side of the *p–n* junction, while the holes move toward the negative side. This movement causes the electric pulse, which is detected by an electronic counter.

Semiconductor diode detectors are applied for $\gamma$-radiation and X-ray spectrometry and as particle detectors.

*The advantages of a semiconductor diode detector:*

- Small physical size.
- Portable.
- Easy to use.
- Arrays are available.
- Real-time measurement.

*The disadvantages of a semiconductor diode detector:*

- Nonlinear radiation dose response.
- Temperature and environmental dependent.
- Limited life span.

## 37.5   THERMOLUMINESCENT DOSIMETER

When certain crystals such as calcium fluoride, lithium fluoride, and others are exposed to ionizing radiation, the electrons in the crystal's atoms are excited to higher energy states, where they stay trapped due to artificially introduced impurities (usually manganese or magnesium). Heating the crystal induces the transfer of the electrons to their ground state. This transfer is accompanied with the emission of a photon. This emission of light by a material that has been exposed to ionizing radiation followed by heating is called *thermoluminescence.*

The amount of emitted light is proportional to the radiation dose received by the material.

This counter can be used for environmental monitoring and inspection of personnel working under radiation exposure.

*The advantages of a thermoluminescent dosimeter:*

- Available in a wide range of materials, sizes, and shapes.
- Point dose measurements are possible.
- Not expensive.
- Easy to use.

*The disadvantages of a thermoluminescent dosimeter:*

- Sensitive to environmental conditions, handling procedures, and heating conditions.
- Repeated use diminishes accuracy.
- Delayed readout.
- Real-time and dose rate measurements not possible.

# INVESTIGATION OF RADIONUCLIDE ACTIVITY AND DETERMINATION OF THE ABSORPTION COEFFICIENT OF GAMMA RADIATION

## 1 OBJECTIVES

To study the attenuation of gamma radiation passing through matter; to measure the activity of the radioactive nuclide; and to determine the absorption coefficient of the gamma radiation.

## 2 THEORY

The *activity* of a radioactive source is the expected number of elementary radioactive disintegrations per unit of time. Activity can be expressed in terms of the number of disintegrations per second (*dps*). Two units of activity are used: the becquerel (*Bq*), equivalent to 1 dps, and the curie (*Ci*), equivalent to $3.7 \times 10^{10}$ dps.

*Radioactive decay* is the spontaneous disintegration of a radionuclide accompanied by the emission of ionizing radiation in the form of $\alpha$-, $\beta$-particles, or $\gamma$-radiation.

Thus, the activity is the rate at which the isotope decays.

The exposure dose rate is defined as

$$ P_{\mathrm{e}} = \frac{dD_{\mathrm{e}}}{dt}. \tag{P13.1} $$

The unit of exposure dose rate is C/kg·s.

*Methods of Measuring Environmental Parameters*, First Edition. Yuriy Posudin.
© 2014 John Wiley & Sons, Inc. Published 2014 by John Wiley & Sons, Inc.

The relationship between the exposure dose rate $\frac{dD_e}{dt}$ ($\mu R/h$) and the activity $A$ of $\gamma$-radiation source can be expressed as follows:

$$\frac{dD_e}{dt} = K_\gamma \frac{A}{r^2}, \tag{P13.2}$$

where $K_\gamma$ is the exposure rate constant that depends on the type of radionuclide (e.g., $K_\gamma = 13.07$ R·cm$^2$/h·mCi for $^{60}$Co and $K_\gamma = 8.25$ R·cm$^2$/h·mCi for $^{226}$Ra); $A$ is the source activity (mCi); $r$ is the distance to the source (cm).

Attenuation of $\gamma$-radiation in matter is described by an exponential law:

$$I = I_0 e^{-\mu d}, \tag{P13.3}$$

where $I$ is the $\gamma$-radiation intensity transmitted through an absorber of thickness $d$; $I_0$ is the $\gamma$-radiation intensity at zero absorber thickness; $d$ is the absorber thickness; and $\mu$ is the attenuation coefficient.

A number of counts $n$ is proportional to the intensity $I$ of gamma radiation and the number of counts that emerge is given by

$$n = n_0 \exp(-\mu d). \tag{P13.4}$$

Taking the logarithm of Equation P13.4 leads to the following relationship:

$$\ln n = \ln n_o - \mu d. \tag{P13.5}$$

From here it is possible to get the absorption coefficient:

$$\mu = \left( \ln \frac{n_0}{n} \right) / d. \tag{P13.6}$$

## 3   EXPERIMENT

1. Connect the radiometer to power supply.
2. Measure the natural background radiation due to cosmic, solar, and terrestrial radiation; calculate the arithmetic mean $\langle P_{eb} \rangle$ and the errors of measurements.
3. Measure the exposure dose rate $P_e$ of radionuclide source by radiometer ranging from 10 to 15 cm, reducing the distance every 2 cm.
4. Determine the exposure dose rate $\langle P_e \rangle$ with regard to the background radiation $\langle P_{eb} \rangle$ by the expression:

$$\langle P_e \rangle = \langle P'_e \rangle - \langle P_{eb} \rangle. \tag{P13.7}$$

5. Using the experimental data construct the graph of $\langle P_e \rangle$ against $r^{-2}$, which characterizes the Equation (P13.2).
6. Find the average value of radionuclide activity $\langle A \rangle = B/K_\gamma$ using the slope

$$B = (y_2 - y_1)/(x_2 - x_1).$$

7. Calculate the relative errors of measurements:

$$\varepsilon_A = \varepsilon_P + \varepsilon_K + 2\varepsilon_r. \tag{P13.8}$$

8. Calculate the confidence interval of the total error:

$$\Delta_A = \langle A \rangle \cdot \varepsilon_A. \tag{P13.9}$$

9. Place a lead absorbing plate between the radionuclide source and the radiometer and measure the average number of counts $\langle n \rangle$; measurements must be hold for a series of lead absorber of total thicknesses $d$.

10. Construct a graph $\ln(n_0/n)$ against $d$, which makes it possible to determine the average absorption coefficient:

$$\mu = B = (y_2 - y_1)/(x_2 - x_1). \tag{P13.10}$$

**Constructive Test**    The unit of the exposure rate constant is $R \cdot cm^2/h \cdot mCi$. However, in the literature or in the internet, it is possible to find such unit as $R \cdot cm^2/mg \cdot h$. Explain the meaning of the unit.

## QUESTIONS AND PROBLEMS

1. What is the physical pollution?

2. What is the vibration level?

3. Explain the hand-arm vibration and whole-body vibration.

4. Name the principal vibration transducers.

5. Give the main definitions of noise.

6. Explain the equivalent sound level and integrating sound level.

7. Name and explain the main types of light pollution.

8. What is the difference between an electric field sensor and magnetic flux density sensor?

9. What device can estimate EMF of cell phone? Your personal computer?

10. What causes radioactive pollution?

11. What causes radioactivity?

12. Name the types of ionizing radiation.

13. Formulate the law of radioactive decay.

14. What is an activity?

15. What is the exposure rate constant? What determines its value?

16. Formulate various doses of ionizing radiation.

**17.** Formulate the exponential law of the attenuation of $\gamma$-radiation in matter.

**18.** What is the unit of absorption coefficient?

**19.** Compare the ionization chamber, proportional counter, and Geiger–Müller counter.

## REFERENCES

Balachandran, B. and Magrab, E.B. 2008. *Vibrations*, 2nd edition. Cengage Learning.

IEEE Standards Coordinating Committee 28. 2002. IEEE standard for safety levels with respect to human exposure to electromagnetic fields, 0–3 kHz. IEEE (IEEE Std C95.6-2002), New York, NY.

Kryter, K.D. 1996. *Handbook of Hearing and the Effects of Noise*. Academic Press, New York.

## FURTHER READING

Beckwith, T.G., Marangoni, R.D., and Lienhard, J.H.V. 2006. *Mechanical Measurements*, 6th edition. Prentice Hall.

Bienkowski, P. and Trzaska, H. 2011. *Electromagnetic Measurements in the Near Field* (Scitech Series on Electromagnetic Compatibility), 2nd edition. Scitech Publishing.

Bogard, P. 2013. *The End of Night: Searching for Natural Darkness in an Age of Artificial Light*. Little, Brown and Company.

Chadwick, J. 1962. *Radioactivity and Radioactive Substances*. Sir Isaac Pitman.

*Chernobyl Catastrophe. 1997.* 3rd edition. Naukova Dumka, Kyiv.

DeVita, S.M. 2000. *Electromagnetic Pollution: A Hidden Stress to Your System*. Wellness Institute of Living and Learning.

Knoll, G.F. 2000. *Radiation Detection and Measurement*. John Wiley and Sons.

McGregor, D.S. 2007. Chapter 8—Detection and measurement of radiation. In: *Fundamentals of Nuclear Science and Engineering* (eds. J. Kenneth Shultis and Richard E. Faw), 2nd edition. CRC, pp. 202–222.

McLean, L. 2011. *The Force: Living Safely in a World of Electromagnetic Pollution*. Scribe Publications Pty Ltd.

Medvedev, Z.A. 1976. Two decades of dissidence. *New Scientist*. November 4.

Mizon, B. 2001. *Light Pollution: Responses and Remedies* (The Patrick Moore Practical Astronomy Series). Springer.

Pfafflin, J.R. and Edward, N.Z. (eds.) 2006. *Encyclopedia of Environmental Science and Engineering*, 5th edition. Volumes 1, 2. CRC Press.

Rich, C. and Longcore, T. 2006. *Ecological Consequences of Artificial Night Lighting*. Island Press.

Romer, A. 1970. *Radiochemistry and the Discovery of Isotopes*. Dover Publications, Inc., New York.

Serway, R.A. 1990. *Physics for Scientists and Engineers*, Parts I and II. Harcourt Brace Jovanovich College Publishers.

## ELECTRONIC REFERENCES

Cobalt-60, http://uwlbrachycourse.wikifoundry.com/page/Cobalt-60 (accessed December 1, 2013).

Comparison of Fukushima and Chernobyl Nuclear Accidents. From Wikipedia, the free encyclopedia, http://en.wikipedia.org/wiki/Comparison_of_Fukushima_and_Chernobyl_nuclear_accidents (accessed December 1, 2013).

Comparisons with Competing Technologies, http://artsensing.com/site/Comparison.html (accessed December 2, 2013).

Computer/VDT Screens, http://hps.org/publicinformation/ate/faqs/computervdtscreen.html (accessed December 2, 2013).

Control of Vibration at Work. Plymouth City Council Health & Safety Performance Standard (HSPS02). 2nd edition. April 2008 revised http://www.plymouth.gov.uk/hsps02_vibration.pdf (accessed December 3, 2013).

den Boer, L.C. (Eelco) and Schroten, A. (Arno). 2007. Traffic Noise Reduction in Europe. Health Effects, Social Costs and Technical and Policy Options to Reduce Road and Rail Traffic Noise, http://www.transportenvironment.org/sites/te/files/media/2008-02_traffic_noise_ce_delft_report.pdf (accessed December 3, 2013).

Ecological Light Pollution. From Wikipedia, the free encyclopedia, http://en.wikipedia.org/wiki/Ecological_light_pollution (accessed December 4, 2013).

Electronic Encyclopedia—Encyclopedia Britannica, http://www.britannica.com/search?query=Vibration (accessed December 4, 2013).

Electromagnetic and Radiofrequency Fields Effect on Human Health, http://www.aaemonline.org/emf_rf_position.html (accessed December 5, 2013).

Electromagnetic Fields and Public Health. Exposure to Extremely Low Frequency Fields, http://www.who.int/peh-emf/publications/facts/fs322/en/index.html (accessed December 5, 2013).

EMF Meters and Detectors, http://emwatch.com/emf-meter-emf-detector.htm (accessed December 6, 2013).

EPA Identifies Noise Levels Affecting Health and Welfare, http://www2.epa.gov/laws-regulations (accessed December 6, 2013).

Extremely Low Frequency (ELF) Radiation, https://www.osha.gov/SLTC/elfradiation/index.html (accessed December 7, 2013).

Federal Office for Radiation Safety, Germany. 1999, http://www.who.int/peh-emf/about/WhatisEMF/en/index3.html (accessed December 7, 2013).

Gaseous Ionization Detectors. From Wikipedia, the free encyclopedia, http://en.wikipedia.org/wiki/Gaseous_ionization_detectors (accessed December 8, 2013).

Grodzinsky, D. General Situation of the Radiological Consequences of the Chernobyl Accident in Ukraine, http://www.rri.kyoto-u.ac.jp/NSRG/reports/kr21/kr21pdf/Grodzinsky.pdf (accessed December 8, 2013).

Hoolihan, D.D. ANSI C63.19: Establishing Compatibility Between Hearing Aids and Cellular Telephones. Compliance Engineering, http://hypertextbook.com/facts/2003/VictorLee.shtml (accessed December 9, 2013).

Hollan, J. 2004. Metabolism-influencing light: measurement by digital cameras. Poster at Cancer and Rhythm, Graz, Austria, Oct. 14–16, 2004, http://amper.ped.muni.cz/noc/english/canc_rhythm/g_camer.pdf (accessed December 9, 2013).

INSAG-7. The Chernobyl Accident: Updating of INSAG-1. A Report by the International Nuclear Safety Advisory Group. Safety Series No. 75-INSAG-7. International Atomic Energy Agency, Vienna, 1992, http://www-pub.iaea.org/MTCD/publications/PDF/Pub913e_web.pdf (accessed December 10, 2013).

Last, W. Electromagnetic-Pollution, http://www.healing-yourself.com/Healing%20the%20Body/Electro-Magnetic-Pollution.html (accessed December 11, 2013).

Last, W. Electromagnetic-Pollution, http://www.livingtreecommunity.com/store2/articles/electromagnetic_pollution.asp (accessed December 11, 2013).

*Magnetic Field Near a Cellular Telephone. The Physics Factbook* (ed. G. Elert), http://hypertextbook.com/facts/2003/VietTran.shtml (accessed December 12, 2013).

Mobile Phone Radiation and Health—Wikipedia, the free…, en.wikipedia.org/wiki/Mobile_phone_radiation_and_health (accessed December 12, 2013).

Negative Effects of Computer Radiation, http://www.ehow.com/list_5883106_negative-effects-computer-radiation.html (accessed December 13, 2013).

Noise and Hearing Loss, http://www.asha.org/public/hearing/disorders/noise.htm (accessed December 13, 2013).

Noise-Induced Hearing Loss, http://www.nidcd.nih.gov/health/hearing/pages/noise.aspx (accessed December 14, 2013).

Noise is Difficult to Define—Part 1, http://www.hearinglossweb.com/Medical/Causes/nihl/diff.htm (accessed December 15, 2013).

NRC: Backgrounder on the Three Mile Island Accident, http://www.nrc.gov/reading-rm/doc-collections/fact-sheets/3mile-isle.html (accessed December 15, 2013).

Nuclear Weapons Testing. From Wikipedia, the free encyclopedia, http://en.wikipedia.org/wiki/Nuclear_weapons_testing (accessed December 16, 2013).

Radiation Effects from the Fukushima Daiichi Nuclear Disaster, http://en.wikipedia.org/wiki/Radiation_effects_from_Fukushima_Daiichi_nuclear_disaster (accessed December 16, 2013).

Rarium-226, http://uwlbrachycourse.wikifoundry.com/page/Radium-226 (accessed December 16, 2013).

Teikari, P. 2007. Light Pollution: Definition, Legislation, Measurement, Modelling and Environmental Effects. Barselona, Catalunya, September 19, 2007, http://users.tkk.fi/~jteikari/Teikari_LightPollution.pdf (accessed December 16, 2013).

TEPCO puts radiation release early in Fukushima Crisis at 900 PBq. Kyodo News. May 24, 2012, http://en.wikipedia.org/wiki/Comparison_of_Fukushima_and_Chernobyl_nuclear_accidents (accessed May 24, 2012).

The Human Consequences of the Chernobyl Nuclear Accident. UNDP and UNICEF, January 22, 2002, http://www.unicef.org/newsline/chernobylreport.pdf (accessed December 17, 2013).

The Physics Factbook™—hypertextbook.com, http://hypertextbook.com/facts/ (accessed December 18, 2013).

What Are Electromagnetic Fields? Current Standards, http://www.who.int/peh-emf/about/WhatisEMF/en/index4.html (accessed December 19, 2013).

放射性物質放出量データの一部誤りについて ("Partially revision of airborne emission estimation of the radioactivity"), http://www.meti.go.jp/press/2011/10/20111020001/20111020001.html (accessed December 20, 2013).

# PART VII

# BIOTIC FACTORS

*Biotic factors* are factors created by living organisms or any living components within an environment in which the action of the organism affects the activity of another organism. Biotic factors in an ecosystem include all living things such as plants, animals, fungi, protists, and bacteria.

Assessment of the biological components of ecosystems is called *biotic monitoring*, the main methods of which are *bioindication* and *biomonitoring*.

# 38

# BIOINDICATION

*Bioindication* involves observing the effect of external factors on ecosystems and their development over long periods of time. The methods utilized include the measurement and identification of pollutants through the use of organisms or communities of organisms (bioindicators), their composition of certain elements or compounds, and their morphological, histological or cellular structure, as well as metabolic and biochemical processes, behavior, and population organization. Collectively this provides information on the *qualitative* aspects of the organism's environment.

A *bioindicator* is an organism that is monitored to estimate the state of health of the environment.

There are several types of bioindicators the selection of which depends on the objectives of the study. For example, they can be lichens, plants, animals, or microorganisms that regularly produce certain molecular signals in response to changes in their environmental conditions.

## 38.1 LICHENS AS BIOINDICATORS

Lichens can be used as bioindicators for air pollution. Lichens are found on a variety of substrates—trees, rock outcroppings, soil, and evergreen leaves. They respond to $SO_2$, $NO_x$, HF, ozone, and metals, since they derive their water and essential nutrients mainly from the atmosphere rather than the soil (Jovan, 2008).

Epiphytic lichens (e.g., *Hypogymnia physodes*) can be used as sensitive bioindicators of atmospheric pollution. A rich flora of lichens indicates good air quality, whereas the absence of lichens indicates $SO_2$ air pollution.

*Methods of Measuring Environmental Parameters*, First Edition. Yuriy Posudin.
© 2014 John Wiley & Sons, Inc. Published 2014 by John Wiley & Sons, Inc.

Pollution-sensitive lichens respond negatively to a wide range of pollutants. These include N-fixing cyanolichens (*Coccocarpia palmicola, Collema furfuraceum, Leptogium cyanescens, Peltigera* spp., *Ramalina* spp., *Tuckermannopsis* spp., and *Usnea* spp.).

Bioindication-utilizing lichens are based on the analysis of the species' presence and abundance, level of settlement and absolute surface area of lichens on the trunks of a trees, species composition, and frequency of occurrence of lichen.

## 38.2   ALGAE AS BIOINDICATORS

*Algae* are a very large and diverse group of simple organisms that inhabit aquatic environments. Algae are able to respond to very low levels of pollutants.

Diatoms are excellent bioindicators of water quality. These algae are characterized by considerable diversity, their sensitivity to the chemical composition of the water, and a narrow range of tolerance to pH, nutrient composition, and the salinity of the aquatic environment.

## 38.3   CLASSIFICATION OF WATER RESERVOIRS

Water reservoirs (lakes, ponds) can be classified based on their productivity or their relative nutrient richness. This type of tropic system is commonly separated into three classes: oligotrophic, mesotrophic, or eutrophic.

*Oligotrophic* indicates water with a high level of oxygen but a low level of dissolved nutrients essential for the growth of algae and plants.

*Eutrophic* water is characterized by a high level of nutrients that result in considerable activity of plant and algae which in turn reduces the dissolved oxygen content.

*Mesotrophic* water has a moderate amount of dissolved nutrients and moderate algae and plant productivity.

## 38.4   WATER QUALITY INDICES

The ability of aquatic organisms to live in water containing different amounts of organic matter, is called *saprobity* (from the Greek word *sapros*—rotten).

The *Saprobic Index* is a numerical estimation of the amount of pollution caused by organic matter which ranges from 0 to 4.

The Saprobic Index $S$ can be calculated as

$$S = \frac{\sum s_i h_i}{\sum h_i},$$

(38.1)

where $S$ is the saprobity index, $s_i$ is individual saprobity index of $i$-th species, and $h_i$ is relative occurrence of $i$-th species in the sample.

The index values for saprobity levels are: for oligosaprobity 0.51–1.50; $\beta$-mesosaprobity 0.51–2.50; $\alpha$-mesosaprobity 2.51–3.50; and polysaprobity 3.51–4.50. Relative occurrence values are: 1—randomly found; 3—typical occurrence; and 5—massive development.

It is possible to convert the saprobity index values into water quality characteristics: water is considered very clean if $S < 1$; clean if $S = 1$–1.5; moderately polluted if $S = 1.51$–2.5; polluted if $S = 2.51$–3.5; dirty if $S = 3.51$–4; and very dirty if $S > 4$.

The saprobic index $S$ for a site was developed by Pantle and Buck (1955) and later modified by Zelinka and Marvan (1961) as follows (Walley et al., 2001):

$$
S = \frac{\displaystyle\sum_{i=1}^{N} s_i w_i h_i}{\displaystyle\sum_{i=1}^{N} w_i h_i}, \tag{38.2}
$$

where $s_i$ is individual saprobity index of $i$-th species, $w_i$ is the indicator weight of the $i$-th species, and $h_i$ is relative occurrence of $i$-th species in a sample.

**Example**     Calculation of saprobic index.

Calculate the saprobic index $S$:

**Use the extract from the list of saprobic indices after Zelinka and Marvan (1961) (bold type)**

| Species of mayflies | Xenosaprobic | Oligosaprobic | $\beta$- mezosaprobic | $\alpha$- mezosaprobic | Indicator weight $w_i$ | Relative occurrence $h_i$ | $w_i \cdot h_i$ | $S_s \cdot w_i \cdot h_i$ |
|---|---|---|---|---|---|---|---|---|
| *Ameletus impinstus* | 10 | | | | 5 | 50 | 250 | 2500 |
| *Baetis gemellus* | 7 | 3 | | | 3 | 26 | 78 | 780 |
| *Baetis pimulus* | 1 | 4 | 4 | 1 | 1 | 32 | 32 | 320 |
| *Baetis rhodani* | 3 | 3 | 3 | 1 | 1 | 40 | 40 | 400 |
| *Ephemera danica* | 1 | 4 | 4 | 1 | 1 | 6 | 6 | 60 |
| $\displaystyle\sum_{i=1}^{N} s_i w_i h_i$ | 3204 | 506 | 272 | 78 | | | | 4060 |
| $\displaystyle\sum_{i=1}^{N} w_i h_i$ | | | | | | | | 406 |
| *Saprobic index S* | 7.89 | 1.25 | 0.67 | 0.19 | | | | |

**Control Exercise**   Calculate saprobic index for the following results of field measurements (after Borchardt, 2007):

| Species | $s_i$ | $h_i$ | $w_i$ | $h_i w_i$ | $s_i h_i w_i$ |
|---|---|---|---|---|---|
| *Ancylus fluviatilis* | 2.0 | 10 | 4 | | |
| *Gammarus fossarum* | 1.6 | 130 | 8 | | |
| *Baetis rhodani* | 2.3 | 50 | 8 | | |
| *Plectocnemia* sp. | 1.5 | 5 | 4 | | |
| Results | | | | $\sum_{i=1}^{N} w_i h_i = ?$ | $\sum_{i=1}^{N} s_i w_i h_i = ?$ |
| Results | | | $S = ?$ | | |

The *Algal Abundance Index* is determined as

$$AAI = [(2N_n + N_3)/N_s]\,100\%, \tag{38.3}$$

where $N_n$—is the number of abundant records; $N_c$—the number of common records; and $N_s$—the number of site visits.

So, water of various classes can be characterized as follows: oligotrophic—AAI < 20; mesotrophic—$20 \leq AAI < 49$; eutrophic—$50 \leq AAI < 69$; hypertrophic—AAI >70 (Marsden, 1997).

The *Generic Diatom Index (GDI)* is based on calculation of sum of the % relative abundances of sensitive genera, the % relative abundances of the tolerant genera, and the % relative abundance of genera used in the selected index:

$$GDI = \frac{\sum \text{sensitive}}{\sum \text{tolerant}}. \tag{38.4}$$

The *Trophic State Index (TSI)* is based on an estimation of the quantities of chlorophyll, phosphorus (both micrograms per liter), Secchi depth (meters), trophic class (Carlson, 1996).

## 38.5   INVERTEBRATES AS BIOINDICATORS

Aquatic invertebrates, particularly benthic macro-invertebrates, inhabit the bottom regions of water reservoirs.

The use of these bioindicators has certain advantages: they differ in their tolerance to the amount and types of pollution, they are easy to collect and identify, and they display limited mobility.

For example, stoneflies (*Plecoptera*) are an indicator of good water quality, while representatives of gull midges (*Itonididae*), apple snails (*Ampullariidae*), and aquatic insects (*Corixidae*) that inhabit ponds and slow moving streams are bioindicators of dirty water. If you find mayflies (*Ephemeroptera*) near water, which are very sensitive to pollution, it means that the water is clean enough for drinking without boiling or treatment.

**Example**   Calculation of macroinvertebrate biotic index.

To calculate the Biotic Index it is necessary to collect in a water basin 100 macro invertebrates and multiply the number of each species by its biotic value. Then you need to add all these products together and divide the sum total by 10. The biotic index above 70 corresponds to the evaluation of water quality as "excellent," 60–69—"good"; 40–59—"fair," and below 40 is poor.

The results of finding of macroinvertebrates are presented in the Table:

| Species | Biotic value B | Number found N | B×N |
|---|---|---|---|
| Dobsonfly larvae | 10 | 14 | 140 |
| Midge fly larvae | 5 | 20 | 100 |
| Fishfly larvae | 6 | 8 | 48 |
| Water penny larvae | 10 | 6 | 60 |
| Mayfly nymph | 10 | 16 | 160 |
| Gilled snail | 4 | 6 | 24 |
| Stonefly larvae | 10 | 18 | 180 |
| Lunged snail | 4 | 4 | 16 |
| Riffle beetle larvae | 10 | 2 | 20 |
| Aquatic worm | 0 | 6 | 0 |
| Total | | **100** | **748** |
| Biotic index total | | 748/10 = 74.8 | |

As soon as the biotic index is above 70, water quality is excellent.

**Control Exercise**   Calculate the biotic index for the water basin in which the following macro invertebrates were found.

| Species | Biotic value B | Number found N | B×N |
|---|---|---|---|
| Leech | 4 | 25 | 100 |
| Aquatic worm | 0 | 10 | 0 |
| Crayfish | 6 | 10 | 60 |
| Water penny larvae | 10 | 20 | 200 |
| Lunged snail | 4 | 20 | 80 |
| Gilled snail | 4 | 15 | 60 |
| Total | | | |
| Biotic index total | | ? | |

*The advantages of a bioindication:*

Methods of bioindication summarize biologically relevant data relating to the environment. These methods are able to respond to:

- short-term and sharp emissions of toxicants; indicate the accumulation and propagation routes of pollutants;
- estimate harmful effects of toxicants on humans and wildlife in the early stages; and
- allow normalizing the permissible load on ecosystems.

*The disadvantages of a bioindication:*

Methods of bioindication:

- Do not give objective information about physical and chemical characteristics of stressors.
- Need more repetition to get statistically significant results.

## REFERENCES

Carlson, R.E. and Simpson, J. 1996. *A Coordinator's Guide to Volunteer Lake Monitoring Methods.* North American Lake Management Society.

Jovan, S. 2008. *Lichen Bioindication of Biodiversity, Air Quality, and Climate*: Baseline Results From Monitoring in Washington, Oregon, and California. United States Department of Agriculture Forest Service Pacific Northwest Research Station General Technical Report PNW-GTR-737 March 2008.

Marsden, M.W., Smith, M.R., and Sargent, R.L. 1997. Trophic status of rivers in the Forth Catchment, Scotland. *Aquat. Conserv. Mar. Freshw. Ecosyst.* 7:211–221.

Pantle, R. and Buck, H. 1955. Die biologische Überwachung der Gewässer und die Darstellung der Ergebnisse. *Bes. Mitt. dt. Gewässerkundl. Jb.* 12:135–143.

Walley, W.J., Grbovic, J., and Dzeroski, S. 2001. A reappraisal of saprobic values and indicator weights based on Slovenian river quality data. *Wat. Res.* 35(18):4285–4292.

Zelinka, M. and Marvan, P. 1961. Zur Präzisierung der biologischen Klassifikation der Reinheit fließender Gewässer. *Arch. Hydrobiol.* 57:389-407.

## ELECTRONIC REFERENCES

Biotic Index—Aquaculture Dictionary, http://www.aquatext.com/tables/bioticind.htm (accessed December 1, 2013).

Borchardt D. Management of Freshwater Resources and Aquatic Ecosystems WS 2007/2008, http://www.comtec.eecs.uni-kassel.de/iwrm-momo/iwrm-momo/uploads/media/Lecture_W_3_Water_Resources_Management_Borchardt_2007.pdf (accessed December 3, 2013).

# 39

# BIOMONITORING

*Biomonitoring* is the use of organisms or communities of organisms (biomonitors), their composition of certain elements or compounds, and their morphological, histological or cellular structure, as well as metabolic and biochemical processes, behavior and population organization that provide information on the *quantitative* aspects of the environment.

## 39.1 TEST-ORGANISMS AND TEST-FUNCTIONS

*Test-organism* (or *biomonitor*) is defined as an organism that provides quantitative information on the quality of the environment around it.

*Test-reaction* or *test-function* is the physiological or behavioral response of an organism to a change in the quality of the environment.

The various types of organisms that have been used as biomonitors include bacteria, protozoa, algae, invertebrates, fungi, and fish.

The following test-reactions can be used during biomonitoring: population growth, amount of living organisms, intensity of reproduction, swimming behavior, immobilization of the cells, rheotaxis, gravitaxis, phototaxis, photosynthetic activity, motility, biochemical effects, activity of oxidative enzymes, histological and morphological changes, respiration, bioluminescence, mechanical strength, electrophysiological activity, membrane permeability, impedance of cell suspension, bioelectrical reaction, energy potential of the cells, and protoplasmic streaming.

*Methods of Measuring Environmental Parameters*, First Edition. Yuriy Posudin.
© 2014 John Wiley & Sons, Inc. Published 2014 by John Wiley & Sons, Inc.

## 39.2 BACTERIA AS TEST-OBJECTS

Bacteria *Bacillus cereus, Beneckea harveyi, Vibrio fischeri,* and *Vibrio harveyi* can be used as test-objects during biomonitoring of aquatic media.

The following test-functions are measured: the intensity of reproduction, bioluminescence, the activity of oxidative enzymes, membrane permeability, and mechanical strength.

Let us consider the principle of operation of the device (The DeltaTox Analyzer and DeltaTox, SDIX) that is based on the registration of bioluminescence. This device includes a highly sensitive analyzer (luminometer), freeze-dried bacterial reagent, and test control and reconstitution solutions. The intensity of luminescence produced by luminescent bacteria after exposure to a test sample is compared with the intensity of optical radiation of a control sample where bacteria are absent. The degree of luminescence loss percent that is related to the metabolic inhibition in the test-organisms indicates the relative toxicity of the sample.

These devices are used for the analysis of wastewater, raw water, drinking water, detection of point sources, early prevention of pollution of surface water and groundwater, and industrial discharges.

The advantages of this method based on bioluminescence registration are the speed and ease of measurement, reliability of measurement results, and analysis of highly toxic environments.

## 39.3 PROTOZOA AS TEST-OBJECTS

Classic representatives of protozoa, which are used as test-objects in biomonitoring, are *Tetrahymena pyriformis, Spirostomun ambiguum,* and *Euplotes* sp.

Test-functions are reproduction intensity, moving activity, morphological changes of the body, respiration, and active transport.

These protozoa are sensitive to the most common ecotoxicants such as heavy metals, petroleum products, detergents, and phenols.

## 39.4 ALGAE AS TEST-OBJECTS

The following algae are used as test-objects: *Scenedesmus quadricauda, Sc. acuminates, Chlorella vulgaris, Euglena gracilis, Dunaliella salina, Desmarestia viridis, Nitella flexilis, Phaeodactylum tricornutum.*

The principal test-functions are: cell death, growth, reproduction, moving activity, photomovement parameters, photosynthetic activity, bioelectric reaction, impedance of suspension, and membrane permeability. Let us consider some of them.

### 39.4.1 Photomovement Parameters of Algae as Text-Functions

To study photomovement in *Dunaliella*, a special experimental videomicrography system was developed (Posudin et al., 1992).

This system of videomicrography utilizes a microscope connected to a light source, monochromator, and videosystem.

It allows the observation and measurement of the photoresponses of either individual organisms or populations as modulated by light stimulus parameters.

Photomovement parameters (e.g., linear and rotational velocity, number of cells moving toward or away from the light source, motility of the cells) modulated by light stimulus parameters (e.g., intensity, polarization, spectral composition) were estimated with this system of videomicrography.

Such a system made it possible to study the effects of abiotic factors (temperature, pH, electric fields, visible, ultraviolet, and ionizing radiation) and various contaminants (heavy metals, surface-active substances, pesticides) on photomovement parameters of algae.

### 39.4.2    Gravitaxis Parameters of Algae as Text-Functions

The gravitational field of the Earth is an important external factor for organisms that are moving in an aquatic medium. The ability of organisms to orient the direction of their movement with respect to the gravitational field is called *gravitaxis*. Gravitaxis is observed in algae such as *E. gracilis*, *Chlamydomonas nivalis*, *Cryptomonas*, *Peridinium gatunense*, *Peridinium faeroense*, *Amphidinium carterae*, *Prorocentrum micans*, and *D. salina*. The direction and level of gravitactic orientation of organisms depends on the type and age of the algae, the presence of heavy metals in the environment, and the effect of solar and in particular ultraviolet radiation. Analysis of angular histograms of cells under the influence of external factors makes it possible to estimate these factors.

An automated system of biomonitoring that is based on the gravitaxis analysis of algae is proposed by Häder and Liu (1990). This cell tracking system consists of a horizontally oriented microscope, a CCD camera, an information processing system, and a monitor for observing the cell population. Such a system makes it possible to record a histogram of the angular distribution of the cells during gravitaxis, to determine the total number of motile cells, the level of gravitaxis, and its direction in normal conditions and under the influence of various toxicants.

It was shown, for example, that such parameters of gravitactic orientation of *E. gracilis* as precision of orientation and swimming velocity of the populations were sensitive to heavy metals (copper, mercury, cadmium, and lead) (Stallwitz and Häder, 1994), pesticides (carbofuran and malathion) (Azizullah et al., 2011), and industrial wastewaters (Azizullah et al., 2012).

### 39.5    INVERTEBRATES AS TEST-OBJECTS

#### 39.5.1    Daphnia as Test-Object

Such representatives of invertebrates as *Daphnia magna*, *Hydra attenuata*, *Hirudo medicinalis*, *Unio tumidus*, *Eulimnogammarus verecosus*, and *Mizuhopecten yessoensis* are commonly used as test-objects during biomonitoring.

The principal test-functions are survival, respiration rate and heart rate, and behavioral response.

### 39.5.2 Daphnia Toximeter

*Daphnia* is a genus of planktonic crustaceans, 1–5 mm (0.04–0.20 in.) in length. Daphnia sometimes called "water fleas" because of their swimming style. These organisms are used for estimation of water quality and detection of hazardous compounds in water from rivers, water treatment plants, distribution systems, and sewers.

The instrument for biomonitoring of aquatic environment is called *Daphnia toximeter* (Environmental Analytical Systems, Canada); it is based on the ability of *Daphnia* to change its behavioral and motor parameters in response to water pollution. These parameters include daphnia size, mortality rate, average speed, speed distribution, average height of swimming, average distance between each daphnia, number of moving organisms, turns and circling movements, and curviness of a path.

Toximeter consists of cross-flow filtration system, daphnia chamber, algae fermenter for daphnia feeding, peltier cooling unit, source of light, and peristaltic pump. The behavior of daphnia (turning, circling) and movement parameters are recorded by a CCD camera which is connected with computer.

Assessment of aquatic toxicity is realized by using a *Toxic Index*, which is based on the evaluation of certain behavioral or motor parameters; sudden simultaneous changes of these parameters cause an acoustic or electronic signal of alarm in Daphnia taximeter.

*Daphnia* toximeter can be applied for the detection of heavy metals such as Zn, Ni, Cr, Fe, Cu, Pb, Hg in aquatic media.

### 39.6 FUNGI AS TEST-OBJECTS

Fungi and actinomycetes such as *Aspergillus niger, Fusarium graminearum,* and *Streptomyces olivaceus* are usually used as test-objects. The principal test-function is growth reaction.

### 39.7 FISH AS TEST-OBJECTS

The following species of fish may be mentioned as examples of test-objects: *Perca fluviatilis, Phoxinus phoxinus, Cyprinus carpio, Danio rerio,* etc.

The main methods of biomonitoring with fish as test-objects are based on the registration of such test-functions as breathing rate and heartbeat, average swim speed, behavior (height, turns, circular motion, curviness), size, number of active fish, location and distribution of fish in chamber, average distance between the test-organisms, and frequency of the electric organ discharges of weak electric fish.

The instrumental versions of biomonitoring systems that are based on the application of fish may be presented by series of Fish Toximeters or an Aquatic Biomonitor (the Intelligent Aquatic Biomonitor System, or iABS) which was developed by the scientists of US Army Center for Environmental Health Research (USACEHR). A typical aquatic biomonitor estimates the behavior of eight bluegrills that are located in the chamber. The pairs of electrodes mounted above and below each of the eight fish allow to register the electrical impulses generated by the fish during their movement in the water, ventilation of gills, and muscle construction. The presence of the pollutant in water causes abnormal behavior of fish, which is accompanied by alarm signal.

The observation chamber has dimensions of 60 cm × 50 cm × 10 cm and equipped with the artificial illumination, CCD camera, software with graphic display of measuring results, and hydraulic supplies.

Such biomonitors can be applied for the detection of toxic materials such as acids, alkalis, heavy metals and earth metals, surface-active substances, pesticides, herbicides, so on. during water-quality and safety analysis.

*The advantages of a biomonitoring:*

• High sensitivity, speed, reliability, and efficiency.
• Ability to create automated systems for collecting and processing information.

*The disadvantage of a biomonitoring:*

• The lack of quantitative assessment of toxic substances in the environment and the possible interaction of individual components of the toxic compounds.

## 39.8   REMOTE WATER-QUALITY MONITORING

A system of remote water-quality monitoring consists of biosensing unit (a fish as test-object) and a system of simultaneous measurement of several water-quality parameters: temperature, hydrogen ion concentration (pH), specific conductance, and stream-stage elevation. Critical water-quality situations result in breathing-rate changes in fish, measurements of which are transmitted by satellite to a data-reception system (Morgan et al., 1989).

## REFERENCES

Azizullah, A., Richter, P., and Häder, D.-P. 2011. Comparative toxicity of the pesticides carbofuran and malathion to the freshwater flagellate *Euglena gracilis. Ecotoxicology,* 20(6):1442–1454.

Azizullah, A., Richter, P., and Häder, D.-P. 2012. Sensitivity of various parameters in *Euglena gracilis* to short-term exposure to industrial wastewaters. *J. Appl. Phycol.* 24(2):187–200.

Häder, D.-P. and Liu, S.-M. 1990. Motility and gravitactic orientation of the Flagellate, *Euglena gracilis,* impaired by artificial and solar UV-B radiation. *Curr. Microbiol.* 21:161–168.

Morgan, T.L., Smith, M.D., and Red, J.T. 1989. Monitoring real-time biological responses to water-quality events from remote data collection platforms and satellite-linkage. In: *Remote Data Transmission* (Proceedings of the Vancouver Workshop, August 1987). Iahs Publ. No. 178.

Posudin, Y.I., Massjuk, N.P., Lilitskaya, G.G., and Radchenko, M.I. 1992. Photomovement of two species of *Dunaliella* Teod. *Algologia,* 2:37–47.

Stallwitz, E. and Häder, D.-P. 1994. Effects of heavy metals on motility and gravitactic orientation of the flagellate, *Euglena gracilis. Eur. J. Protistol.,* 30(1):18–24.

# PRACTICAL EXERCISE **14**

# PHOTOMOVEMENT PARAMETERS AS TEST-FUNCTIONS DURING BIOMONITORING

## 1 SIMULTANEOUS USE OF SEVERAL TEST-FUNCTIONS DURING BIOMONITORING

For the majority of biomonitoring methods, the impact of chemicals on aquatic environments is assessed using only one parameter that is modulated by the compound of interest. Assessment of only one test-function significantly limits the effectiveness of biomonitoring, in that other chemicals in an aquatic medium may produce the same effect.

Consider, for example, the effects of salts of copper and cadmium on the linear velocity of green alga *Dunaliella*. Suppose that the salt concentration is $10^{-4}$ M (Posudin et al., 1996a). The ratio of the velocity $v$ of the cells in the sample with a salt to the velocity $v_c$ of cells in the control sample (without salt) is $0.92 \pm 0.05$ for copper and $0.93 \pm 0.05$ for cadmium, that is, values of relative velocities are reliably indistinguishable.

The use of simultaneous measurements of several photomovement parameters is proposed in that it allows increasing the sensitivity of biomonitoring. The vector method for biomonitoring is recommended for estimating the effect of toxicant concentration in aquatic environments using simultaneous measurement of two or more movement parameters. This method facilitates processing of data from large-scale measurements, allows the more precise quantitative estimation of the effect of toxicant concentration, and can also facilitate toxicant identification.

*Methods of Measuring Environmental Parameters*, First Edition. Yuriy Posudin.
© 2014 John Wiley & Sons, Inc. Published 2014 by John Wiley & Sons, Inc.

## 2   VECTOR METHOD OF BIOMONITORING

The vector method is based on determining the value $r$ and direction $\theta$ of the vector $\vec{R}$ that has the following projections on the axis of a $N$-dimensional system of coordinates: $Xi/X_{1c}, X_2/X_{2c}, \ldots, X_N/X_{Nc}$ (where $X_1, X_2, \ldots, X_N$ and $Xi_c, X_{2c}, \ldots, X_{Nc}$ are the photomovement parameters for the microorganisms in the pollutant and control samples, respectively).

In a two-dimensional system of coordinates ($N = 2$), the value $r$ and direction $\theta$ are defined as follows:

$$r = \sqrt{(X_1/X_{1c})^2 + (X_2/X_{2c})^2}, \tag{P14.1}$$

$$\theta = \text{arctg}\,[(X_1/X_{1c})/(X_2/X_{2c})], \tag{P14.2}$$

where $X_i$ and $X_{ic}$ are movement parameters of the test-objects in experimental and control samples, respectively ($i = 1, 2$).

The vector method for quantitative estimation of the toxic effects of pollutants on photomovement parameters of motile microorganisms was initially proposed by Posudin et al. (1996b). Test-objects are placed into treatment (with toxicant) and control cuvettes in a videomicrography system that simultaneously records several photomovement parameters (e.g., linear velocity $V$, rotational velocity $n$, phototopotaxis $F$, and the number of immobile cells $N_{im}$ in relation to the total number $N_0$).

**Example**   Investigate the effect of the same salts of copper and cadmium on linear velocity $v$ and phototaxis $F$ of *Dunaliella*.

**Solution**   Using Equations P14.1 and P14.2 (where $X_1 = v$; $X_{1c} = v_c$ and $X_2 = F$; $X_{2c} = F_c$), we can obtain the following values:

$$r = |\vec{R}| = 0.97; \theta = 71^0 \text{ for copper salt};$$

$$r = |\vec{R}| = 1.02; \theta = 79^0 \text{ for cadmium salt}.$$

Figure P14.1 displays the values and directions of vector $\vec{R}$ in response to both salts at a $10^{-4}$ M concentration. It is evident that the use of simultaneous measurements of two photomovement parameters allows increasing the resolution of toxicants and sensitivity of biomonitoring.

**Control Exercise**   In a three-dimensional system of coordinates ($N = 3$), the value $r$ and directions ($\theta_1$ and $\theta_2$) are determined as follows:

$$r = \sqrt{(X_1/X_{2c})^2 + (X_2/X_{2c})^2 + (X_3/X_{3c})^2}; \tag{P14.3}$$

$$\theta_1 = \arccos[(X_1/X_{1c})/r], \tag{P14.4}$$

$$\theta_2 = \text{arctg}[(X_3/X_{3c})/(X_1/X_{1c})]. \tag{P14.5}$$

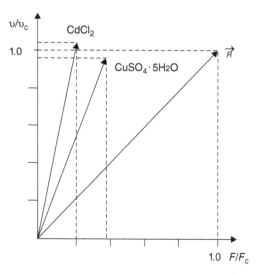

**FIGURE P14.1**   Dependence of the value $r$ and direction $\theta$ of vector $\vec{R}$, which is constructed in two-dimensional system of coordinates, on the type of the salts of heavy metals ($CuSO_4 \cdot 5H_2O$ and $CdCl_2$) at the same ($10^{-4}$ M) concentration.

Determine the value $r$ and direction ($\theta_1$ and $\theta_2$) of vector $\vec{R}$ in three-dimensional system of coordinates $(V/V_c;\ F/F_c;\ (N_{im}/N_0)/(N_{im}/N_0)_c$ for such photomovement parameters of *Dunaliella*: $X_1 = v = 23.1 \pm 0.2$ µm/s; $X_{1c} = v_c = 25.1 \pm 0.6$ µm/s; $X_2 = F = 0.09 \pm 0.04$; $X_{2c} = F_c = 0.28 \pm 0.05$; $X_3 = (N_{im}/N_0) = 0.27 \pm 0.08$; $X_{2c} = (N_{im}/N_0)_c = 0.10 \pm 0.005$.
Concentration of copper salt is $10^{-4}$ M.

## QUESTIONS AND PROBLEMS

1. What is bioindication?

2. What is bioindicator?

3. What is saprobity?

4. Explain the trophic system of classification.

5. Determine the Algal Abundance Index.

6. What is the Generic Diatom Index?

7. Give the definition of biomonitoring.

8. Define the test-object and test-function.

9. What is the principal difference between bioindication and biomonitoring?

# REFERENCES

Posudin, Y.I., Massjuk, N.P., and Lilitskaya, G.G. 1996a. Photomovement parameters of algae as test-functions during biomonitoring. *Polish J. Env. Sci.*, 5(3):15–57.

Posudin, Y.I., Massjuk, N.P., and Lilitskaya, G.G. 1996b. Vector method of biotesting aquatic media. *Algologia*, 1(1):15–25.


# FURTHER READING

Clausen, J., van Wijk, R., and Albrecht, H. 2012. Weakly electric fish for biomonitoring water quality. *Environ. Technol.*, 33(10–12):1089–1099.

Lebert, M. and Häder, D.-P. 1999. Image analysis: a versatile tool for numerous applications. *G.I.T. Special Edition Imaging Microscopy*, 1:5–6.

Lee, C. 2006. *Bluegill on Guard in Region's Water Supply*. Washington Post Staff Writer. Monday, September 18, 2006.

Posudin, Y.I., Massjuk, N.P., and Lilitskaya, G.G. 2010. *Photomovement of Dunaliella Teod.* Vieweg+Teubner Research.


# ELECTRONIC REFERENCES

The BBE Fish Toximeter. A Powerful Instrument for Water Toxicity Monitoring, http://www.ppsystems.com/Literature/Fish%20Toximeter.pdf (accessed December 12, 2013).

DeltaTox User's Manual, http://www.sdix.com/uploadedFiles/Content/Products/Water_Quality_Tests/Deltatox-Analyzer-UserGuide.pdf (accessed December 5, 2013).

Test Method: Generic Diatom Indices. Revised: May 30, 2007. http://www.deltaenvironmental.com.au/management/Lab_methods/Generic_indices.htm (accessed December 4, 2013).

Trader, D. USACEHR Employs Fish to Monitor Water Supplies. February 1, 2012. http://www.army.mil/article/72938/USACEHR_employs_fish_to_monitor_water_supplies/ (accessed December 16, 2013).

# APPENDIX

The SI prefixes for multiples of units and submultiples of units

| Factor | Prefix Name | Prefix Symbol | Factor | Prefix Name | Prefix Symbol |
|--------|-------------|---------------|--------|-------------|---------------|
| $10^{24}$ | Yotta | Y | $10^{-1}$ | Deci | d |
| $10^{21}$ | Zetta | Z | $10^{-2}$ | Centi | c |
| $10^{18}$ | Exa | E | $10^{-3}$ | Milli | m |
| $10^{15}$ | Peta | P | $10^{-6}$ | Micro | $\mu$ |
| $10^{12}$ | Tera | T | $10^{-9}$ | Nano | n |
| $10^{9}$ | Giga | G | $10^{-12}$ | Pico | p |
| $10^{6}$ | Mega | M | $10^{-15}$ | Femto | f |
| $10^{3}$ | Kilo | k | $10^{-18}$ | Atto | a |
| $10^{2}$ | Hecto | h | $10^{-21}$ | Zepto | z |
| $10^{1}$ | Deca | d | $10^{-24}$ | Yocto | y |

## THE SEVEN BASE QUANTITIES OF THE INTERNATIONAL SYSTEM OF QUANTITIES

*International System of Units (SI)* is the system of units, based on the International System of Quantities, their names and symbols, including a series of prefixes and their names and symbols, together with rules for their use, adopted by the General Conference on Weights and Measures (CGPM).

*Methods of Measuring Environmental Parameters*, First Edition. Yuriy Posudin.
© 2014 John Wiley & Sons, Inc. Published 2014 by John Wiley & Sons, Inc.

*Base units* are the building blocks of SI—all other units of measure can be derived from the base units.

(International System of Units, From Wikipedia, the free encyclopedia. http://en.wikipedia.org/wiki/International_System_of_Units)

| Base Quantity | Base Unit | |
|---|---|---|
| Name | Name | Symbol |
| Length | Meter | m |
| Mass | Kilogram | kg |
| Time | Second | s |
| Electric current | Ampere | A |
| Thermodynamic temperature | Kelvin | K |
| Amount of substance | Mole | mol |
| Luminous intensity | Candela | cd |

## PSYCHROMETRIC TABLES

### 1. Vapor Pressure, mm Hg

| Air Temperature, °C | 0.0 | 0.1 | 0.2 | 0.3 | 0.4 | 0.5 | 0.6 | 0.7 | 0.8 | 0.9 |
|---|---|---|---|---|---|---|---|---|---|---|
| 0 | 4.58 | 4.61 | 4.65 | 4.68 | 4.71 | 4.75 | 4.79 | 4.82 | 4.85 | 4.89 |
| 1 | 4.92 | 4.96 | 5.00 | 5.03 | 5.07 | 5.10 | 5.15 | 5.18 | 5.21 | 5.25 |
| 2 | 5.29 | 5.33 | 5.37 | 5.41 | 5.45 | 5.48 | 5.52 | 5.57 | 5.60 | 5.64 |
| 3 | 5.69 | 5.72 | 5.76 | 5.84 | 5.89 | 5.93 | 5.93 | 5.97 | 6.02 | 6.05 |
| 4 | 6.10 | 6.14 | 6.18 | 6.23 | 6.27 | 6.32 | 6.36 | 6.41 | 6.45 | 6.50 |
| 5 | 6.54 | 6.59 | 6.63 | 6.68 | 6.73 | 6.77 | 6.82 | 6.87 | 6.92 | 6.96 |
| 6 | 7.01 | 7.06 | 7.11 | 7.16 | 7.21 | 7.26 | 7.31 | 7.36 | 7.41 | 7.46 |
| 7 | 7.52 | 7.54 | 7.61 | 7.67 | 7.72 | 7.77 | 7.83 | 7.88 | 7.94 | 7.99 |
| 8 | 8.04 | 8.10 | 8.15 | 8.21 | 8.27 | 8.33 | 8.38 | 8.44 | 8.49 | 8.55 |
| 9 | 8.61 | 8.67 | 8.72 | 8.78 | 8.84 | 8.90 | 8.96 | 9.02 | 9.08 | 9.15 |
| 10 | 9.21 | 9.27 | 9.33 | 9.40 | 9.46 | 9.53 | 9.59 | 9.65 | 9.71 | 9.78 |
| 11 | 9.85 | 9.91 | 9.98 | 10.04 | 10.11 | 10.18 | 10.25 | 10.32 | 10.38 | 10.45 |
| 12 | 10.52 | 10.59 | 10.66 | 10.74 | 10.80 | 10.88 | 10.95 | 11.02 | 11.09 | 11.16 |
| 13 | 11.24 | 11.31 | 11.39 | 11.46 | 11.54 | 11.61 | 11.69 | 11.76 | 11.84 | 11.91 |
| 14 | 11.99 | 12.07 | 12.15 | 12.23 | 12.31 | 12.39 | 12.47 | 12.55 | 12.63 | 12.72 |
| 15 | 12.78 | 12.87 | 12.95 | 13.03 | 13.12 | 13.20 | 13.29 | 13.38 | 13.46 | 13.55 |
| 16 | 13.64 | 13.72 | 13.80 | 13.89 | 13.98 | 14.07 | 14.16 | 14.25 | 14.34 | 14.43 |
| 17 | 14.53 | 14.62 | 14.71 | 14.81 | 14.90 | 15.00 | 15.09 | 15.18 | 15.27 | 15.38 |
| 18 | 15.48 | 15.57 | 15.67 | 15.77 | 15.87 | 15.97 | 16.07 | 16.17 | 16.27 | 16.38 |
| 19 | 16.47 | 16.58 | 16.68 | 16.79 | 16.89 | 17.00 | 17.10 | 17.21 | 17.31 | 17.43 |
| 20 | 17.53 | 17.64 | 17.75 | 17.86 | 17.97 | 18.08 | 18.19 | 18.30 | 18.42 | 18.54 |
| 21 | T8.65 | 18.76 | 18.88 | 18.99 | 19.11 | 19.23 | 19.35 | 19.47 | 19.59 | 19.71 |

| Air Temperature, °C | 0.0 | 0.1 | 0.2 | 0.3 | 0.4 | 0.5 | 0.6 | 0.7 | 0.8 | 0.9 |
|---|---|---|---|---|---|---|---|---|---|---|
| 22 | 19.83 | 19.95 | 20.07 | 20.20 | 20.32 | 20.44 | 20.56 | 20.69 | 20.82 | 20.94 |
| 23 | 21.07 | 21.19 | 21.33 | 21.46 | 21.58 | 21.71 | 21.83 | 21.98 | 22.11 | 22.24 |
| 24 | 22.38 | 22.51 | 22.65 | 22.78 | 22.92 | 23.06 | 23.20 | 23.33 | 23.47 | 23.62 |
| 25 | 23.76 | 23.90 | 24.04 | 24.19 | 24.33 | 24.47 | 24.62 | 24.76 | 24.91 | 25.06 |
| 26 | 25.21 | 25.36 | 25.51 | 25.66 | 25.81 | 25.96 | 26.12 | 26.27 | 26.43 | 26.59 |
| 27 | 26.74 | 26.90 | 27.06 | 27.21 | 27.37 | 27.54 | 27.70 | 27.86 | 28.18 | 28.18 |
| 28 | 28.34 | 28.51 | 28.69 | 28.85 | 29.02 | 29.19 | 29.36 | 29.53 | 29.70 | 29.87 |
| 29 | 30.04 | 30.22 | 30.40 | 30.57 | 30.75 | 30.92 | 31.10 | 31.28 | 31.46 | 31.64 |
| 30 | 31.83 | 32.01 | 32.20 | 32.38 | 32.57 | 32.75 | 32.94 | 33.13 | 33.32 | 33.51 |
| 31 | 33.70 | 33.89 | 34.09 | 34.28 | 34.49 | 34.67 | 34.87 | 35.07 | 35.27 | 35.47 |
| 32 | 35.67 | 35.87 | 36.07 | 36.28 | 36.48 | 36.69 | 36.89 | 37.10 | 37.31 | 37.52 |
| 33 | 37.73 | 37.95 | 38.16 | 38.38 | 38.59 | 38.81 | 39.02 | 39.24 | 39.47 | 39.68 |
| 34 | 39.91 | 40.13 | 40.35 | 40.58 | 40.80 | 41.03 | 41.26 | 41.49 | 41.72 | 41.95 |
| 35 | 42.18 | 42.42 | 42.66 | 42.89 | 43.13 | 43.37 | 43.61 | 43.85 | 44.09 | 44.33 |
| 36 | 44.58 | 44.82 | 45.06 | 45.31 | 45.57 | 45.81 | 46.06 | 46.32 | 46.57 | 46.83 |
| 37 | 47.08 | 47.34 | 47.60 | 47.85 | 48.12 | 48.38 | 48.64 | 48.90 | 49.17 | 49.44 |
| 38 | 49.71 | 49.98 | 50.25 | 50.52 | 50.80 | 51.07 | 51.34 | 51.62 | 51.09 | 52.18 |
| 39 | 52.46 | 52.61 | 53.02 | 53.31 | 53.60 | 53.89 | 54.17 | 54.46 | 54.76 | 55.05 |

## 2. Vapor Pressure, Pa

| Air Temperature, °C | 0.0 | 0.1 | 0.2 | 0.3 | 0.4 | 0.5 | 0.6 | 0.7 | 0.8 | 0.9 |
|---|---|---|---|---|---|---|---|---|---|---|
| 0 | 611 | 615 | 620 | 624 | 629 | 633 | 638 | 642 | 647 | 652 |
| 1 | 656 | 661 | 666 | 671 | 676 | 680 | 686 | 690 | 695 | 700 |
| 2 | 705 | 710 | 716 | 721 | 726 | 731 | 736 | 742 | 747 | 752 |
| 3 | 758 | 763 | 768 | 779 | 785 | 790 | 790 | 796 | 802 | 80? |
| 4 | 813 | 819 | 824 | 830 | 836 | 842 | 848 | 854 | 860 | 866 |
| 5 | 872 | 878 | 884 | 891 | a 97 | 903 | 909 | 916 | 922 | 928 |
| 6 | 935 | 941 | 948 | 954 | 961 | 968 | 974 | 981 | 988 | 995 |
| 7 | 1002 | 1008 | 1015 | 1022 | 1029 | 1036 | 1044 | 1051 | 1058 | 1065 |
| 8 | 1072 | 1080 | 1087 | 1095 | 1102 | 1110 | 1117 | 1125 | 1132 | 1140 |
| 9 | 1148 | 1156 | 1163 | 1171 | 1179 | 1187 | 1195 | 1203 | 1211 | 1220 |
| 10 | 1228 | 1236 | 1244 | 1253 | 1261 | 1270 | 1278 | 1287 | 1295 | 1304 |
| 11 | 1313 | 1321 | 1330 | 1339 | 1348 | 1357 | 1366 | 1375 | 1384 | 1393 |
| 12 | 1403 | 1412 | 1421 | 1431 | 1440 | 1450 | 1459 | 1469 | 1478 | 1488 |
| 13 | 1498 | 1508 | 1518 | 1528 | 1538 | 1548 | 1558 | 1568 | 1578 | 1588 |
| 14 | 1599 | 1609 | 1620 | 1630 | 1641 | 1051 | 1662 | 1673 | 1684 | 1695 |
| 15 | 1704 | 1715 | 1726 | 1737 | 1749 | 1760 | 1771 | 1783 | 1794 | 1806 |
| 16 | 1817 | 1829 | 1840 | 1852 | 1664 | 1 876 | 1888 | 1900 | 1912 | 1924 |
| 17 | 1937 | 1949 | 1961 | 1974 | 1986 | 2000 | 2011 | 2024 | 2036 | 2050 |

*(continued)*

| Air Temperature, °C | 0.0 | 0.1 | 0.2 | 0.3 | 0.4 | 0.5 | 0.6 | 0.7 | 0.8 | 0.9 |
|---|---|---|---|---|---|---|---|---|---|---|
| 18 | 2063 | 2076 | 2089 | 2102 | 2115 | 2129 | 2142 | 2155 | 2169 | 2183 |
| 19 | 2196 | 2210 | 2224 | 2238 | 2252 | 2266 | 2280 | 2294 | 2308 | 2323 |
| 20 | 2337 | 2352 | 2366 | 2381 | 2396 | 2410 | 2425 | 2440 | 2455 | 2471 |
| 21 | 2486 | 2501 | 2517 | 2532 | 2548 | 2563 | 2579 | 2595 | 2611 | 2627 |
| 22 | 2643 | 2659 | 2675 | 2692 | 2708 | 2724 | 2741 | 2758 | 2775 | 2791 |
| 23 | 2808 | 2825 | 2843 | 2860 | 2877 | 2894 | 2912 | 2930 | 2947 | 2965 |
| 24 | 2983 | 3001 | 3019 | 3037 | 3055 | 3074 | 3092 | 3110 | 3129 | 3148 |
| 25 | 3167 | 3186 | 3205 | 3224 | 3243 | 3262 | 3282 | 3301 | 3321 | 3341 |
| 26 | 336161 | 3381 | 3401 | 3421 | 3441 | 3461 | 3482 | 3502 | 3523 | 3544 |
| 27 | 3565 | 3586 | 3007 | 3628 | 3649 | 3671 | 3692 | 3714 | 3735 | 3757 |
| 28 | 3779 | 3801 | 3824 | 3846 | 3868 | 3891 | 3913 | 3936 | 3959 | 3982 |
| 29 | 4005 | 4028 | 4052 | 4075 | 4099 | 4122 | 4146 | 4170 | 4194 | 4218 |
| 30 | 4243 | 4267 | 4292 | 4316 | 4341 | 4366 | 4391 | 4416 | 4441 | 4467 |
| 31 | 4492 | 4518 | 4544 | 4570 | 4596 | 4622 | 4648 | 4675 | 4701 | 4728 |
| 32 | 4755 | 4782 | 4809 | 4836 | 4863 | 4891 | 4918 | 4946 | 4974 | 5002 |
| 33 | 5030 | 5059 | 5087 | 5116 | 5144 | 5173 | 5202 | 5231 | 5261 | 5290 |
| 34 | 5320 | 5349 | 5379 | 5409 | 5439 | 5470 | 5500 | 5531 | 5561 | 5592 |
| 35 | 5623 | 5654 | 5686 | 5717 | 5749 | 5781 | 5813 | 5845 | 5877 | 5909 |
| 36 | 5942 | 5975 | 6007 | 6040 | 6074 | 6107 | 6140 | 6174 | 6208 | 6242 |
| 37 | 6276 | 6310 | 6345 | 6379 | 6414 | 6449 | 6484 | 6519 | 6555 | 6590 |
| 38 | 6626 | 6662 | 6698 | 6734 | 6771 | 6807 | 6844 | 6881 | 6918 | 6955 |
| 39 | 6993 | 7031 | 7068 | 7106 | 7145 | 7183 | 7221 | 7260 | 7299 | 7338 |

## METROLOGICAL CHARACTERISTICS OF MEASURING INSTRUMENTS

*Accuracy* is the degree of the difference between a measured quantity value and the true quantity value.

*Precision* is the ability of a measurement to be consistently reproduced.

*Sensitivity* can be defined as the ratio of the change in the signal from the measuring system to the change in the value of the parameter being measured.

*Resolution* is the smallest change in the parameter being measured (e.g., division or figure of the scale).

*Linearity of Response* corresponds to a uniform scale device.

*Zero Drift* characterizes the zero instability in the absence of signal, caused by climatic and other conditions.

*Response Time* describes the rate at which the device responds to the changes of the input signal; it corresponds to the time interval between the change of the parameter and the moment of the measurement of this change.

*Detection Limit* is the smallest amount or concentration of a particular substance in a sample that can be reliably detected during a measurement process.

# INDEX

*Methods of Measuring Environmental Parameters*, First Edition. Yuriy Posudin.
© 2014 John Wiley & Sons, Inc. Published 2014 by John Wiley & Sons, Inc.